Adhesives in Biomedical Applications

Scrivener Publishing
100 Cummings Center, Suite 541J
Beverly, MA 01915-6106

Adhesion and Adhesives: Fundamental and Applied Aspects

The topics to be covered include, but not limited to, basic and theoretical aspects of adhesion; modeling of adhesion phenomena; mechanisms of adhesion; surface and interfacial analysis and characterization; unraveling of events at interfaces; characterization of interphases; adhesion of thin films and coatings; adhesion aspects in reinforced composites; formation, characterization and durability of adhesive joints; surface preparation methods; polymer surface modification; biological adhesion; particle adhesion; adhesion of metallized plastics; adhesion of diamond-like films; adhesion promoters; contact angle, wettability and adhesion; superhydrophobicity and superhydrophilicity. With regards to adhesives, the Series will include, but not limited to, green adhesives; novel and high-performance adhesives; and medical adhesive applications.

Series Editor: Dr. K.L. Mittal
Email: raareviews@gmail.com

Publishers at Scrivener
Martin Scrivener (martin@scrivenerpublishing.com)
Phillip Carmical (pcarmical@scrivenerpublishing.com)

Adhesives in Biomedical Applications

Edited by
K.L. Mittal
and
S. Neogi

Scrivener Publishing

WILEY

This edition first published 2023 by John Wiley & Sons, Inc., 111 River Street, Hoboken, NJ 07030, USA and Scrivener Publishing LLC, 100 Cummings Center, Suite 541J, Beverly, MA 01915, USA

For more information about Scrivener publications please visit www.scrivenerpublishing.com.

Wiley Global Headquarters
111 River Street, Hoboken, NJ 07030, USA

For details of our global editorial offices, customer services, and more information about Wiley products visit us at www.wiley.com.

Library of Congress Cataloging-in-Publication Data

ISBN 978-1-394-20920-0

Cover image: Pixabay.Com
Cover design by Russell Richardson

Set in size of 11pt and Minion Pro by Manila Typesetting Company, Makati, Philippines

Printed in the USA

10 9 8 7 6 5 4 3 2 1

Contents

Preface

Adhesives are gaining popularity in many and varied biomedical applications nowadays. Adhesives are being used as a replacement for sutures and staples. Sutures and staples are being replaced because they have disadvantages such as scarring, infection, keloid formation, poor skin healing, or hernia in the case of abdominal sutures. On the other hand, adhesives dramatically reduce healthcare costs, significantly reduce time spent in surgery, curb the risks of bleeding, and are generally easy to use. Adhesives also find their use in diagnostic imaging and theranostics, various biomedical devices, dental adhesives, dermal adhesives, etc.

An ideal adhesive to be used for biomedical applications must have the following properties. It must be loaded substantially with the active compound, must bind to specific receptor sites on the cell or mucus surface, must interact with mucus or its components for adequate adhesion, must allow controlled release of the active compound when swollen, must be excreted unaltered or biologically degraded to inactive, nontoxic oligomers. It must possess a sufficient amount of hydrogen-bonding chemical groups, must possess high molecular weight, must possess high chain flexibility, and must have the surface tension that may induce spreading onto the mucous layer. Commonly used adhesives for biomedical applications originate from cellulose, gelatin, chitosan, collagen, poly(lactic-co-glycolic acid), poly(ethylene glycol), poly(acrylic acid), etc.

This book, containing eleven chapters, is divided into two parts: Part 1: General Topics; and Part 2: Specific Adhesives, Characteristics, and Applications. Topics covered include: historical developments of various adhesives for biomedical applications; global industry development and analysis of adhesives for biomedical applications; biomedical adhesives; bioadhesion: fundamentals and mechanisms; fibrin glue; herbal bioactives-based mucoadhesive drug delivery systems; adhesive hydrogels; adhesives in dermal patches; medical adhesives from extracted mussel adhesive proteins; dental adhesives; and the role of adhesive-based systems for diagnostic imaging and theranostic applications.

The chapters are written by renowned researchers actively involved in the arena of adhesives in biomedical applications. As current adhesives are ameliorated, or new, more effective adhesives are developed, new application vistas will emerge. Also currently there is a high tempo of interest and research in developing "green" adhesives based on bioresources (e.g., mussel proteins). Moreover, the use of nanoparticles to improve and enhance adhesives for biomedical applications is actively pursued.

The book is profusely referenced and copiously illustrated. It should be of great appeal and interest to medical doctors, surgeons, cosmetologists, people in the pharmaceutical industry, adhesionists, adhesive technologists, those involved with biomedical devices and bioimplants, polymer scientists, and materials scientists.

Now it gives us great pleasure to acknowledge all those who played essential roles in giving this book a body form. Naturally, first and foremost, our profound thanks go to the authors for their keen interest, sustained enthusiasm, unwavering cooperation, and sharing their valuable research experience in the form of written accounts (which essentially provided the grist for this book), without which this book could not be materialized. Also, the steadfast interest and whole-hearted support of Martin Scrivener (publisher) in this book endeavor is highly appreciated.

Kash Mittal
Hopewell Jct., NY, USA
E-mail: raareviews@gmail.com
Sudarsan Neogi
Indian Institute of Technology Kharagpur, India
July 2023

Part 1
GENERAL TOPICS

1

Historical Developments of Various Adhesives for Biomedical Applications

Nagavendra Kommineni[1], Raju Saka[2], Vaskuri G. S. Sainaga Jyothi[3], Arun Butreddy[4], Jyotsna G. Vitore[5] and Wahid Khan[6*]

[1]*Center for Biomedical Research, Population Council, New York, NY, USA*
[2]*Department of Pharmaceutics, National Institute of Pharmaceutical Education and Research, Hyderabad, Telangana, India*
[3]*Department of Pharmaceutical Sciences, University of Tennessee Health Science Center, Memphis, USA*
[4]*School of Pharmacy, Department of Pharmaceutics & Drug Delivery, University of Mississippi, Oxford, MS, USA*
[5]*Department of Pharmaceutics, National Institute of Pharmaceutical Education and Research, Ahmedabad, Gujarat, India*
[6]*Natco Research Centre, NATCO Pharma Limited, Hyderabad, India*

Abstract
Adhesives are versatile and have been widely researched in recent decades due to their significant effect on health care with applications in various areas. Biomaterials have grown into an attractive research area where the need for exploration of biocompatible, efficient and economical bioadhesives has increased. Adhesives have been in use since ancient times by our ancestors for wound healing. There were many efficient materials derived from plant, animal and micro-organism sources which were used as adhesives. These adhesives have played a major role in surgery and wound healing. First-generation adhesives were found to have drawbacks with regard to adhesion in wet conditions and also toxicity associated with their synthetic nature. However, over time much advancement has taken place in this area and various biocompatible adhesives have been identified and isolated for surgical applications. Even the applications of adhesives have extended to unexplored areas like drug delivery and tissue engineering. In current decades, new bioadhesive materials have come into picture to meet the needs in drug delivery and wound healing. This chapter is aimed to cover the historical development

**Corresponding author*: mail4wahid@gmail.com

K.L. Mittal and S. Neogi (eds.) Adhesives in Biomedical Applications, (3–24) © 2023 Scrivener Publishing LLC

of various adhesives for medical applications. In addition, various development phases of these adhesives along with the historical milestones have been covered. This chapter also enlightens the reader about most recent and updated applications of bioadhesives in biomedical sciences and finally highlights the important points to consider while exploring this field of adhesives for biomedical applications.

***Keywords*:** Adhesives, drug delivery, biomedical application, tissue engineering, wound healing

1.1 Origin of Adhesives

Adhesive bonding has influenced day-to-day life of individuals. It has long history and versatile applications. There are many industries and technologies which contribute to the recent advancement of adhesives. Our early ancestors were unaware of the concept of an adhesive material. Historically, discovery of adhesive materials started with nature. Most of the adhesives were originated from plant and animal sources [1, 2]. In addition, a few reports have suggested that microorganisms such as algae, bacteria and other insects, amphibians including echinoderms, carnivorous, orchids, frogs, spiders are also prominent sources of adhesives. Many micro-organisms are reported to produce sticky material on the surface of their bodies [3].

However, there are numerous historical evidences in literature regarding adhesives to suggest that humans had used adhesives for thousands of years for making objects more useful, attractive and decorous. In addition, adhesives were used in making ornaments and weapons. Adhesives have evolved gradually. The understanding of adhesiveness had a long way from sticky material obtained in cooking to the various advanced adhesives available in recent times. The past transformation period in the history of adhesives was well recognized where the suggestion made by Roman author and scientist Pliny the Elder that Daedalus was credited for inventing glue. However, based on the technical consideration and archaeological evidence the Pliny's suggestion was denied and it became necessary to find out the origin of man-made adhesives [4].

Adhesives are used in wound healing, dressing or prevention of bleeding in the medical field. In the current developed era of medical technology, various applications have been found for adhesives [5]. Numerous theories and hypotheses have been postulated to explain and demonstrate the process of adhesion including van der Waals forces, chemical bonding and electrostatic attraction [6–11]. Historically, human ancestors including primitive people and homosapiens used adhesives in treating wounds and healing injury. They used grass and leaves to cover the wound area.

Primitive people used to close wound by sewing it together with muscle fibers and shreds of tendon. Large wounds were cleaned by inserting spikes along the edges of the wounds and then tying these together with clothes and tendons. Around 1100BC, the oldest sutures were first described in ancient Egyptian literature. Historically, sutures were recognized as blessed combination of preventing infection and wound healing. 4000 years back, plant derived gums were used in the form of adhesives [12]. In addition, tree resin and beeswax were applied as sealants or adhesives. However, the ancient Greeks, Egyptians and Indians preferred suturing to heal the wound. Perhaps, the major challenge is to find alternative to suture the wound.

In the middle of revolutionary era, many scientists tried to refine the sutures for getting to current advanced status. In the 19th century, the modified and advanced adhesives were explored, and rubber cements appeared. Still, the advancement in this field had to wait for the 21St century when the natural adhesives were transformed into novel compounds. In the past 40 years, paradigm shift has been observed in the development and advancement of adhesives in the medical field. New adhesives are also continuously developed taking inspiration from natural adhesive materials. In 1970, U.S. Clean Air Act and the formation of the Environmental Protection Agency and Occupational Safety and Health Administration were established. Even today, there are several types of adhesive materials available with good safety profile and sustainability. Furthermore, water-based adhesives, ultraviolent and electron beam cured adhesives, hot-melt adhesives have been developed. There are certain chemicals which are reported to be carcinogenic and toxic in nature, hence the approaches for removing residual organic solvents with minimal risk of toxicity are preferred. In the recent era scientists have developed pressure-sensitive adhesives (PSAs) in order to absorb a significant amount of fluid, and these are rubber based for ease of utility. Following this, the revolutionary change occurred where exploration of silicone-based adhesives came in picture, which provides comfort and promotes healing. In this digital era, the needs of patients have changed from conventional to frictionless and wearable adhesives in wound closure.

1.2 Prominence of Biomedical Adhesives in Wound Healing and Drug Delivery

The use of adhesives in wound healing and drug delivery gained importance as adhesives offered various advantages.

1.2.1 Wound Healing

Interest in tissue adhesives has gained prominence in the past half century. They were introduced as an alternative to adhesive tapes or bandages. Adhesive tapes and bandages are applicable for topical wound therapy, while tissue adhesives were aimed to control bleeding irrespective of site of injury, i.e., the major advantage is that these could also be used for internal applications [13, 14]. This paved the way for advances in surgery and wound healing. Various classes of adhesives were introduced with varied functionality and applications. Fibrin sealants were the first tissue adhesives with their first use dating back to early 20th century.

Cyanoacrylates were one of the earliest adhesives approved for wound healing. These are synthetic derivatives of acrylic acid widely used for surgical and traumatic wound repair. These have advantages like higher mechanical strength compared to other adhesives [15, 16]. Moreover, cyanoacrylate adhesives are cheaper compared to fibrin or collagen-based sealants. This led to their wide use in surgical applications [17]. Their development started in 1950s for medical use. However, the approval was obtained only in early 1980s for their use only as topical skin adhesive [18]. N-butyl-2-cyanoacrylate marketed as Histoacryl Blue (TISSUESEAL, USA) was the first cyanoacrylate used for tissue adhesive application [19]. The first U.S. Food and Drug Administration (FDA) approval of cyanoacrylate adhesive for internal use was received in early 21st century. This product was approved for topical application to close easily approximated skin edges of minimum-tension wounds of surgical incisions and for simple, well-cleansed, trauma-induced lacerations. Histoacryl Blue and Histoacryl Flexible may be used in conjunction with, but not in place of, deep dermal sutures [20]. However, cyanoacrylates suffer from drawbacks such as poor elasticity and significant toxicity due to their synthetic nature. This led to the development of safer biocompatible alternatives.

Fibrin sealants are another class of popular tissue adhesives. The use of fibrin sealants dates back to early 20th century with first use of fibrin reported in 1909 [21]. However, the popularity of fibrin sealants rose in 1940s [22, 23]. The sealant is of biological origin and is derived from human plasma. The sealant is formed as a crosslink of fibrinogen and fibrin in the presence of factor XIII and calcium as catalysts. The crosslink prevents blood loss causing effective wound healing [24]. The fibrin sealants are prepared from human blood [25]. The major advantage of fibrin sealants is that they are entirely of biological origin and hence there is no risk of toxicity. Moreover, the crosslink formed degrades slowly without any abnormal degradants formation. Thus, these became a popular alternative for wound healing applications. These can also be used internally for sensitive areas thus increasing their spectrum of applications. However, these suffer from drawbacks like poor adhesion strength and reduction in efficiency under wet condition and long sealing time [26]. Moreover, sealants derived from different human sources vary in fibrin and thrombin contents thus affecting degree of adhesion. This can be avoided by pooling the plasma from various sources [27]. However, the pooled sources have risk of viral transmission and also bacterial growth when used without antibiotics [28]. Nevertheless, fibrin sealants are most popular adhesives with wide use in surgical interventions.

Gelatin-resorcinol-formaldehyde/glutaraldehyde glues are another earlier class of adhesives that have been continuously used since 1960s. Adhesives are also made with a combination of different components. Gelatin in the adhesive imparts elasticity to the adhesive. Formaldehyde-based adhesives have strong bonding while those based on glutaraldehyde have better *in vivo* stability. The mechanism of adhesion is due to crosslinking of polymer chains. The advantage of these adhesives is that they have higher adhesion strength compared to fibrin sealants which is similar to that of cyanoacrylate adhesives. However, poor adhesion in wet conditions and carcinogenicity due to aldehyde content limited their clinical applications [29–31].

Collagen-based tissue adhesives are one of the newer classes of adhesives. Collagen is also a biological component which forms a major part of extracellular matrix. It acts by triggering an intrinsic pathway for blood coagulation. The mechanism of adhesion is crosslinking. As it is derived from mammalian source the risk of infection is less likely. Thus, these have found wide use in wound healing and surgery. Few collagen-based products were approved like Costasis™/Dynastat™ (Cohesion Technologies) and FloSeal Matrix® (Baxter International Inc.) [32]. Like fibrin-based sealants, collagen sealants have slower curing rate for adhesive applications. Thus, the adhesion strength is lower compared to synthetic adhesives. Moreover, the collagen can swell upon tissue compression making it unsuitable for ophthalmic or neurologic surgeries [33]. Approaches to increase adhesion strength are being evaluated to make these ideal for wound healing.

Albumin-based adhesives are another class of tissue adhesives which rose to prominence in late 1990s. Albumin-glutaraldehyde adhesive marketed as BioGlue® Surgical Adhesive (CryoLife Inc.), is a two-component system consisting of bovine serum albumin and glutaraldehyde in stoichiometric ratio. Compared to other tissue adhesives from biological sources, these adhesives have higher tensile and shear strengths. Moreover, albumin-based adhesives are not derived from pooled sources and hence risk of infection is very low. However, these require bloodless environment thus preventing their use in active bleeding. Similarly, presence of bovine serum albumin may cause hypersensitivity reaction in some individuals. Even glutaraldehyde was found to be toxic for cells which limits its uses [34]. More clinical studies are required to establish their use in surgery.

Biomimetic adhesives are the newest class of adhesives which are mostly nature derived. These were inspired by nature where these were used for protection from predators, self-healing and attachment to surfaces. Various bioinspired adhesives are being evaluated for application in wound healing and surgeries. Examples of bioinspired adhesives include mussel-derived adhesive proteins, barnacles, sandcastle worms, prolamine, gecko adhesives, etc. Unlike other adhesives which cause adhesion due to crosslinking, the bioinspired adhesives undergo chemical adhesion thus providing highest strength even under wet conditions [35]. Studies were performed using hydrogels along with these adhesive materials to improve adhesion.

1.2.2 Drug Delivery

Application of bioadhesives in drug delivery is a newer aspect in adhesive technology though these were evaluated much earlier. One of the first methods of drug delivery using adhesives is antibiotic delivery with fibrin sealants. Here the researchers used fibrin sealant as a matrix to control the drug release. Drugs evaluated ranged from ampicillin, ciprofloxacin to tetracycline [36]. Similarly, the application was expanded to include the delivery of chemotherapeutic agents. Even cyanoacrylate adhesives were successfully applied in delivery of antibiotic like vancomycin [37, 38].

Site-specific drug delivery gained importance to achieve either localized delivery (ophthalmic drug delivery) or enhanced systemic bioavailability (colonic and transdermal drug delivery). Various adhesive polymers were used for controlled delivery of drugs. These included chitosan, cellulose derivatives, methacrylate polymers, alginates, lectins and thiolated derivatives of these polymers. These were selected based on the purpose of application.

1.3 Generations of Bioadhesives

The term "bioadhesive" implies the adhesion of a synthetic or natural molecule to a biological surface which could be either a mucus surface or epithelial membrane [39]. The most extensively explored bioadhesives are a group of macromolecules that are hydrophilic in nature.

These macromolecules contain functional groups which are proficient to form hydrogen bonds with the biological surface. These functional groups include hydroxyl groups, carboxyl groups and amine groups. However, they are non-specific in nature in terms of adhesion and adhere to any surface but require moisture for the activation. Hence, they are often referred as wet adhesives. Nevertheless, once they are activated, they exhibit a strong adhesion to the surface. The bioadhesives may overhydrate and form slippery mucilage unless the water uptake is restricted [40]. However, if the adhesive joint is dried then it results in a strong adhesive bond. They are considered safe by regulatory agencies and are defined as "off-the-shelf" adhesives. These bioadhesives are referred as first-generation bioadhesives [41].

The most commonly used bioadhesives which fall under the category of first-generation bioadhesives are carbopol [42], sodium alginate [43], polycarbophil [44] and chitosan [45]. These bioadhesives are categorized as synthetic or natural bioadhesives. Carbopol, carbophil and poly(acrylic acid) are widely used synthetic bioadhesives. Natural or semi-synthetic bioadhesives include chitosan, xanthan gum, alginate, etc. Cellulose based bioadhesives are also explored widely including methyl cellulose, sodium carboxymethyl cellulose and hydroxypropylmethyl cellulose [46].

The bioadhesion of macromolecules to the biological surface occurs via interactions which include ionic bonds, covalent bonds, hydrogen bonds, hydrophobic bonds and van der Waals bonds. Ionic bonds are formed between the oppositely charged ions and covalent bonds are formed by sharing a pair of electrons. Hydrogen bonds are formed between electronegative atoms and hydrogen atoms [47], van der Waals forces are formed between the polar molecules, and hydrophobic bonds are formed as a result of hydrophobic interactions. Covalent and ionic interactions result in strong adhesion; however, hydrogen bonds, hydrophobic bonds and van der Waals bonds are quite weaker bonds [48]. The type of interaction manifested between the macromolecule and the biological surface is based on the property of the bioadhesive molecules, for instance, whether anionic, non-ionic and cationic in nature.

Interaction between the macromolecules and the biological surface is explained by the chemical bonds. However, to demonstrate the mechanism of formation of bioadhesion numerous theories have been proposed [49]. The transfer of electrons between the biological surface and macromolecules results as a consequence of difference in the electronic structures. This manifests adhesion between the adhering surfaces as proposed in electronic theory [50]. Wetting theory is applicable to liquid systems where the affinity of liquid to spread on the biological surface is calculated as spreading coefficient considering the surface and interfacial tensions [51, 52]. Adsorption theory describes the adhesion between the biological surface and macromolecules resulting via hydrogen bonding or van der Waals forces where a bond is formed between the biological surface and the bioadhesive macromolecule [53, 54]. Diffusion theory explains the diffusion of chains of macromolecules into the biological surface. The diffusion of chains is driven by the difference in concentration across the biological interface, mobility of chains and the diffusion coefficient [55]. Mechanical theory suggests the interlocking of bioadhesive molecules into the rough surface of the biological membrane. Also, rough surfaces offer an enhanced surface area accessible for interaction along with an increased viscoelastic and plastic energy release during bond failure, which are considered to be more imperative in the process of adhesion than a mechanical effect [56]. Fracture theory is quite different from the rest of the theories where it relates the force required to detach two surfaces involved in adhesion to the adhesion strength [57]. These theories are proposed to explain the mechanism behind establishing adhesion. However, numerous factors affect the strength of adhesion which include molecular weight of bioadhesives [58], concentration of bioadhesive molecules/polymer [59], chain length and its flexibility [60], interpenetration of polymer into mucus gel layer [61], degree of hydration [62], charge and functional group on the macromolecule [63, 64] and pH of the environment [65]. Charge and functional group present on the macromolecule affect the degree of bioadhesion and result in nonspecific attachment to the site. Bioadhesion of macromolecules to the biological membrane is also influenced by hydration where the hydration promotes interpenetration of polymer chains into the biological surface. However, the excess hydration leads to the formation of slippery mucilage hampering the bioadhesive property. This resulted in exploring new generation of bioadhesives which are specific in adhesion to the desired site. The widely explored new generation bioadhesives are lectins and thiolated macromolecules.

Lectins structurally belong to the class of proteins or glycoproteins where they bind specifically to the carbohydrate moiety, for instance, bind to the glycosylated membrane components [66]. Nevertheless, lectins also mediate receptor adhesion leading to internalization via endocytosis or transcytosis. Thiolated macromolecules are the derivatives of polyacrylates or chitosan. Thiol groups are present in their structure which results in the formation of covalent bonds between the thiol groups and the cysteine groups in the mucus membrane [67]. The derived thiolated macromolecules are chitosan–thioglycolic acid [68] and poly(acrylic acid)–cysteine [69].

Thus, these new generation bioadhesives offer specific and strong bioadhesive property compared to the old-generation bioadhesives.

1.4 Timeline of Developments and Advances

The scientific community has made great progress in the creation of tissue regenerative bioadhesives for musculoskeletal tissues in the last few decades. Co-polymerization, cross-linking, mixing, and surface modification are only a few of the chemical modifications that have resulted in significant improvements in the basic physical properties. Despite potential improvements in adhesion strength and bulk modulus, existing bioadhesives have yet to attain mechanical qualities required to restore the functional properties of the majority of musculoskeletal tissues in a timely manner. However, there are only a limited number of *in vivo* studies on bioadhesives efficacy in tissue repair and regeneration that require further research.

According to a PubMed literature search, the number of papers with the term "tissue adhesives *in vivo*" has remained relatively constant in the last decade, however it only accounts for 6.7 percent of all papers searched for "tissue adhesives." In the last 20 years, the number of publications with the key word bioadhesive has steadily increased and the number of papers with bioadhesives *in vivo* is also on the rise, accounting for more than 20% of all papers containing bioadhesives. These findings from the literature search would indicate a growing research interest in adhesive biomaterials with a biological purpose other than their typical role as tissue adhesives providing physical bonding, hemostasis, or sealing.

Since their FDA approval in 1998, fibrin tissue adhesives or tissue sealants have been the most extensively utilized bioadhesives in the United States. Fibrin is the only substance approved by the FDA for use as a hemostat, tissue glue, and tissue sealant at this time [70, 71]. In the past few decades, aldehyde-based adhesives, i.e., gelatin - resorcinol cross-linked with formaldehyde (GRF) and GRFG (GRF with glutaraldehyde) have been widely used for vascular, thoracoscopic, gastrointestinal, liver, and urinary track surgeries in Europe and Japan. BioGlue® 24 (CryoLife, Kennesaw, GA) is a glutaraldehyde-based protein-aldehyde system (PAS) that is commercially available. It is made up of two components: bovine serum albumin (BSA) and glutaraldehyde, and its bonding action is identical to that of GRF/GRFG. In 1999, the FDA approved iBoGlue® for use in the United States as an adjuvant to suturing or stapling for acute thoracic aortic dissection and cardiac surgery [72, 73].

The use of bioadhesives as a delivery vehicle is a relatively new concept [74]. Bioadhesives with high target efficiency are highly implemented as delivery vehicles even if the adhesive property is low as it is likely useful for the localized distribution of cells, growth factors, and small molecules on the target location. It would be more feasible to achieve high adhesion strength and mechanical properties that would be an effective treatment option for degenerative musculoskeletal diseases such as tendinopathy if a bioadhesive's sole purpose was to provide controlled delivery rather than to support cell and tissue ingrowth [75]. Furthermore, the development and application of bioadhesives as controlled delivery vehicles with significant clinical impact is gaining popularity [75, 76].

The *in vivo* degradation of bioadhesives is an important but understudied research area. The *in vivo* functioning of a bioadhesive is directly linked to its degradation rate, whether for mechanical support, cell and tissue ingrowth, or regulated distribution. Bioadhesives used in securing tissue grafts or filling tissue defects must degrade in a way that is balanced with the development and remodeling of new tissue. Similarly, the release kinetics of a bioactive molecule loaded into a bioadhesive as a delivery vehicle is heavily influenced by the degradation of the vehicle. Because of the inherent differences between *in vitro* and *in vivo* in terms of biochemical and mechanical environments, degradation of biodegradable materials *in vitro* often resulted in *in vivo* degradation [77–80].

1.5 Current and Future Applications

1.5.1 Wound Healing

Wound closure was the first application of tissue adhesives and hence much development has been carried out in this area compared to other applications. Cyanoacrylate derivatives have been widely used in endoscopy, skin closure, esophageal variceal bleeding, peptic ulcer bleeding, and internal applications [81]. Histoacryl (Braun, Melsungen, Germany) is the most famous of all cyanoacrylates that has been in use since long time. Similarly, Dermabond® (Ethicon, Inc., Somerville, NJ) is a 2-octylcyanoacrylate which is most famous for skin closure where the patients reported less pain and faster healing compared to other products [34]. Nevertheless, its use has declined due to the need for more biocompatible alternatives. Fibrin glues have been widely used for surgical applications. Fibrin glues have been utilized for hemostasis in cardiac surgery, vascular graft sealing, vascular surgery, thoracic surgery, hepatobiliary, and pancreatic surgery [82–85].

Similarly, fibrin glue was found to be useful in preventing bleeding from gastro-intestinal ulcers, fistulas and perforations [86–88]. It is also used in special cases like neurosurgery where fibrin glue has been used to seal cerebrospinal fluid (CSF) leakage. Albumin adhesives have been used for haemostasis in vascular and cardiac anastomoses. Marketed albumin glues like BioGlue have been effectively used in preventing pulmonary air leakage [89–91]. They are even widely used for treatment of fistulas. However, the presence of glutaraldehyde reduced its usage due to the risk of cytotoxicity. Poly(ethylene glycol)-based adhesives have been used as an adjuvant for suturing and grafting and also in hemostasis. Though they are effective surgical adhesives, their swelling and rapid degradation limited their use as an adjuvant for surgical applications [92]. Mussel-inspired adhesives being the latest entry in the arena of biological adhesives are being explored for tissue adhesive applications. Since isolation of mussel proteins from nature is a costly and inefficient process, the researchers have identified the critical component of mussel protein called dopamine (DOPA) which is essential for its strong adhesion [93]. Based on this finding, the researchers have synthesized various DOPA-based polymer derivatives which have been found to have higher adhesion strength compared to available tissue adhesives. For example, poly (γ-glutamic acid) hydrogel formed by combining with dopamine was found to have 10-fold higher adhesion strength in wet condition compared to the marketed fibrin sealant [94]. These adhesives were also widely researched for bone adhesion in combination with (an)other adhesive polymers like chitosan and dextran which led to multifold increment in adhesion strength [95, 96]. Good biocompatibility also enabled mussel-inspired adhesives to diversify their applications. Gecko-inspired adhesives were also applied for tissue adhesion. Based on the pillared structure of gecko, the researchers have synthesized polyimide hairs which showed excellent adhesion to the site of application [97]. However, the inability of the pillars to retain adhesion for long time and also in wet condition limited their use. To address this issue, novel modification techniques like coating the pillars with a synthetic adhesive material to mimic mussel adhesive have been implemented. This led to improved adhesion even in wet condition with adhesion maintained up to 1000 contact cycles [98]. Similarly, Mahdavi et al also developed similar nanopillar structure with variation in ratio of tip to base diameter and also tip to pitch diameter. This resulted in strong adhesion demonstrated in *in vivo* study [99]. Though, multitude of advantages can be explored through biomimetic adhesives like mussel and gecko, still numerous clinical studies need to be performed to establish the effectiveness of these new generation adhesives.

1.5.2 Drug Delivery

The adhesive scaffold can also serve as a rate-controlling matrix for various synthetic and biological agents [100]. This property was utilized by formulation scientists for an effective local delivery of drugs for increasing therapeutic effect and also to facilitate tissue healing [101]. Fibrin glues, being the most utilized and researched class of adhesives, were also evaluated extensively for drug delivery applications. The ease of altering the mechanical properties of fibrin glue structure through simple compositional changes provided an avenue for research in drug delivery. The tunable properties can be achieved by varying crosslinking time, fibrinogen content, calcium content, and thrombin content [81]. Even anti-fibrinolytic agents can also be added in the glue to delay the fibrinolysis, thus providing long lasting matrix for controlled release. These factors enabled loading of various types of therapeutic agents. Fibrin glues were found to be useful as temporary storage depots for growth factors like vasculoendothelial growth factor (VEGF), nerve growth factor, bone morphogenic protein and platelet derived growth factor (PDGF) to aid in tissue engineering. Similarly, drugs like vancomycin, lidocaine, sisomicin, doxorubicin and 5-fluoro uracil were effectively delivered from the matrix of fibrin to elicit local effects. Recently, gene delivery vectors were also used in fibrin matrix for delivery [36]. However, the major drawback using fibrin glue is that it cannot serve as a depot for prolonged drug delivery as it degrades in days. This limits its wider application. Delivery of vitamins was also done with adhesives. Lazovic *et al.* prepared vitamin A and C soaked collagen dressing which led to improved healing compared with adhesive sheets without vitamins [102]. Similarly, elements like zinc were also loaded in wound dressings to promote healing. Also, poly(n-butylcyanoacrylate) based electrospun meshes were evaluated for delivery of chemotherapeutic agents like 5-fluorouracil where it was observed that the adhesive matrix sustained drug release for about 96 days [103]. These factors enable the tissue adhesives to be excellent drug delivery vehicles. Newer materials like mussel-derived molecules were also used for drug delivery applications. Talebian *et al.* prepared mussel-inspired biofibres which were able to control the release of doxorubicin drug and were also found to be better in treating cancer than an injectable doxorubicin [104]. Sunoqrot *et al.* [105] prepared polydopamine coated methoxy-poly(ethylene glycol)-b-poly(ε-caprolactone) nanoparticles loaded with rifampicin for gastro-retentive drug delivery. The coated nanoparticels showed higher mucin adsorption compared to the uncoated nanoparticles indicating enhanced adhesion due to polydopamine. *Ex-vivo* wash-off test also showed that the polydopamine coated nanoparticles were able to reside longer on the stomach mucosa compared to uncoated ones. This indicated a potential gastro-retentive drug delivery application but needs to be explored extensively [105]. Zhang *et al.* prepared multifunctional mussel-inspired microneedles for transdermal drug delivery of glucocorticoid. The designed system was able to improve the adhesion of the device followed by controlled release over time to reduce osteoarthritis in an animal model [106]. Though these drug delivery systems possess multitude of advantages, much clinical data needs to be generated to prove that these systems would be ideal for further exploration in medical applications.

1.5.3 Tissue Engineering

Biological adhesives were assumed as candidates for tissue engineering owing to their excellent biocompatibility, low toxicity and biodegradability. These scaffolds were found to have milder immune responses compared to their synthetic counterparts. Tissue adhesives are widely used as scaffolds for tissue regeneration. The tissue engineering is done in two ways: either by injecting the scaffold and cells into the body or by culturing cells on the scaffold *in vitro* and then introducing at the target site. Adhesives like fibrin glues were effectively used in tissue engineering of cartilage tissue, cardiac tissue and bone growth [107]. Research studies found that the fibrin matrix was an excellent scaffold for growth of osteoblasts, myoblasts and chondrocytes. Collagen was also used as a scaffold for cartilage repair [108]. However, collagen-based scaffolds suffer from rapid degradation. Similarly, researchers have developed bovine serum albumin-based hydrogel scaffold for cardiomyocytes for cardiac tissue repair. Polymerized albumin scaffolds were found to be excellent candidates for growth and differentiation of mesenchymal stem cells [109]. The albumin scaffolds were also found to enhance bone formation in various model studies by various researchers making albumin one of first choices for tissue engineering applications. Unlike collagen-based scaffolds, albumin-based scaffolds were found to have longer life due to slow degradation which will aid in effective tissue regeneration.

Biomimetic adhesives have also been being effectively used in tissue engineering applications. DOPA from mussel adhesive protein is a widely employed material in tissue engineering applications. Polydopamine, a bioinspired component from mussel adhesive protein is widely used for tissue engineering application. These polydopamine based scaffolds were used for skin, cartilage and nerve tissue repair. In a study the hydrogel consisting of polydopamine complexed with sodium alginate and polyacrylamide facilitated skin fibroblast and keratinocyte adhesion, proliferation and expression [110]. In another study, a polyacrylamide hydrogel containing polydopamine-chondroitin sulphate facilitated increased cell adhesion thus improving cartilage repair [111]. Similarly the application of polydopamine also extended to neural regeneration where the scaffold systems employing polydopamine were found to promote nerve regeneration through improved adhesion of Schwann cells and neuronal stem cells [112]. Nevertheless, this application needs to be evaluated extensively to establish adhesives as an integral part of tissue engineering applications.

1.6 Summary

The earliest application of bioadhesives can be traced back to the 20th century. Since then, bioadhesives have been explored widely all over the world. However, the extraction of fibrin from human blood plasma and its use as a bioadhesive was initiated in 1940s where it was used as an adhesive for peripheral nerves. Fibrin was commercialized and approved in Europe in 1972 and by US FDA in 1998. The prominent use of bioadhesives has been in drug delivery and wound dressing and has resulted in exploration of new generations of bioadhesives and derivatization of existing bioadhesive molecules. The feasibility of tailoring the bioadhesives as per the requirements in the biomedical field is attracting attention from the scientific community. Hence, bioadhesives are being explored to meet the unresolved needs in the pharmaceutical and biomedical arenas.

References

1. A.M. Smith, The biochemistry and mechanics of gastropod adhesive gels, in: *Biological Adhesives*. A.M. Smith, and J.A. Callow, (Eds.) pp. 167-182, Springer, Berlin, Heidelberg (2006).
2. J. Byern and I. Grunwald (Eds.), *Biological Adhesive Systems: From Nature to Technical and Medical Application*, Springer Science & Business Media (2011).
3. M. Hansell, *Built by Animals: The Natural History of Animal Architecture*. Oxford University Press, UK (2007).
4. P.A. Fay, A history of adhesive bonding, in *Adhesive Bonding*, second edition, R. D. Adams (Ed.), pp. 3-40, Woodhead Publishing, United Kingdom (2021).
5. R.D. Adams (Ed.), *Adhesive Bonding: Science, Technology and Applications*. Elsevier (2005).

6. J. Schulz and M. Nardin, Theories and mechanisms of adhesion, in: *Handbook of Adhesive Technology.* A. Pizzi and K.L. Mittal (Eds.) pp. 19-34, Marcel Dekker, New York (1994).
7. D.E. Packham, The mechanical theory of adhesion, in: *Handbook of Adhesive Technology*, Second edition, A. Pizzi and K.L. Mittal (Eds.) pp. 69-93, CRC Press, Boca Raton, FL (2003).
8. M.M. Chehimi, A. Azioune, and E. Cabet-Deliry, Acid–base interactions: relevance to adhesion and adhesive bonding, in: *Handbook of Adhesive Technology*, second edition, A. Pizzi and K.L. Mittal (Eds.) pp. 95-144, CRC Press, Boca Raton FL (2003).
9. F.M. Fowkes, Acid-base interactions in polymer adhesion, in: *Physicochemical Aspects of Polymer Surfaces*, Vol. 2, K.L. Mittal (Ed.) pp. 583-603, Plenum Press, New York (1983).
10. K.L. Mittal (Ed.), *Acid-Base Interactions: Relevance to Adhesion Science and Technology*. Vol. 2, CRC Press, Boca Raton, FL (2000).
11. K.L. Mittal and H.R. Anderson, Jr. (Eds.), *Acid-Base Interactions: Relevance to Adhesion Science and Technology*, CRC Press, Boca Raton, FL (1991).
12. E. Dinte and B. Sylvester, Adhesives: applications and recent advances. in: *Applied Adhesive Bonding in Science and Technology*, H. Özer (Ed.) pp. 119, InTech, London, United Kingdom (2018).
13. M. Glickman, A. Gheissari, S. Money, J. Martin and J.L Ballard, CoSeal Multicenter Vascular Surgery Study Group. A polymeric sealant inhibits anastomotic suture hole bleeding more rapidly than gelfoam/thrombin: results of a randomized controlled trial. *Archives Surgery*, 137, 326-331 (2002).
14. A. Ritchie and L. Rocke, Staples versus sutures in the closure of scalp wounds: a prospective, double-blind, randomized trial. *Injury*, 20, 217-218 (1989).
15. A.L. Singer, L. Berrutti, H.C Thode Jr and S.A. McClain, Octylcyanoacrylate for the treatment of partial-thickness burns in swine: a randomized, controlled experiment. *Academic Emergency Medicine*, 6, 688-692 (1999).
16. B. Burns, Cyanoacrylates: Towards high temperature resistant instant adhesives. A critical review. *Rev. Adhesion Adhesives*, 5, 361-390 (2017).
17. J. Quinn, J. Maw, K. Ramotar, G. Wenckebach and G. Wells, Octylcyanoacrylate tissue adhesive versus suture wound repair in a contaminated wound model. *Surgery*, 122, 69-72 (1997).
18. J.V. Quinn, *Tissue Adhesives in Wound Care*. BC Decker, Hamilton, Ontario, Canada (1998).
19. J.P. Noordzij, P.A. Foresman, G.T. Rodeheaver, J.V. Quinn and R.F. Edlich. Tissue adhesive wound repair revisited. *J. Emergency Medicine*, 12, 645-649 (1994).
20. G.J. Mattamal, US FDA perspective on the regulation of cyanoacrylate polymer tissue adhesives in clinical applications. *Materials Science Forum* 539, 692-697 (2007).
21. S. Bergel, The effect of the fibrin. *Dtsch Med Wochenschr*, 35, 663-665 (1909).

22. J. Young and P. Medawar, Fibrin suture of peripheral nerves: measurement of the rate of regeneration. *Lancet*, 236, 126-128 (1940).
23. E.P. Cronkite, E.L. Lozner, and J.M. Deaver, Use of thrombin and fibrinogen in skin grafting: preliminary report. *J. Amer. Med. Assoc*, 124, 976-978 (1944).
24. D.H. Sierra, Fibrin sealant adhesive systems: a review of their chemistry, material properties and clinical applications. *J. Biomaterials Applications*, 7, 309-352 (1993).
25. W.D. Spotnitz, P.D. Mintz, N. Avery, T.C. Bithell, S. Kaul and S.P. Nolan. Fibrin glue from stored human plasma. An inexpensive and efficient method for local blood bank preparation. *Amer. Surgeon*, 53, 460-462 (1987).
26. D.H. Sierra, D.S. Feldman, R. Saltz and S. Huang. A method to determine shear adhesive strength of fibrin sealants. *J. Appl. Biomaterials*, 3, 147-151 (1992).
27. G.F. Gestring and R. Lerner, Autologous fibrinogen for tissue-adhesion, hemostasis and embolization. *Vascular Surgery* 17, 294-304 (1983).
28. S.I. Rapaport, A. Zivelin, R.A. Minow, C.S. Hunter and K. Donnelly. Clinical significance of antibodies to bovine and human thrombin and factor V after surgical use of bovine thrombin. *Amer. J. Clinical Pathology*, 97, 84-91 (1992).
29. C. Lin and J. Ritter, Effect of synthesis pH on the structure of carbon xerogels. *Carbon*, 35, 1271-1278 (1997).
30. J.M. Albes, C. Krettek, B. Hausen, R. Rohde, A. Haverich and H.G. Borst. Biophysical properties of the gelatin-resorcinformaldehyde/glutaraldehyde adhesive. *Annals Thoracic Surgery*, 56, 910-915 (1993).
31. L.I. Bonchek and N.S. Braunwald, Experimental evaluation of a cross-linked gelatin adhesive in gastrointestinal surgery. *Annals Surgery*, 165, 420-424 (1967).
32. L.J.R. Foster, Bioadhesives as surgical sealants: A review. in *Bioadhesion and Biomimetics: From Nature to Applications*, H. Bianco-Peled. and M. Davidovich-Pinhas (Eds.), pp. 203-234, Jenny Stanford Publishing, New York, (2015).
33. W.D. Spotnitz, and S. Burks, Hemostats, sealants, and adhesives: components of the surgical toolbox. *Transfusion*, 48, 1502-1516 (2008).
34. L. Ge and S. Chen, Recent advances in tissue adhesives for clinical medicine. *Polymers*, 12, 939 (2020).
35. S. Rathi, R. Saka, AJ. Domb and W. Khan. Protein-based bioadhesives and bioglues. *Polym. Advanced Technol.*, 30, 217-234 (2019).
36. P.P. Spicer and A.G. Mikos, Fibrin glue as a drug delivery system. *J. Controlled Release*, 148, 49-55 (2010).
37. M.M. Eskandari, O.G. Ozturk, H.G. Eskandari, E. Balli and C. Yilmaz. Cyanoacrylate adhesive provides efficient local drug delivery. *Clinical Orthopaedics Related Research*. 451, 242-250 (2006).
38. K.L. Mittal, I.S. Bakshi, and J.K. Narang (Eds.), *Bioadhesives in Drug Delivery*. Wiley-Scrivener, Beverly, MA (2020).

39. J. Woodley, Bioadhesion. *Clinical Pharmacokinetics,* 40, 77-84 (2001).
40. J.W. Lee, J.H. Park, and J.R. Robinson, Bioadhesive-based dosage forms: The next generation. *J. Pharm. Sci,* 89, 850-866 (2000).
41. T.D. Bui, S. Huerta, and I.L. Gordon, Negative pressure wound therapy with off-the-shelf components. *Amer. J. Surgery,* 192, 235-237 (2006).
42. W. Zou, G. Cao, Y. Xi and N. Zhang, New approach for local delivery of rapamycin by bioadhesive PLGA-carbopol nanoparticles. *Drug Delivery,* 16, 15-23 (2009).
43. J.C. Richardson, P.W. Dettmar, F.C. Hampson and C.D. Melia, Oesophageal bioadhesion of sodium alginate suspensions: particle swelling and mucosal retention. *European J. Pharm. Sci* , 23, 49-56 (2004).
44. Z. Zhu, Y. Zhai, N. Zhang, D. Leng and P. Ding. The development of polycarbophil as a bioadhesive material in pharmacy. *Asian J. Pharm. Sci,* 8, 218-227 (2013).
45. T.A. Khan, K.K. Peh, and H.S. Ch'ng, Mechanical, bioadhesive strength and biological evaluations of chitosan films for wound dressing. *J Pharm Pharmaceut Sci,* 3, 303-311 (2000).
46. G.R de Araujo, L. de Oliveira Porfírio, L.A. Silva, D.G. Santana, P.F. Barbosa, C.P. Dos Santos, N. Narain, V.H. Sarmento, R. de Souza Nunes, E. Ting and A.A. Lira. In situ microemulsion-gel obtained from bioadhesive hydroxypropyl methylcellulose films for transdermal administration of zidovudine. *Colloids Surfaces. B,* 188, 110739 (2020).
47. L. Teng, Y. Chen, M. Jin, Y. Jia, Y. Wang and L. Ren, Weak hydrogen bonds lead to self-healable and bioadhesive hybrid polymeric hydrogels with mineralization-active functions. *Biomacromolecules,* 19, 1939-1949 (2018).
48. J.K. Vasir, K. Tambwekar, and S. Garg, Bioadhesive microspheres as a controlled drug delivery system. *Int. J. Pharmaceutics,* 255, 13-32 (2003).
49. K. Pathak and R. Malviya, Introduction, theories and mechanisms of bioadhesion, in: *Bioadhesives in Drug Delivery,* K.L. Mittal, I.S. Bakshi, and J.K. Narang (Eds.) pp. 3-28, Wiley-Scrivener, Beverly, MA (2020).
50. B.V. Derjaguin, Y.P. Toporov, V.M. Muller and I.N. Aleinikova, On the relationship between the electrostatic and the molecular component of the adhesion of elastic particles to a solid surface. *J. Colloid Interface Sci,* 58, 528-533 (1997).
51. E. Tagami, Theory of the interface between immiscible polymers. *J. Chem. Phys.,* 57, 1812-1813 (1972).
52. K.L. Mittal, The role of the interface in adhesion phenomena. *Polym. Eng. Sci.,* 17, 467-473 (1977).
53. W.C. Wake, *Adhesion and the Formulation of Adhesives,* pp. 303-313, Applied Science, London, UK (1976).
54. D. Kaelble and J. Moacanin, A surface energy analysis of bioadhesion. *Polymer,* 18, 475-482 (1977).
55. D. Campion, The influence of structure on autohesion (self-tack) and other forms of diffusion into polymers. *J. Adhesion,* 7, 1-23 (1975).

56. N.A. Peppas and J.J. Sahlin, Hydrogels as mucoadhesive and bioadhesive materials: a review. *Biomaterials*, 17, 1553-1561 (1996).
57. G. Ponchel, F. Touchard, D. Duchêne and N.A. Peppas. Bioadhesive analysis of controlled-release systems. I. Fracture and interpenetration analysis in poly (acrylic acid)-containing systems. *J. Controlled Release*, 5, 129-141 (1987).
58. Y. Huang, W. Leobandung, A. Foss and N.A. Peppas, Molecular aspects of muco-and bioadhesion: Tethered structures and site-specific surfaces. *J. Controlled Release*, 65, 63-71 (2000).
59. M.I. Ugwoke, R.U. Agu, N. Verbeke and R. Kinget, Nasal mucoadhesive drug delivery: background, applications, trends and future perspectives. *Advanced Drug Delivery Reviews*, 57, 1640-1665 (2005).
60. SA. Mortazavi and J.D. Smart, Factors influencing gel-strengthening at the mucoadhesive-mucus interface. *J. Pharm. Pharmacol*, 46, 86-90 (1994).
61. M.E. Imam, M. Hornof, C. Valenta, G. Reznicek and A. Bernkop-Schnürch. Evidence for the interpenetration of mucoadhesive polymers into the mucous gel layer. *STP Pharma Sci.*, 13, 171-176 (2003).
62. H.H. Sigurdsson, T. Loftsson, and C.-M. Lehr, Assessment of mucoadhesion by a resonant mirror biosensor. *Int. J. Pharmaceutics*, 325, 75-81 (2006).
63. F. Madsen, K. Eberth, and J.D. Smart, A rheological assessment of the nature of interactions between mucoadhesive polymers and a homogenised mucus gel. *Biomaterials*, 19, 1083-1092 (1998).
64. C.M. Lehr, J.A. Bouwstra, E.H. Schacht and H.E. Junginger, *In vitro* evaluation of mucoadhesive properties of chitosan and some other natural polymers. *Int. J. Pharmaceutics*, 78, 43-48 (1992).
65. Y. Sudhakar, K. Kuotsu, and A. Bandyopadhyay, Buccal bioadhesive drug delivery—a promising option for orally less efficient drugs. *J. Controlled Release*, 114, 15-40 (2006).
66. C.M. Lehr, Lectin-mediated drug delivery: The second generation of bioadhesives. *J. Controlled Release*, 65, 19-29 (2000).
67. A. Bernkop-Schnürch, Thiomers: a new generation of mucoadhesive polymers. *Advanced Drug Delivery Reviews*, 57, 1569-1582 (2005).
68. L.Q. Wang and F.D. Cui, Chitosan-thioglycolic acid conjugate microspheres: their use for oral delivery of rEPO. *J. Drug Delivery Sci. Technol*, 16, 421-426 (2006).
69. V. Leitner, M. Marschütz, and A. Bernkop-Schnürch, Mucoadhesive and cohesive properties of poly (acrylic acid)-cysteine conjugates with regard to their molecular mass. *Eur. J. Pharm. Sci*, 18, 89-96 (2003).
70. N.V. Shah and R. Meislin, Current state and use of biological adhesives in orthopedic surgery. *Orthopedics*, 36, 945-956 (2013).
71. W.D. Spotnitz, Fibrin sealant: the only approved hemostat, sealant, and adhesive—a laboratory and clinical perspective. *Int. Scholarly Res. Notices* 2014 (2014).

72. S. Tarafder, G.Y. Park, J. Felix and C.H. Lee, Bioadhesives for musculoskeletal tissue regeneration. *Acta Biomaterialia.* 117, 77-92 (2020).
73. M. Mehdizadeh and J. Yang, Design strategies and applications of tissue bioadhesives. *Macromolecular Biosci.*, 13, 271-288 (2013).
74. M. Zhu, K. Wei, S. Lin, X. Chen, C.C. Wu, G. Li and L. Bian, Bioadhesive polymersome for localized and sustained drug delivery at pathological sites with harsh enzymatic and fluidic environment via supramolecular host-guest complexation. *Small*, 14, 1702288 (2018).
75. J. Li and D.J. Mooney, Designing hydrogels for controlled drug delivery. *Nature Reviews Materials*, 1,1-17 (2016).
76. V.A. Kumar, N.L. Taylor, S. Shi, N.C. Wickremasinghe, R.N. D'Souza and J.D. Hartgerink, Self-assembling multidomain peptides tailor biological responses through biphasic release. *Biomaterials* 52, 71-78 (2015).
77. C.H. Lee, J.L. Cook, A. Mendelson, E.K. Moioli, H. Yao and J.J. Mao, Regeneration of the articular surface of the rabbit synovial joint by cell homing: a proof of concept study. *Lancet*, 376, 440-448 (2010).
78. C.H. Lee, J. Hajibandeh, T. Suzuki, A. Fan, P. Shang and J.J. Mao, Three-dimensional printed multiphase scaffolds for regeneration of periodontium complex. *Tissue Eng. Part A*, 20, 1342-1351 (2014).
79. C.H. Lee, S.A. Rodeo, L.A. Fortier, C. Lu, C. Erisken and J.J. Mao, Protein-releasing polymeric scaffolds induce fibrochondrocytic differentiation of endogenous cells for knee meniscus regeneration in sheep. *Sci Transl Med.*, 6, 266ra171 (2014).
80. C.H. Lee, B. Shah, E.K. Moioli and J.J. Mao, CTGF directs fibroblast differentiation from human mesenchymal) stem/stromal cells and defines connective tissue healing in a rodent injury model. *J. Clinical Investigation.*, 120, 3340-3349 (2010).
81. V. Bhagat and M.L. Becker, Degradable adhesives for surgery and tissue engineering. *Biomacromolecules*, 18, 3009-3039 (2017).
82. C.I. Shaffrey, W.D. Spotnitz, M.E. Shaffrey and J.A. Jane, Neurosurgical applications of fibrin glue: augmentation of dural closure in 134 patients. *Neurosurgery*, 26, 207-210 (1990).
83. O. Staindl, Tissue adhesion with highly concentrated human fibrinogen in otolaryngology. *Annals Otology Rhinology Laryngology.*, 88, 413-418 (1979).
84. J.W. Baker, W.D. Spotnitz, and S.P. Nolan, A technique for spray application of fibrin glue during cardiac operations. *Annals Thoracic Surgery*, 43, 564-565 (1987).
85. T.L. Matthew, W.D. Spotnitz, I.L. Kron, T.M. Daniel, C.G. Tribble and S.P. Nolan, Four years' experience with fibrin sealant in thoracic and cardiovascular surgery. *Annals Thoracic Surgery*, 50, 40-43 (1990).
86. Z. Gnjidic, D. Tomac, L. Negovetic and T. Gjinolli, Fibrin sealants in the management of cerebrospinal fistulae. *Biomed Prog* 7, 39-42 (1994).
87. P. Rutgeerts, E. Rauws, P. Wara, P. Swain, A. Hoos, E. Solleder, J. Halttunen, G. Dobrilla, G. Richter and R. Prassler, Randomised trial of single and repeated

fibrin glue compared with injection of polidocanol in treatment of bleeding peptic ulcer. *Lancet*, 350, 692-696 (1997).
88. L. Laine. Endoscopic therapy for bleeding ulcers: room for improvement? *Gastrointestinal Endoscopy*, 57, 557-560 (2003).
89. J. Passage, H. Jalali, R.K. Tam, S. Harrocks and M.F. O'Brien. BioGlue surgical adhesive—an appraisal of its indications in cardiac surgery. *Annals Thoracic Surgery*, 74, 432-437 (2002).
90. T.B. Reece, T.S. Maxey, and I.L. Kron. A prospectus on tissue adhesives. *Amer. J. Surgery*, 182, S40-S44 (2001).
91. J. Passage, R. Tam, M. Windsor and M. O'Brien. Bioglue: a review of the use of this new surgical adhesive in thoracic surgery. *ANZ. J. Surgery*, 75, 315-318 (2005).
92. J.C. Wheat and J.S. Wolf. Advances in bioadhesives, tissue sealants, and hemostatic agents. *Urologic Clinics*, 36, 265-275 (2009).
93. B.P. Lee, P.B. Messersmith, J.N. Israelachvili and J.H. Waite. Mussel-inspired adhesives and coatings. *Annual. Review. Mater. Res* 41, 99-132 (2011).
94. W. Chen, R. Wang, T. Xu, X. Ma, Z. Yao, Chi B.O. and H. Xu. A mussel-inspired poly (γ-glutamic acid) tissue adhesive with high wet strength for wound closure. *J. Mater. Chem. B*, 5, 5668-5678 (2017).
95. B. Hoffmann, E. Volkmer, A. Kokott, P. Augat, M. Ohnmacht, N. Sedlmayr, M. Schieker, L. Claes, W. Mutschler and G. Ziegler. Characterisation of a new bioadhesive system based on polysaccharides with the potential to be used as bone glue. *J Mater Sci Mater Med*, 20, 2001-2009 (2009).
96. A. Dey, P. Bhattacharya, and S. Neogi. Bioadhesives in biomedical applications: A critical review. *Rev. Adhesion Adhesives*, 8, 130-152 (2020).
97. A.K. Geim, S.V. Dubonos, IV. Grigorieva, K..S. Novoselov, A.A. Zhukov and S.Y. Shapoval. Microfabricated adhesive mimicking gecko foot-hair. *Nature Materials*, 2, 461-463 (2003).
98. H. Lee, B.P. Lee, and P.B. Messersmith. A reversible wet/dry adhesive inspired by mussels and geckos. *Nature*, 448, 338-341 (2007).
99. A. Mahdavi, L. Ferreira, C. Sundback, J.W. Nichol, E.P. Chan, D.J. Carter, C.J. Bettinger, S. Patanavanich, L. Chignozha, E. Ben-Joseph and A. Galakatos. A biodegradable and biocompatible gecko-inspired tissue adhesive. *Proc. Natl. Acad. Sci (USA)*, 105, 2307-2312 (2008).
100. I. Singh, G. Devi, B.A. Barik, A. Sharma and L. Kaur. Mucoadhesive pellets for drug delivery applications: A critical review. *Rev. Adhesion Adhesives*, 8, 153-167 (2020).
101. M. Joshi, R. Shankar, and K. Pathak, Bioadhesive nanoformulations—concepts and preclinical studies: A critical review. *Rev. Adhesion Adhesives,* 7, 295-330 (2019).
102. G. Lazovic, M. Colic, M. Grubor and M. Jovanovic. The application of collagen sheet in open wound healing. *Annals Burns Fire Disasters*, 18, 151-156 (2005).

103. Y. R. Carriles, M. Suetel, S. Henze, R.Á. Brito and W.D. Mueller. Electrospun meshes of poly (n-butyl cyanoacrylate) and their potential applications for drug delivery and tissue engineering. *Int. J. Pharm*, 606, 120735 (2021).
104. S. Talebian, I.K. Shim, S.C. Kim, G.M. Spinks, K.L. Vine and J. Foroughi. Coaxial mussel-inspired biofibers: Making of a robust and efficacious depot for cancer drug delivery. *J. Mater. Chem. B*, 8, 5064-5079 (2020).
105. S. Sunoqrot, L. Hasan, A. Alsadi, R. Hamed and O. Tarawneh. Interactions of mussel-inspired polymeric nanoparticles with gastric mucin: Implications for gastro-retentive drug delivery. *Colloids Surfaces. B*, 156, 1-8 (2017).
106. X. Zhang, G. Chen, Y. Yu, L. Sun and Y. Zhao. Bioinspired adhesive and antibacterial microneedles for versatile transdermal drug delivery. *Research*, 1-9 (2020).
107. Y. Li, H. Meng, Y. Liu and B.P. Lee. Fibrin gel as an injectable biodegradable scaffold and cell carrier for tissue engineering. *Sci. World J*, 1-10 (2015).
108. C. Levinson, E. Cavalli, B. von Rechenberg, M. Zenobi-Wong, and S.E. Darwiche. Combination of a collagen scaffold and an adhesive hyaluronan-based hydrogel for cartilage regeneration: A proof of concept in an ovine model. *Cartilage*, 13, 636S-649S (2021).
109. D.B, Horváthy, M. Simon, C.M. Schwarz, M. Masteling, G. Vácz, I. Hornyák and Z. Lacza. Serum albumin as a local therapeutic agent in cell therapy and tissue engineering *Biofactors*, 43, 315-330 (2017).
110. M. Suneetha, K.M. Rao, and S.S. Han, Mussel-inspired cell/tissue-adhesive, hemostatic hydrogels for tissue engineering applications. *ACS Omega*, 4, 12647-12656 (2019).
111. L. Han, M. Wang, P. Li, D. Gan, L. Yan, J. Xu, K. Wang, L. Fang, C.W. Chan, H. Zhang and H. Yuan. Mussel-inspired tissue-adhesive hydrogel based on the polydopamine–chondroitin sulfate complex for growth-factor-free cartilage regeneration. *ACS Appl. Mater Interfaces*, 10, 28015-28026 (2018).
112. H. Cheong, J. Kim, B.J. Kim, E. Kim, H.Y. Park, B.H. Choi, K.I. Joo, M.L. Cho, J.W. Rhie, J.I. Lee and H.J. Cha. Multi-dimensional bioinspired tactics using an engineered mussel protein glue-based nanofiber conduit for accelerated functional nerve regeneration. *Acta Biomaterialia*, 90, 87-99 (2019).

2

Global Industry Development and Analysis of Adhesives for Biomedical Applications

Muhammed Yusuf Kandur[1], R. Hemamalini[2] and Ebru Toksoy Öner[1*]

[1]*Industrial Biotechnology and Systems Biology Group, Department of Bioengineering, Marmara University, Istanbul, Turkey*
[2]*R&D Wing, Industrial Research & Development Unit, Indian Institute of Technology Delhi, Hauz Khas, New Delhi, India*

Abstract
The know-how gained on adhesives together with the rising awareness and preference for environmentally friendly adhesives have allowed bioadhesive research to gain immense momentum in the last two decades. Adhesives produced from living organisms or those compatible with living tissues fall within the ambit of bioadhesives. As per Scopus, of the 4900 patents filed in the realm of 'bioadhesives' in the last four decades, 4200 have been filed in the last two decades alone. As a result, a plethora of new adhesives have been developed. Some of these adhesives are obtained from natural sources, while others are synthesized from natural or synthetic raw materials. Adhesives so produced have applications in various industries including, but not limited to, paper and pulp, cosmetics, packaging, and healthcare. North America leads the global bioadhesives market in terms of production and consumption.

Keywords: Adhesives, biomedicine, biomedical adhesive market value, biomedical adhesive market trends

2.1 Introduction

Adhesion is a physical phenomenon in which there is an attractive force between dissimilar molecules, unlike cohesion where similar molecules

**Corresponding author*: ebru.toksoy@marmara.edu.tr

K.L. Mittal and S. Neogi (eds.) Adhesives in Biomedical Applications, (25–46) © 2023 Scrivener Publishing LLC

attract each other. An adhesive is usually a polymeric substance that helps bond together two different materials by virtue of an attractive force that develops between the adhesive and the materials being bonded. These surfaces may be held together by any of the four types of forces: surface and field forces, material bridges, mechanical interlocking, or suction forces [1]. The attractive forces, in turn, depend on the chemistry of interacting surfaces which may be ionic interactions, dipole-dipole forces, hydrogen bonding, surface chelate complexes, or covalent bonds depending on the chemical composition of the surfaces [2].

Adhesion had existed in nature even before mankind recognized the phenomenon or its underlying mechanism and although it has been studied for years, what is known for sure is that the exact mechanism of adhesion varies depending on the surface and/or adhesive involved. For instance, a gecko or frog that can stick to dry or wet surfaces respectively, the mechanical interlocking of hooked fruits to animal bodies for seed dispersal, wings interlocked on the body of beetles, the suckers of an octopus, or a climbing fish illustrate adhesion in nature. Even the presence of dew drops on leaf surfaces is an example of adhesion which is crucial for influencing the rainwater interception by leaves before reaching the soil [1, 3–5]. Water-repellent lotus leaves and water-walking aquatic insects have inspired the development of self-healing adhesives and other technologies such as aquatic robots and small boats [6, 7].

The common natural sources of adhesives include plant resins, animal glues, and minerals [8]. Adhesives have been used in shoe-making,

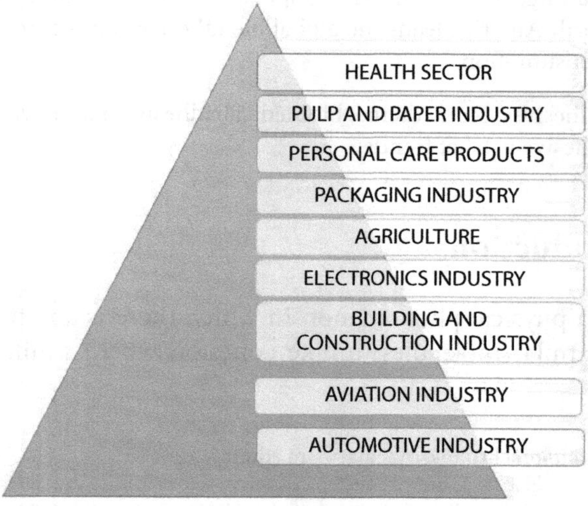

Figure 2.1 Applications of adhesives [9].

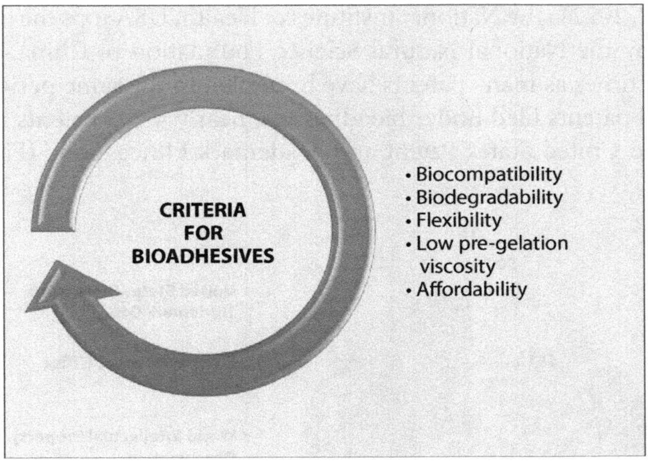

Figure 2.2 Criteria that need to be met by bioadhesives for biomedical applications [13].

packaging, furniture, plastics, automotive, electronics, and construction industries for many decades. Medicine and aerospace are the latest entrants to this list. In aerospace, adhesives afford a significant advantage over welding and bolting [9] [Figure 2.1].

In medicine, bioadhesives are now preferred over conventional surgical procedures such as sutures and staples. Currently, bioadhesives in the market include synthetic, natural, and biomimetic adhesives and are used in wound healing, tissue regeneration, smart drug delivery systems, cancer treatment, and wearable electronics [10–14]. For clinical applications, various aspects need to be considered for the selection of an adhesive. These include biocompatibility with living tissues, easy biodegradability, and most importantly affordability, among others. The complete list of criteria is shown in Figure 2.2 [13].

This chapter provides an overview of developments that have taken place in the field of bioadhesives globally over the years with respect to research, sources, specific applications, products, and technological developments. An attempt has been made to cover the historical aspects leading to the latest accomplishments under each of these headings for bioadhesives.

2.2 Research Landscape of Bioadhesives

A Scopus search for the keyword 'bioadhesives' yielded over five hundred articles between 1978 and 2022 (as per keyword search done on

January 17, 2022). The National Institutes of Health, USA tops the list, closely followed by the National Natural Science Foundation of China. However, nearly ten times as many patents have been filed in the same period. Of the total 4,985 patents filed under bioadhesives, nearly 4,200 patents have been filed in the United States Patent and Trademark Office alone [Figure 2.3].

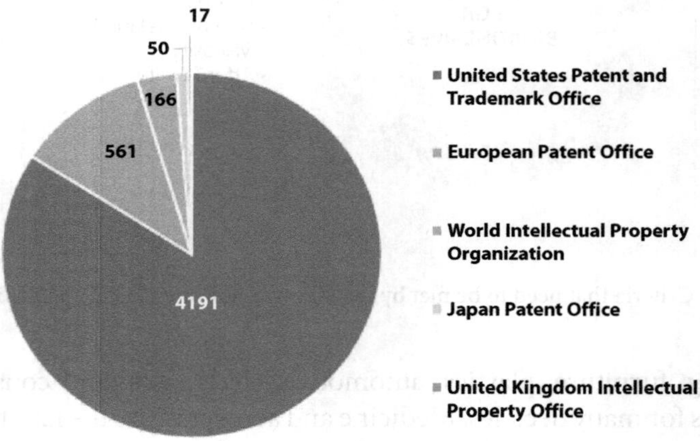

Figure 2.3 Agency-wise distribution of the number of patents filed under 'bioadhesives' between 1973-2022 as per Scopus (https://www.scopus.com/) Accessed on January 17, 2022 for the keyword 'bioadhesives'.

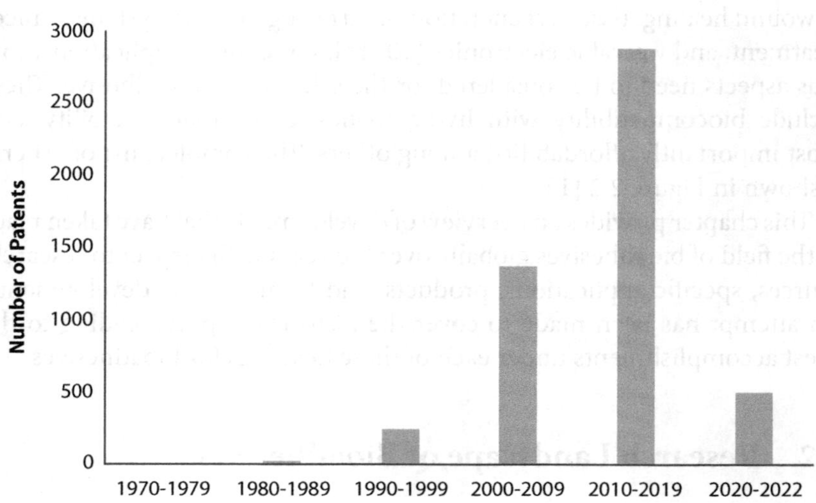

Figure 2.4 Number of patents filed under 'bioadhesives' in each decade. Scopus (https://www.scopus.com/) Accessed on January 17, 2022 for the keyword 'bioadhesives'.

In the current decade, 491 patents have already been filed at the time of writing this chapter since 2020 [Figure 2.4] [Scopus (https://www.scopus.com/) search on January 17, 2022, for the keyword, 'bioadhesives']. Aside from the filed patents, market forecasts can also give hints about the importance of such adhesive materials and trends the manufacturers tend to follow. According to a research report performed by IMARC Group, bioadhesive market value has already surpassed US$ 2 billion and is expected to reach US$ 4.81 billion [14].

2.3 Sources of Bioadhesives for Biomedical Applications

The most common application of bioadhesives in the biomedical field is in surgery. As surgical sealants, bioadhesives offer better alternatives to the conventional surgical closing methods such as sutures, wires, and staples which are invasive, often cause tissue damage resulting in inflammation, infection and even mortality. In this section, the conventional sources of bioadhesives, as well as the latest additions, have been covered under the broad headings of natural, synthetic, and biomimetic adhesives. While different authors approach the classification of bioadhesives in different ways, this approach seemed the simplest while affording maximum flexibility for covering the latest developments in the field.

2.3.1 Natural Bioadhesives

Substances derived from natural proteins or carbohydrates having adhesive property are classified under natural adhesives. These adhesives are prepared by chemical cross-linking of natural polymers and are the most biocompatible adhesives.

2.3.1.1 Protein Bioadhesives

2.3.1.1.1 Fibrin

A fibrin-based bioadhesive contains human pooled plasma liquid or autogenous blood of the patient and has been effectively used as a sealant, haemostat as well as adhesive. The human pooled plasma is the source of fibrin and thrombin. It is especially effective in treating capillary and venous bleedings (which tend to be slower than arterial bleeding). This type of adhesives hasten the clotting process under dry conditions as it can take some time for the fibrin to form a clot. In cases where heavy bleeding is encountered, fibrin sealants are applied in combination with a carrier sponge to prevent the washing away of the liquid fibrin [15]. Crossseal, TachoSil, Tisseel, Evicel and Hemaseel are some commercial human fibrin-based bioadhesives [11]. Fibrin-based sealants meet the biocompatibility and biodegradability criteria but suffer from disadvantages of poor mechanical strength, poor adhesion under wet conditions, risk of viral transmission, and complex production. These problems are overcome by Vitagel from Stryker and the CryoSeal system from Thermogen, which produce autologous fibrin glue from the patient's own plasma [13]. Although, fibrin-based adhesives tend to dry and adhere slower on the surface, they can also cause swelling and tissue compression so they should be used with care. FloSeal™ and Proceed™ are two collagen-based adhesives approved for use in the US. Condress® is currently the most widely used wound dressing; it is available either in gel form or in spray form [11, 16]. In 2018, the global fibrin sealants market value exceeded US$ 700 million and is expected to pass US$ 1,150 million by 2026 [17].

2.3.1.1.2 Gelatin

Gelatin is a relatively new bioadhesive obtained through chemical or physical degradation of collagen. It is both highly biocompatible and biodegradable and it exhibits low antigenicity. This adhesive traps coagulating agents in its naturally occurring interstices within its helix structure as it adheres to the wound. It naturally induces platelet adhesion and aggregation and activates coagulation factors. Gelatin has advantages of cost-effectiveness, ease-of-preparation, modification, and being completely resorbable and its derivatives have been widely used in biomedicine, primarily as adhesive films during corneal surgeries. Gelatin hydrogels have also been found to integrate well with cartilage tissue and found to be effective in osteoarthritis. Addition of gelatin tends to shorten healing time and process and enhances elastic modulus and the crosslink density of adhesive network. When used with microgels, gelatin improves the bulk mechanical properties, one of which is adhesiveness. Recent trials with gelatin involve on-site forming gels and trigger-sensitive adhesives. The adhesive strength of this polymer can even be manipulated by altering its substitution degree alone meaning its adhesive property can be altered or additional benefits can be implemented easily [18–20]. Gelatin-based adhesives had a market value of US$ 193 million in 2019 and is expected to surpass US$ 310 million within the decade [21].

2.3.1.1.3 Albumin

Albumin is abundant, intrinsically biocompatible, biodegradable, non-immunogenic and non-toxic. Because of these properties and its unique capability to bind to a variety of compounds, it is one of the most ideal natural adhesives for use in biomedical applications. It can also reduce non-specific protein adsorption while enhancing cell adhesion and proliferation. A product called BioGlue® contains bovine albumin with glutaraldehyde which reacts with the lysine residues of albumin allowing it to form a strong sealant within 20-30 seconds which requires the use of gauze pads to prevent unwanted contact [11]. However, this comes at a cost. The gel may compress the applied tissue which can limit the blood vessel growth and cause nerve damage, adhesive embolism, pseudoaneurysm and myonecrosis. A few of these side-effects have been offset in a product called ProGel® by substituting bovine albumin with human albumin and adding modified poly (ethylene glycol) to the polymer. There have been numerous wound healing studies that used albumin to replace sutures, prevent air and/or liquid leakages. PREVELEAK® is a surgical adhesive comprised of a mixture of albumin and polyaldehyde. Currently, this double-barrelled syringe system becomes a gel within 10-15 seconds after application. It is efficient in sealing leakages but can cause minor discomfort. It has been developed for cardiovascular surgical procedures and its effectiveness in sealing has been proven [22, 23]. The market value of global albumin was around US$ 4.3 billion in 2018 and is expected to pass US$ 6.5 billion by 2026 [24].

2.3.1.2 Carbohydrate-Based Bioadhesives

2.3.1.2.1 Chitosan

Chitosan, a natural polysaccharide derived from chitin found in crustacean shell and fungal cell wall, is a widely used natural bioadhesive. It is inexpensive, biocompatible, non-toxic, biodegradable, accelerates wound healing and is naturally mucoadhesive and antimicrobial. Its natural positive charge allows chitosan to attract blood cells which hastens the clotting process. These properties have allowed chitosan to be used in oral, ocular, nasal, parenteral and transdermal drug delivery, nano-drug delivery and haemostasis. Chitosan-based bioadhesives are also used in combination with other bioadhesives resulting in better adhesiveness and improved anti-microbial property, which makes chitosan-based polysaccharides sought-after bioadhesives. Films, sponges and hydrogels obtained from chitosan exhibit excellent bioavailability and improved mucoadhesive property. Once placed on an open wound, thin chitosan membranes can adhere to fibroblasts which allows them to enhance the proliferation process along with tissue regeneration. Chitosan adhesives can be used on both soft and hard tissues making them good candidates for a wide array of possible applications. Currently, Hem Con™ and Celox™ are the most common chitosan-based products used for sealing. Hem Con™ is a highly effective chitosan-based bandage used by military personnel. Celox™ is a powder formulation used to seal intra-oral lacerations. Surgilux is a more sophisticated, light-induced, chitosan-based biomedical product. The mixture is activated with a laser beam allowing it to bind target tissues together [18, 25, 26].

2.3.1.2.2 Dextran

Due to its glucose-based structure, polymers such as dextran are biodegradable, biocompatible, and exhibit low cytotoxicity making them highly important in biomedicine. Dextran alone has been used in medicine for more than a half-century as an anti-thrombolytic agent and as a carrier agent. Numerous studies have not only shown that dextran can be a good bioadhesive, but its properties can be altered with the addition of certain chemicals. Though dextran-based adhesives can be used as skin adhesives that have more desirable properties than fibrin-based adhesives, dextran has been mostly studied in the field of smart drug delivery systems. It not only degrades controllably but also can be shaped into microspheres, nanoparticles, and injectable hydrogels with inherent antibacterial property. The main reason why dextran-based adhesives do not dominate the market is their instability in aqueous environments [11, 27].

2.3.1.2.3 Cellulose

Cellulose is not only one of the most widely found and studied polymers, but it is also one of the most adhesive. This renewable and natural polymer can be found in plant cells as well as various microorganisms and can be used in different forms to obtain a biomedical adhesive. Bacterial cellulose from *Acetobacter xylnium* is a promising polymer that has been already used in several biomedical applications for both soft and hard tissues. Cellulose nanofibers, which are basically cellulose broken down into nanoscales, are becoming more widely investigated than cellulose itself. Due to its high hydroxyl surface groups, microbial cellulose tends to be more hydrophilic, biodegradable, and is more prone to modifications. The bacterial cellulose synthesis pathway also allows to work with higher purity in larger amounts of cellulose compared to cellulose obtained from plants. Bacterial cellulose exhibits unique nanostructure, high water holding capacity, high crystallinity, high mechanical strength, and high degree of polymerization. By physically breaking down cellulose into smaller fragments, one can improve physical properties such as adhesive characteristic, physical stability, and adsorption. This process is marginally easier for bacterial cellulose since bacteria tend to produce this polymer in nano-fibrillar form [28]. Once the cellulose is obtained, it can be modified through different pathways which can give the polymer additional beneficial properties. Cellulose acetate, for example, can be used in bone and dental tissues to ease the integration process of an implant. Cellulose sulfate, carboxymethyl cellulose, and cellulose nitrate are some of the studied cellulose modifications in the biomedical area, each with its own unique properties. A recent study used bacterial cellulose to obtain an adhesive for oral cavity applications by combining it with photo-activated carbine-based bioadhesive. This novel composite material seemed to improve the adhesive strength as well as the stability of the adhesive in aqueous envionment. The fact that this mucoadhesive can be triggered through UV radiation makes it a highly functional product. Synaero™ is a surgical hemostatic gel that not only stops bleeding but also allows air passage post-surgery which is critical to bypass the complications encountered with other bioadhesive gels such as albumin [11, 29].

2.3.2 Synthetic Adhesives

2.3.2.1 Poly(ethylene glycol)

Poly(ethylene glycol) (PEG) is a hydrophilic, mucoadhesive, non-ionic and biocompatible synthetic polymer. The fact that it does not trigger a response from the immune system makes PEG a preferred compound in biomedicine. It has been so widely studied and in fact PEG is currently used in numerous commercial biomedical adhesives such as DuraSeal™, FocalSeal®, CoSeal®, AdvaSeal and Systagenix both as the primary adhesive and as a complementary one. Adhesives obtained from PEG have a wide range of applications. These can be tailored to be trigger sensitive, various molecules can be bound to PEG or they can be applied as a combination of different components so that they can have specified properties. This increases the versatility of PEG adhesives in biomedicine for leakage prevention, drug delivery, tissue regeneration, and structural support for cell growth [16, 30]. The market value of PEG biomedical adhesives is estimated to be around US$ 3.5 million and is expected to be more than US$ 5 million by the year 2028.

2.3.2.2 Polycyanoacrylate

Polycyanoacrylate-based bioadhesives are commonly used in the cosmetics industry, vascular, retinal and tendon repairs, and haemostasis owing to their strong bonding. They constitute a majority of the Food and Drug Administration (FDA) USA approved bioadhesives. The liquid monomers of this bioadhesive undergo catalyst-independent polymerization. Despite its prevalence, this adhesive is mainly used for exterior applications due to the potential toxicity encountered during its hydrolysis. Formaldehyde and alkyl cyanoacetate are some of the toxic by-products that can cause harmful side effects such as histological alterations, necrosis, inflammation, asthma, and dermatitis [28, 31]. Dermabond® is a widely used commercial polycyanoacrylate bioadhesive used as surgical sutures and as dressing for superficial wounds. Approved at the end of the 20th century, this product has been successfully used for surgical wound closures ever since. Histoacryl®, Indermil®, Omnex® and Surgiseal® are a few other commercial cyanoacrylate-based adhesives. A clear advantage offered by polycyanoacrylates is that they can be self-healing because the monomers polymerize through anionic polymerization in presence of hydroxide. Though they may seem a good candidate for bioadhesive applications their adhesive property is not the strongest. In some instances, these adhesives were found to be actually weaker than sutures. Once dried, cyanoacrylate-based adhesives tend to form a brittle residue which is not ideal for wound dressing. If these obstacles can be removed, polycyanoacrylates have a good potential to be commercially used [18, 32–34].

2.3.2.3 Chondroitin Sulfate

Chondroitin sulfate is an interesting polysaccharide that is already present in tissues such as cartilage. It consists of alternating N-acetyl-galactosamine and glucuronic acid monomers which are sulfated at different positions. Moreover, this polymer is amenable to additional modifications resulting in even better adhesives. Though this adhesive is relatively new and there are not many commercial adhesives using this polymer, scientific papers published recently suggest that this adhesive has a promising potential. Reacting chondroitin sulfate with glycidyl methacrylate allows the production of photo-inducible hydrogel, whereas modifying this natural biopolymer with sodium periodate results in reactive aldehyde groups which can form cross-linked networks with amine groups. Using chondroitin sulfate hydrogels as an adhesive layer between the native cartilage tissues and implemented biomaterials tends to result in strong adhesion over a long period of time

[35, 37, 38]. In a recent study performed by a Chinese research group, chondroitin sulfate was used to obtain a porous wound dressing which turned to soft hydrogel to seal wounds after contacting water [36].

If compared with each other, it can be seen that synthetic adhesives, which are easier to obtain in bulk processes, tend to present a stronger adhesion to the surface they interact with and are more susceptible to modification. However, they tend to lack the inherent biocompatibility which can result in low biosorption and cytotoxic reactions. Natural adhesives, on the other hand, are advantageous compared to the synthetic adhesives since they mostly are biocompatible and biodegradable. On the other hand, they can be difficult to work with in terms of quantity and purity not to mention the possibility to act as a vector depending on the source [38]. The value of Chondroitin Sulfate in the global market size was estimated as US$ 1.25 billion in the year 2022. This was partly attributable to the rise in cases of osteoarthritis globally. According to one estimate, this market has been projected to have a CAGR of 3.6% and is expected to attain a US$ 1.72 billion by the year 2031 [39].

2.4 Biomedical Applications of Bioadhesives

Adhesives can be utilized in numerous different industries, but biomedical industry may be one of the most promising. The body of complex organisms such as humans is a mixture of soft and hard tissues; some of which are stationary, and some need to contract and release constantly. This means that a perfect adhesive for one application may not be applicable at all for another and that is why an extensive knowledge about not only the regional properties of the body but also about the nature of the potential adhesive materials and precursors is required; making bioadhesives a multidisciplinary research area where experts from different areas of science as well as manufacturers and physicians come together to obtain a novel product. In the context of biomedical applications, the most popular use of adhesives is sealants, an alternative to surgical sutures. These materials not only decrease the infection rate as also provide an ease of practice when it comes to control bleeding and wound closing as well as improving the healing process, reducing the invasiveness and trauma [2, 29]. Along with providing structural support and enhancing the healing process, adhesives are gaining notice in more modern applications in areas such as smart drug delivery systems and diagnostics. The most common applications of bioadhesives in medicine and clinical practices are summarised in Figure 2.5. Based on years of research performed, different adhesives with

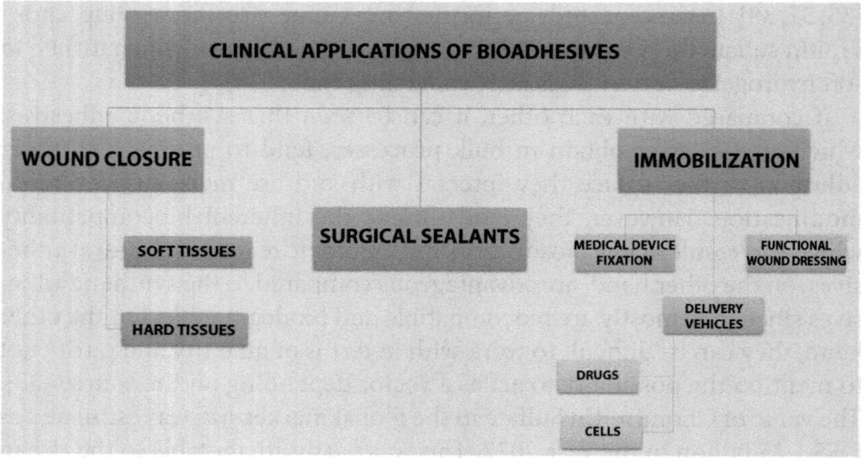

Figure 2.5 Bioadhesives in clinical applications [32].

additional benefits can be designed; such materials not only bond tissues together, they can also improve the healing process and deliver pharmaceutical active molecules. They can heal themselves or alternatively can carry healing factors and/or cellular structures whilst being stimuli-responsive allowing the production of multifunctional adhesives. For obvious reasons, many recent bioadhesive products have focused on obtaining multifunctional bioadhesives rather than traditional routes. These adhesives may inherently have additional beneficial properties due to the chemical composition or by design of the adhesive. Alternatively additional pharmaceutical molecules or chemicals can be integrated into the formulation. Although they can be encountered in a myriad of application areas, adhesives in biomedicine are mainly used for sealing leakage, wound healing and for targeted drug and/or cell delivery, where tissue adhesion becomes as important as the carried material itself [32, 40].

Surgical adhesives have a wide range of applications. A few typical applications are listed in Figure 2.6. It is easy to imagine the impact of polymeric bioadhesives on soft tissues and wound closure, but they are just as prevalent and impactful on hard tissues such as bone and dental tissues. Dentures, for example, are an application area where adhesives have been used since the early 20th century. It is known that the correct adhesive should not only allow the patient to eat, chew and bite properly but it should not cause any type of irritation or discomfort making the adhesive selection a crucial step. Although adhesives are not the only method for securing an implant, it is generally accepted as the most ideal [41, 42]. Bone fractures, similar to denture applications, is a fascinating area for application of bioadhesives since

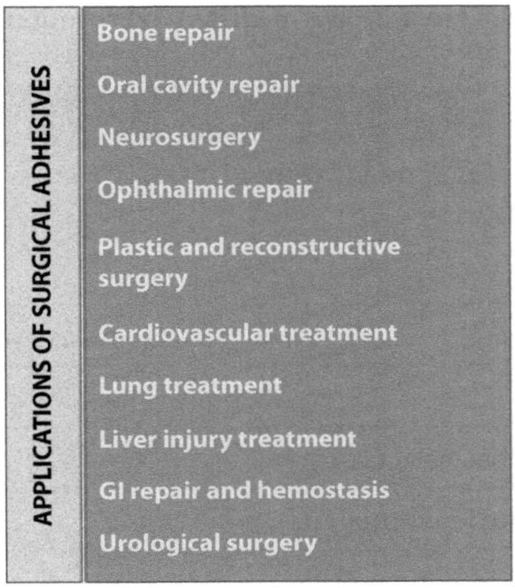

Figure 2.6 Applications of surgical adhesives [10].

it will reduce the operation and healing duration for broken and/or fractured bone tissue through a less invasive method compared to traditional operations. Bone fractures are among the most common injuries and their treatment can be considered somewhat outdated. The healing process of a bone tissue consists of four phases: I) Hematoma Formation, II) Early Inflammation, III) Repair, and finally IV) Remodelling phase. The aim of surgical intervention is to bring the fractured pieces together to hasten these processes. After the pieces are brought together the region in question is stabilised through casting or implants. Though it is widely practiced, it is nowhere near practical or cheap. There is also the possibility for an infection or an imperfect healing even with the guidance of stabilisers such as casting, making bone adhesives even more patient-friendly. These materials potentially can bypass the risk of infection, the physical stress applied on the damaged area and high costs due to secondary interventions if they can be biocompatible, stabilize the fractured bone tissue and improve healing [18].

Aside from bringing hard and soft tissues together to hasten the healing process, bioadhesives can be used to seal leakages and carry important loads such as immobilised cells and therapeutic agents. Vast application areas of biomedicine combined with latest developments such as wearable technology or trigger-sensitive systems allow bioadhesives to be used in a plethora of novel applications such as triggerable stem cell delivery which happens to be extremely effective on mucus covered tissues [29, 40]. Figure 2.6 lists applications of surgical adhesives.

Lately multifunctional bioadhesives that can bring a soft and/or hard tissue together whilst providing additional beneficial properties to the patient

Table 2.1 Advantages of bioadhesive-mediated interventions over conventional methods [41].

Bioadhesives	Conventional methods
Natural and synthetic compounds with tissue adhesion property	Staples, sutures and wires
Non-invasive	Invasive
Less damaging to tissues	Damage to tissues is inevitable
Easy to use	Requires skill
Possess inherent antibacterial property	-
Possess inherent antioxidant property	-
Possess inherent anti-inflammatory property	-
Are self-healing	-
Injectable	-
Lesser risk of leakage	High incidence of post-surgical leakage

have been gaining interest. These properties usually are either drug delivery or self-healing to improve the healing process all together, but other functionalities such as providing structural support can be implemented by adding the appropriate components. The fact that these multifunctional bioadhesives can be designed to be a specific trigger sensitive (i.e. pH, temperature, etc.) makes them even more attractive [40]. A more comprehensive comparison of bioadhesives with conventional methods can be found in Table 2.1.

2.5 Latest Industrial Developments

The biomedical adhesives market has come a long way from the first time they were introduced as a concept in the 20th century. Market value of global bioadhesives was US$ 2.18 billion in 2021 and is expected to have a compound annual growth rate of nearly 14% during 2022-2027. Some of the companies that dominate the market include Henkel Corporation, Ashland Inc., EcoSynthetix, Dow Chemical Company, and Adhesives

Research Inc. Biomedical applications is only a fraction of the true capabilities of bioadhesives, in fact most of the adhesives developed are used in paper and packaging industries. A recent trend in adhesive market is to produce bio-adhesives since they tend to be more environmentally-friendly and sustainable [14].

Smart drug delivery is an exciting field of application for adhesives where they can be used to deliver specific drugs, cells or pharmaceutical active molecules to specific sites of interest. Tissues with mucous membrane particularly are great candidates for biomedical applications of adhesives (since adhesives not only tend to mimic the natural state of these tissues but these tissues also tend to be rich with vessels that can carry excess material outside the area) [43].

Table 2.2 includes a few examples of novel bioadhesives that major producers tend to focus on and lists the most widely used bioadhesives.

2.5.1 Novel Adhesives

Tsang *et al.* reported the case of a bacterium, *Caulobacter crescentus* which exudes a glue that helps bind it very strongly to different surfaces. Interestingly, this bacterium forms biofilms inside catheters, ship hulls, water pipes and other submerged surfaces [44].

Although bioadhesives are usually classified into two different categories, recently a new type of bioadhesives has emerged: biomimetic adhesives. These adhesives are considered to be the next generation in adhesive technology. As the name implies, these adhesives take advantage of the adhesive systems already found in nature which have been perfected after thousands of years of evolution. Most common types of biomimetic adhesives used are inspired from mussels, geckos, different species of frogs and worms. These adhesives are not only "field-tested" over millions of years of evolution, but they also can be obtained from wet environments such as damp caves or subaquatic environments which is exactly one of the challenges some of the modern adhesives have trouble overcoming. Most modern adhesives either break down or perform poorly within soft tissues where high water content is present. Natural aquatic adhesives, on the other hand, are built to perform effectivelyin high water containing environments [30, 38, 45].

Mussel adhesive proteins (MAPs) are one of the most extensively studied materials. When mussels anchor themselves to surfaces, they secrete a thread-like adhesive called byssus, which anchors them to the target surface even under mechanical stress. Byssus is composed of 20-30 different types of MAPs that can adhere to many surfaces, including stone, metal, and glass. Tyrosine, which is one of the proteins comprising MAPs, goes through modifications to form 3,4-dihydroxy-phenylalanine (DOPA). The catechol found in DOPA plays a crucial role in the adhesive property of MAPs because it can utilize hydrogen bonding, hydrophobic interactions and covalent bonding; however, it cannot be mass produced or extracted in high yields. Although, mussel-inspired adhesives are one of the most studied sealants not on its own but rather as a modification agent which gives different properties to different adhesives. These adhesives are not only biocompatible, biodegradable, elastic but they can also be reversible. There have even been some studies that used mussel-inspired injectable adhesives for gastrointestinal tract [11, 37].

Gecko-mimetic adhesives are another widely practiced and well-known class of biomimetic sealants. These small lizards are able to climb vertical surfaces and walk and rest upside down due to the millions of tiny hairs located on their feet. These nanoscaled hairs provide enough adhesion to allow geckos attach and detach from surfaces with ease and control. Inspired by this property scientists were able to obtain polyimide nano-pillars which proved to be a highly effective adhesive, as long as it stayed dry. However, combining this adhesive system with the above-mentioned DOPA not only increases its adhesion strength but also allows it to be effective under wet conditions [16, 36].

Over the past few years, numerous biomedical adhesives with varying properties have been developed. Whether these are naturally sourced or synthesised, their adhesive property needs to be verified through different trials. These trials can be tensile test where the adhesion strength is measured through a pull-off mechanism, or it can be a lap shear test. In lap shear test the adhesion strength of the product is tested using of two sheets with the adhesive in between where one of them is pulled apart parallel to the interface. Alternatively, the adherends can be pulled off from each other. Biocompatibility and biodegradability are also two important parameters to be tested when dealing with bioadhesives since these materials will most likely interact with live tissues whether the environment is considered wet or dry. Also, different tissues can exert different mechanical and chemical stresses making it impossible to obtain an all-purpose adhesive/sealant for biomedical applications. Table 2.2 lists some of the commercially available biomedical adhesives [46, 47].

Various adhesives are being tested for different types of applications every year, not just for medicine but also for construction, chemical applications, etc. Considering their wide range of applications, finesse, and easily modifiable properties, bioadhesives have a potential to dominate multiple markets. Because different applications require different

Table 2.2 List of commercially available bioadhesives [23, 37, 48].

Commercial name	Supplier	Main adhesive
Crosseal	OMRIX	Fibrinogen
Vitagel	Stryker	Collagen
ProGel	NeoMend	Albumin
BioGlue	CryoLife	Albumin
Dermabond	Ethicon	Cyanoacrylate
Actamax	Actamax Surgical Material LLC	Dextran
Focal Seal-L	Focal Inc.	PEG-co-poly(lactic acid)/ poly(trimethylene carbonate)
HemCon	Bandage Pro	Chitosan
Adherus	Hyperbranch	PEG/branched polyethylenimine
PeriAcryl	GluStitch Inc.	Cyanoacrylate
DuraSeal	Intergralife	PEG
FloSeal	Baxter	Gelatin/Thrombin
Gelfoam	Pfizer	Gelatin
TISSEL	Baxter Healthcare Corp.	Human Plasma
EVICEL	OMRIX	Human Plasma
Artiss	Baxter Healthcare Corp.	Human Plasma
TachoSil	Baxter Healthcare Corp.	Fibrinogen; Thrombin coated collagen
Evithrom	Johnson & Johnson	Thrombin
EVARREST	Ethicon, Johnson & Johnson	Fibrinogen; Thrombin
PreveLeak	Mallinckrodt Pharma IP Trading DAC	Albumin
Silvercel	Systagenix	Alginate; Carboxymethyl cellulose; Silver coated nylon fibers
Sorbsan	Aspen Medical	Calcium salt of alginic acid
CoStasis	Haemacure	Collagen; Calcium chloride; Thrombin
Surgiflo	Ethicon, Johnson & Johnson	Gelatin; Thrombin

properties from bioadhesives, there cannot be a "perfect bioadhesive" that can be used for every application. This may seem like a disadvantage at first sight, however it allows different bioadhesives to expand into different markets. Irrespective of whether they are obtained from natural polymers, synthetic polymers or proteins, bioadhesives are sure to be critical in scientific breakthroughs for different industries. Adhesives obtained from natural sources (directly or by mimicking) are attracting more and more interest from both industry and academia. Recently industrial producers have started to focus on adhesives obtained from or mimicking the adhesive structures found in nature. Marine organisms are particularly rich in terms of adhesives suitable for biomedical applications since their adhesives are already biocompatible and suitable for wet environments. Mussels, sponges, urchins, squids, starfish and jelly fish are only some of the aquatic organisms with different adhesive precursors, adhesive materials and adhesion mechanisms that can be utilized by the market [11, 49]. The fact that some regulatory bodies such as FDA or US Environmental Protection Agency (EPA) favour naturally sourced adhesive materials over synthetic ones is another driving force in this shift [50].

2.6 Summary

As more and more adhesives are being discovered and researched it is not difficult to foresee how pivotal they will become in many industries. Considering how widespread adhesive options are available, it is easy to assume that many of the present commercial adhesives will be irreplicable in various markets. Currently some of the most widespread industries that adhesives have become irreplaceable are furniture, construction, clothing, and packaging. With the recent applications of bioadhesives as an alternative to surgical sutures and tissue bonding capability more and more alternatives are being investigated. Considering how much tissues can vary in terms of characteristics it is important to select the correct material in tissue engineering applications. Though biomedical applications of adhesives are somewhat limited to wound closure and minor tissue engineering applications for the time being, it is not unrealistic to assume that they will shape the future of medicine by affecting the way scientists and physicians look at smart drug delivery systems, cosmetics, and tissue engineering.

Acknowledgment

The authors wish to thank Marmara University for the financial sources provided.

References

1. Z. Gu, S. Li, F. Zhang, S. Wang, Z. Gu, S. Li, S. Wang and F. Zhang, Understanding surface adhesion in nature: a peeling model, *Advanced Science*, 3, 1-13 (2016).
2. S. Magalhães, L. Alves, B. Medronho, A.C. Fonseca, A. Romano, J.F.J. Coelho and M. Norgren, Brief overview on bio-based adhesives and sealants, *Polymers*, 11, 1685 (2019).
3. H. Wang, H. Shi, Y. Li and Y. Wang, The effects of leaf roughness, surface free energy and work of adhesion on leaf water drop adhesion, *PLoS One*. 9, 1-10 (2014).
4. D.J. Gardner, Theories and mechanisms of adhesion, in: *Handbook of Adhesive Technology, 3rd Edition*, A.Pizzi and K.L. Mittal (Eds.), pp. 3-18, CRC Press, Boca Raton (2018).
5. D.E. Packham, The mechanical theory of adhesion, in: *Handbook of Adhesive Technology*, second edition, A.Pizzi and K.L. Mittal (Eds.), pp. 69-93, CRC Press, Boca Raton, FL (2003).
6. Q.F. Guan, Z.M. Han, Y. Zhu, W.L. Xu, H. Bin Yang, Z.C. Ling, B.B. Yan, K.P. Yang, C.H. Yin, H. Wu and S.H. Yu, Bio-inspired lotus-fiber-like spiral hydrogel bacterial cellulose fibers, *Nano Letters* 21952–21958 (2021). https://doi.org/10.1021/acs.nanolett.0c03707
7. X.Q. Kong, J.L. Liu, W.J. Zhang and Y.D. Qu, Load-bearing ability of the mosquito tarsus on water surfaces arising from its flexibility, *AIP Advances* 5, 037101-1-037101-9 (2015).
8. R. Ronquillo, *Overview of Adhesives - Types, Pros and Cons, and Selection Considerations*, (n.d.) https://www.thomasnet.com/articles/adhesives-sealants/overview-of-adhesives/ (accessed January 30, 2022).
9. Dinte, E. and Sylvester, B., Adhesives: *Applications and Recent Advances, in: Applied Adhesive Bonding in Science and Technology*, Halil Özer (Ed.), IntechOpen (2018). https://www.intechopen.com/chapters/57763
10. D.S. Shah, B.K. Lawson and M. Yaszemski, Description and definition of adhesives, and related terminology, in: *Biomaterials Science*, fourth edition, W.R. Wagner, S. E. Sakiyama-Elbert, G. Zhang and M. J. Yaszemski (Eds.) pp. 1181–1198, Academic Press, Cambridge, Massachusetts, USA (2020).
11. L. Ge and S. Chen, Recent advances in tissue adhesives for clinical medicine, *Polymers (Basel)*. 12, 1-22 (2020).

12. Z. Ma, G. Bao and J. Li, Multifaceted design and emerging applications of tissue adhesives, *Advanced Materials* 33, 1-29 (2021).
13. K. Zheng, Q. Gu, D. Zhou, M. Zhou and L. Zhang, Recent progress in surgical adhesives for biomedical applications, *Smart Mater. Medicine*, 3, 41–65 (2022). https://doi.org/10.1016/j.smaim.2021.11.004.
14. IMARC Group, Bioadhesive Market: Global Industry Trends, Share, Size, Growth, *Opportunity and Forecast* 2022-2027 (2022) https://www.researchandmarkets.com/reports/5530404/bioadhesive-market-global-industry-trends.
15. W.D Spotnitz, Tissue adhesives: science, products, and clinical use, in: *Musculoskeletal Tissue Regeneration*, W.S. Pietrzak (Ed.), pp. 531-546, Humana Press (2008).
16. Chivers, R.A., Adhesion in medicine, in: *Handbook in Adhesion Technology, Second edition*, L.F. M. da Silva, A. Öchsner and R. D. Adams (Eds.), pp. 1729-1749, Springer, Cham (2018).
17. Fortune Business Insights, Fibrin Sealants Market Size, Share, Trends Industry Report 2026, 2020. https://www.fortunebusinessinsights.com/industry-reports/fibrin-sealants-market-100316
18. J. Il Kang and K.M. Park, Advances in gelatin-based hydrogels for wound management, *J. Mater. Chem. B.*, 9, 1503–1520 (2021).
19. A. Dey, P. Bhattacharya and S. Neogi, Bioadhesives in biomedical applications: A critical review, *Rev. Adhesion Adhesives*, 8, 130-152(2020).
20. S. Sharifi, M.M. Islam, H. Sharifi, R. Islam, D. Koza, F. Reyes-Ortega, D. Alba-Molina, P.H. Nilsson, C.H. Dohlman, T.E. Mollnes, J. Chodosh and M. Gonzalez-Andrades, Tuning gelatin-based hydrogel towards bioadhesive ocular tissue engineering applications, *Bioactive Materials* 6, 3947–3961 (2021).
21. Coherent Market Insights, Europe gelatin and bone glue market size and forecast to 2027, (2020). https://www.coherentmarketinsights.com/market-insight/europe-gelatin-and-bone-glue-market-3793 (accessed March 3, 2022).
22. A.H.J. Gowda, Y. Bu, O. Kudina, K.V. Krishna, R.A. Bohara, D. Eglin and A. Pandit, Design of tunable gelatin-dopamine based bioadhesives, *Int. J. Biol. Macromol.* 164, 1384–1391 (2020).
23. Summary of Safety and Effectiveness Data (SSED) Report, US FDA P100030/S008, assigned to Mallinckrodt Pharma IP Trading DAC (2017)
24. Fortune Business Insights, Global Albumin Market Size, Share Report [2019-2026], https://www.fortunebusinessinsights.com/albumin-market-102637 (accessed March 3, 2022).
25. L. Kaur and I. Singh, Chitosan-catechol conjugates–A novel class of bioadhesive polymers: A critical review, *Rev. Adhesion Adhesives*, 7, 51–67 (2019).
26. D.A. Gyles, L.D. Castro, J.O.C. Silva and R.M. Ribeiro-Costa, A review of the designs and prominent biomedical advances of natural and synthetic hydrogel formulations, *Eur. Polym. J.* 88 373–392 (2017).

27. N. Mati-Baouche, P.H. Elchinger, H. De Baynast, G. Pierre, C. Delattre and P. Michaud, Chitosan as an adhesive, *Eur. Polym. J.* 60, 198–212 (2014).
28. M.C. Giano, Z. Ibrahim, S.H. Medina, K.A. Sarhane, J.M. Christensen, Y. Yamada, G. Brandacher,and J.P. Schneider, Injectable bioadhesive hydrogels with innate antibacterial properties, *Nature Communications*, 5, 1–9 (2014).
29. T.D. Gomes, S.G. Caridade, M.P. Sousa, S. Azevedo, M.Y. Kandur, E.T. Öner, N.M. Alves and J.F. Mano, Adhesive free-standing multilayer films containing sulfated levan for biomedical applications, *Acta Biomater.* 69, 183-195 (2018).
30. M. Oprea and S.I. Voicu, Recent advances in composites based on cellulose derivatives for biomedical applications, *Carbohydrate Polym.* 247, 116683 (2020).
31. P. Rajesh Raja, Cyanoacrylate adhesives: A critical review, *Rev. Adhesion Adhesives*, 4, 398–416 (2016).
32. N. Saha, N. Saha, T. Sáha, E. Toksoy Öner, U.V. Brodnjak, H. Redl, J. von Byern and P. Sáha, Polymer based bioadhesive biomaterials for medical application—a perspective of redefining healthcare system management, *Polymers (Basel).* 12, 1- 19 (2020).
33. E.M. Petrie, Cyanoacrylate adhesives in surgical applications: A critical review, *Rev. Adhesion Adhesives*, 2, 253–310 (2014).
34. A.J. Vernengo, Adhesive materials for biomedical applications, in: *Adhesives-Applications and Properties*, A. Rudawska (Ed.), pp. 99-136, IntechOpen, London, (2016).
35. G. Trujillo-de Santiago, R. Sharifi, K. Yue, E.S. Sani, S.S. Kashaf, M.M. Alvarez, J. Leijten, A. Khademhosseini, R. Dana and N. Annabi, Ocular adhesives: Design, chemistry, crosslinking mechanisms, and applications, *Biomaterials.* 197, 345–367 (2019).
36. X. Pei, J. Wang, Y. Cong and J. Fu, Recent progress in polymer hydrogel bioadhesives, *J. Polym. Sci.* 59, 1312–1337 (2021).
37. N. Wanasingha, N.K. Dutta and N.R. Choudhury, Emerging bioadhesives: from traditional bioactive and bioinert to a new biomimetic protein-based approach, *Adv. Colloid Interface Sci.* 296, 102521 (2021).
38. S. Wang, X. Ji, S. Chen, C. Zhang, Y. Wang, H. Lin and L. Zhao, Study of double-bonded carboxymethyl chitosan/cysteamine-modified chondroitin sulfate composite dressing for hemostatic application, *Eur. Polym. J.* 162, 110875 (2022).
39. Global Chondroitin Sulfate Market Share, Size, Growth, Research, Forecast: By Product: Food Grade, Pharmaceutical Grade; By Source: Bovine, Porcine, Poultry, Fish, Sharks, Crocodiles, Others; By Form: Powder, Tablets, Capsules; By Application: Pharmaceutical and Nutraceutical, Others; Regional Analysis; Supplier Landscape: 2023-2031, https://www.expertmarketresearch.com/reports/chondroitin-sulfate-market (accessed June 13, 2023).

40. R. Pinnaratip, M.S.A. Bhuiyan, K. Meyers, R.M. Rajachar and B.P. Lee, Multifunctional biomedical adhesives, *Advanced Healthare. Mater.* 8 1801568 (2019).
41. Frost & Sullivan, Global Biodegradable Pressure Sensitive Adhesives Technology Market Report 2021-2026, Technology Advances in Biodegradable Pressure Sensitive Adhesives, https://www.researchandmarkets.com/reports/5309957/technology-advances-in-biodegradable-pressure?utm_source=CI&utm_medium=PressRelease&utm_code=g2vxtq&utm_campaign=1591991+-+Global+Biodegradable+Pressure+Sensitive+Adhesives+Technology+Market+Report+202-2026&utm_exec=chdo54prd (accessed June 13, 2023).
42 W. Duan, X. Bian and Y. Bu, Applications of bioadhesives: a mini review, *Frontiers Bioeng. Biotechnol.*, 9, 702-714 (2021).
43. K.C. Ugoeze, Bioadhesive polymers for drug delivery applications, in: *Bioadhesives in Drug Delivery*, K.L. Mittal, I.S. Bakshi and J.K. Narang (Eds.), pp: 29-90, Wiley-Scrivener, Beverly, MA, (2020).
44. P.H. Tsang, G. Li, Y. V. Brun, L. Ben Freund and J.X. Tang, Adhesion of single bacterial cells in the micronewton range, *Proc. Natl. Acad. Sci. USA*, 103, 5764–5768 (2006).
45. W. Zhu, Y.J. Chuah and D.A. Wang, Bioadhesives for internal medical applications: A review, *Acta Biomater.* 74, 1–16 (2018).
46. J. Singh, N.C.S. Tan, U.R. Mahadevaswamy, N. Chanchareonsook, T.W.J. Steele and S. Lim, Bacterial cellulose adhesive composites for oral cavity applications, *Carbohydrate Polym.* 274, 118403 (2021).
47. K.H. Park, K.Y. Seong, S.Y. Yang, and S. Seo, Advances in medical adhesives inspired by aquatic organisms' adhesion, *Biomater. Res.* 21, 1–9 (2017).
48. V. Dhandapani, P. Saseedharan, D. Groleau and P. Vermette, Overview of approval procedures for bioadhesives in the United States of America and Canada, *J. Biomed. Mater. Res. Part B*, 10, 950-966 (2021).
49. J. Li, X. Yu, E.E. Martinez, J. Zhu, T. Wang, S. Shi, S.R. Shin, S. Hassan and C. Guo, Emerging biopolymer-based bioadhesives, *Macromolecular Biosci.* 2100340 (2021).
50. Markets and Markets, Bioadhesives Market Global Forecast to 2025. https://www.marketsandmarkets.com/Market-Reports/bioadhesive-market-16386893.html (accessed March 6, 2022).

3

Biomedical Adhesives

Jaehun Mun and Sungbaek Seo*

Department of Biomaterials Science (BK21 FOUR Program), College of Natural Resources and Life Science/Life and Industry Convergence Research Institute, Pusan National University, Miryang, Republic of Korea

Abstract

Lately, there has been a growing interest in high-quality healthcare and rehabilitative services. In this context, various medical devices and related adhesives have been developed. This chapter provides an overview of biomedical adhesives and their mechanisms of adhesion. Here, the components of both soft and hard tissue adhesives, which are classified based on the region of their application on tissues, are described. Additionally, a brief description regarding the functions and development of adhesives such as BioGlue®, used for medical surgery and wound dressing, is also provided.

Keywords**:** Biomedical adhesives, soft tissue, hard tissue, suture, dental cement, bone cement, Bioglue, wound dressing

3.1 Introduction

Based on data from 38 OECD countries in 2020, people aged 65 or older account for 9.3% of the population, and the aging index has gradually increased from 90 in 2010 to 100.4 in 2015 and 109.2 in 2020 [1]. In societies with a high aging index, the incidences of various diseases, including vascular and digestive diseases, have increased. Therefore, there has been an increasing interest in developing advanced medical technologies and therapies for improving the quality of health and recovery after medical treatment. The increasing demand for noninvasive and scar-resistant

**Corresponding author*: sbseo81@pusan.ac.kr

K.L. Mittal and S. Neogi (eds.) Adhesives in Biomedical Applications, (47–70) © 2023 Scrivener Publishing LLC

medical practices has paved the way for the development and use of effective medical adhesives in medical devices.

3.1.1 Definition of a Biomedical Adhesive

A biomedical adhesive refers to a high-tech medical material that forms bonds to damaged tissues by simple manipulation instead of medical sutures [2]. Medical adhesives can be applied not only for surgical suturing but also for various biomedical devices such as slip-coating or drug coating on catheters, hard contact lenses, and as skin adhesive wound dressings. The adhesive can be used on both soft and hard tissues in surgical or therapeutic procedures. Depending on the number of ingredients, these adhesives can be classified as single- or multi-component systems [3, 4].

3.1.2 Proposed Mechanisms of Adhesion

The mechanisms of adhesion involving biomedical adhesives can be described by six theories: 1) mechanical interlocking, 2) electrostatic attraction, 3) adsorption, 4) diffusion, 5) chemical bonding, and 6) weak boundary layers.

3.1.2.1 Mechanical Interlocking

Mechanical engagement, also known as mechanical interlocking, proposed by McBain and Hopkins in 1925, is a process in which the adhesive penetrates the voids, holes, and irregularities on the surface of the material to which it is applied; the strength of an adhesive is dictated by the extent of its penetration in the material [5]. The most representative example of mechanical interlocking is the anchoring effect proposed by Borroff and Wake, who studied the adhesion between rubber and fibers [6]. In this phenomenon, one end of the fiber penetrates the rubber, which is an adhesive material [7–9].

3.1.2.2 Electrostatic Attraction

The electrostatic theory was first proposed in 1948 by Deryaguin. According to this theory, an electrical double layer forms at the interface where adhesives exist with different electronic band structures, as the electrons move to match their Fermi levels. When the interface is destroyed in the system and the two materials serving as capacitors are split, energy is required for the separation. Therefore, the adhesion force between the two materials is generated by the electrostatic attraction between the electrical double layers [10–14].

3.1.2.3 *Adsorption*

The adsorption (thermodynamic) theory of adhesion, introduced by Sharpe and Schonhorn, states that bonding of the adherend material to the adhesive occurs due to forces between atoms or molecules at the interface, most commonly van der Waals forces and Lewis acid-base interactions, which are mostly attributed to thermodynamic forces related to surface free energy of the adhesive [15].

3.1.2.4 *Diffusion*

The diffusion theory assumes that adhesion is induced by mutual infiltration of adhesives between macromolecules at the interface or between polymers or interconnections, thereby forming a uniform phase. This mechanism is mainly based on the theory proposed by Voyutskii, which states that large molecular chains or chain segments are sufficiently fluid or are compatible with each other; these characteristics are vital in processes such as bonding and welding. Therefore, assuming that mutual diffusion occurs, the strength of the adhesion force varies depending on the contact time, temperature, characteristics, and molecular weight of the polymer [16].

3.1.2.5 *Chemical Bonding*

This theory states that the chemical bond formed at the adhesive-adherend interface dictates the adhesion force between the two materials. Unlike other bonding forces, such as van der Waals forces, the adhesion force is considered a primary chemical bonding force. Primary and secondary forces are classified based on the difference in the relative force or bonding energy of the interaction, where the covalent bonding energy is approximately 100–1000 kJ/mol and the energy due to van der Waals interactions and hydrogen bonding does not exceed 50 kJ/mol. The formation of a chemical bond is strongly related to the reactivity of adhesives. Other forms of primary bonding include ionic and covalent bondings, which have also been reported. In addition, the most important treatment to enhance interfacial chemical bonding is to use an adhesion promoter, called a coupling agent, which improves the bonding force between adhesive and adherend. The adhesion promoter chemically reacts with the adhesive and adherend on opposite sides, thereby serving as a chemical bridge at the interface [17–19].

3.1.2.6 *Weak Boundary Layers*

Bikerman proposed the initial concept of the weak boundary layers theory, which states that the cohesive force of weak boundary layers determines the adhesion force when an adhesion interface is destroyed. This method assumes that when the adhesive is removed, adhesive adjacent to an interface fails cohesively rather than separation at the interface. For example, small molecules or impurities with short chains which exist in the interior may move to the surface, thereby forming a weak interface layer, which is the first to be destroyed [20].

3.1.3 Considerations for Development of Biomedical Adhesives

Biomedical adhesives have been commonly used instead of sutures to bond skin or other biological tissues; however, damage to the sutures such as loosening of sutures or accidental thread cuts, can cause hygiene problems, inflammatory reactions, and additional trauma to tissues. Therefore, biomedical adhesives in medical devices should be developed in such a way that these have sufficient mechanical properties, biodegradability, and homeostatic effects and can be quickly applied and sterilized at room temperature under moist conditions.

3.2 Types of Biomedical Adhesives and their Components

Biological tissue adhesives can be divided into those used for soft tissues, such as internal organs, or for hard tissues, such as bones or teeth (Table 3.1). The corresponding adhesives can be categorized into single- and multi-component systems, depending on the ingredients [3, 4].

3.2.1 Soft Tissue Adhesives

The material components mainly involved in soft tissue adhesives are: cyanoacrylate, fibrin, gelatin, polyurethane, and mussel adhesive proteins.

Table 3.1 Categories of biomedical adhesives and components involved.

<table>
<tr><th>Type of tissue where adhesives are applied</th><th colspan="2">Components involved in biomedical adhesives</th></tr>
<tr><td rowspan="6">Soft tissue</td><td colspan="2">Cyanoacrylate</td></tr>
<tr><td colspan="2">Fibrin glue</td></tr>
<tr><td colspan="2">Protein glue</td></tr>
<tr><td colspan="2">Polyurethane</td></tr>
<tr><td colspan="2">Poly(ethylene glycol) (PEG)</td></tr>
<tr><td colspan="2">Mussel-derived adhesive proteins</td></tr>
<tr><td rowspan="7">Hard tissue</td><td rowspan="4">Dental cement</td><td>Zinc phosphate</td></tr>
<tr><td>Zinc oxide Eugenol</td></tr>
<tr><td>Glass ionomers</td></tr>
<tr><td>Resins</td></tr>
<tr><td rowspan="3">Bone cement</td><td>Poly(methyl methacrylate) (PMMA)</td></tr>
<tr><td>Calcium phosphate</td></tr>
<tr><td>Bispenol-A-glycidylmethacrylate (Bis-GMA)</td></tr>
</table>

3.2.1.1 Cyanoacrylate-Based Adhesives

Cyanoacrylate-based adhesives are cured by moisture, without an initiator, at room temperature in a short time. These adhesives are transparent and have high adhesion strength, but are vulnerable to impact and have poor heat and water resistance [3, 4]. As the number of carbon atoms in the side chain increases, the adhesion strength decreases. However, in polymers, as the number of carbon atoms increases, the hydrolysis rate decreases; thus, less formaldehyde is generated, resulting in decreased tissue toxicity [3, 4]. Methyl and ethyl cyanoacrylate adhesives were initially used as suture adhesives, but are currently rarely used because of severe tissue toxicity. Therefore, cyanoacrylates with butyl and octyl groups have been developed and used [3, 4]. The use of cyanoacrylate is increasing in pediatric patients and plastic surgery because it does not require removal of the suture and there is no pain.

3.2.1.1.1 Methyl Cyanoacrylate-Based Adhesives

Methyl cyanoacrylate is an organic compound comprised of methyl ester, nitrile, and alkene which are low-viscosity, colorless liquids. It is mainly used as a raw material for instant adhesives and is rapidly polymerized in the presence of moisture. It is not susceptible to blooming and has minimal odor. Although its adhesion strength is relatively weak compared to ethyl cyanoacrylate, it can adhere to most metal substrates (Figure 3.1).

3.2.1.1.2 Ethyl Cyanoacrylate-Based Adhesives

Ethyl cyanoacrylate (EC) is a low-viscosity liquid. It is used as the main ingredient in instant adhesives (Figure 3.2) and can be dissolved in acetone and methyl ethyl ketone. It also rapidly polymerizes in the presence of moisture. This is the most commonly used type of instant adhesive. However, it causes bleaching and has an irritating odor, requiring careful handling.

3.2.1.1.3 Butyl Cyanoacrylate-Based Adhesives

Butyl cyanoacrylate is a pungent, transparent colorless liquid. It is insoluble in water and is used as an instant medical adhesive (Figure 3.3).

3.2.1.1.4 Octyl Cyanoacrylate-Based Adhesives

Octyl cyanoacrylate was developed to address toxicity concerns and to reduce skin irritation and allergic reactions. It is primarily used in medical, veterinary, and emergency care (Figure 3.4).

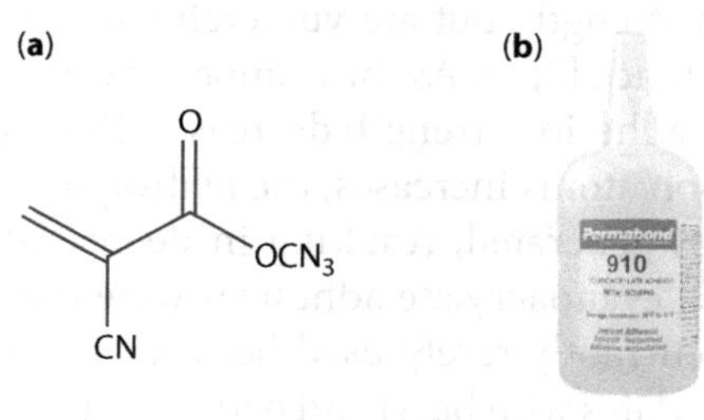

Figure 3.1 (a) Chemical structure of methyl 2-cyanoacrylate. (b) Permabond 910: used for bonding metal surfaces and bonds well to a variety of materials, including plastics and rubber.

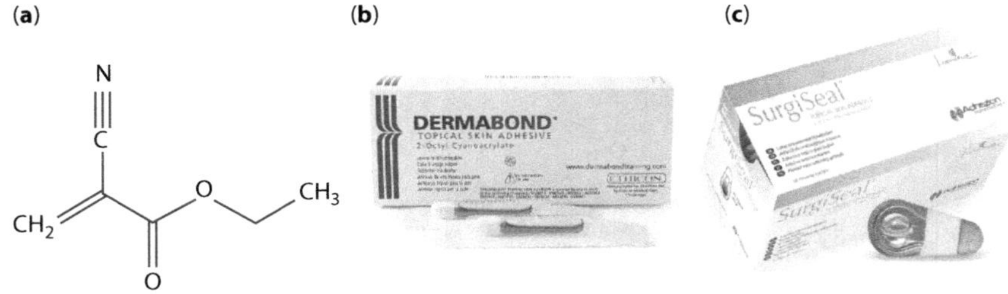

Figure 3.2 (a) Chemical structure of ethyl-2-cyanoacrylate. (b) Dermabond (Ethicon, Somerville, NJ, USA): It is a liquid topical skin adhesive used to suture wounds. It has excellent strength, microbial protection and single-layer application. Compared to traditional materials, the strength of wound suturing is enhanced by 75%. (c) SurgiSeal (Adhezion Biomedical, Hudson, VA, USA): It causes less secondary damage and is painless to remove. It also forms a protective film which prevents bacterial infection or foreign substances from penetrating, allows the patient to bathe, and is easily applied.

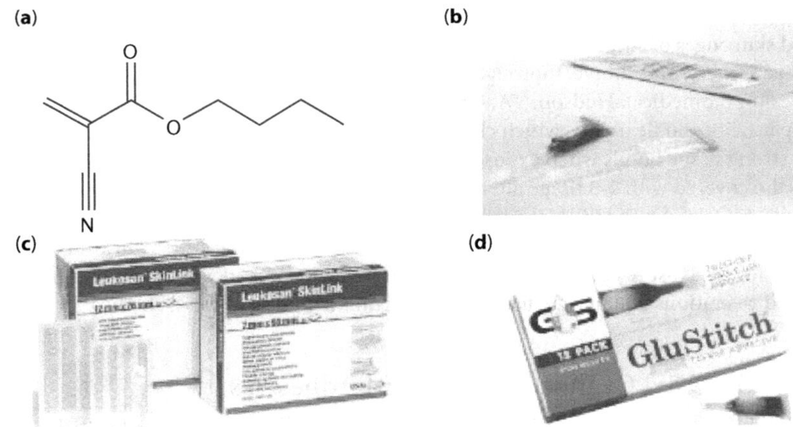

Figure 3.3 (a) Chemical structure of n-butyl cyanoacrylate. (b) Histoacryl® (B. Braun Melsungen AG, Tuttlingen, Germany): The Histoacryl® Flexible Pack, a tissue adhesive and microbial blocking product that can be locally applied, consists of n-butyl 2-cyanoacrylate and a softener. Histoacryl® Flexible Pack is displayed in blue. (c) SkinLink (Smith & Nephew, Sydney, Australia): This creates an anchor between the skin and adhesive, maintaining the flexibility of the peripheral skin, while securing the integrity of closure. The adhesive diffuses into the edges of the perforations, reinforcing the structure and strength of the closures. The protective membrane prevents the adhesive from drenching the entire closure while preserving its flexibility. (d) GluStitch: GluStitch® Twist is a CE marked, class IIa, sterile, topical tissue adhesive, composed of monomeric n-butyl-2-cyanoacrylate. GluStitch® Twist is in liquid form until it is applied to the skin and spread into a thin layer. GluStitch® Twist is intended for topical closure of surgical skin incisions and trauma-induced lacerations.

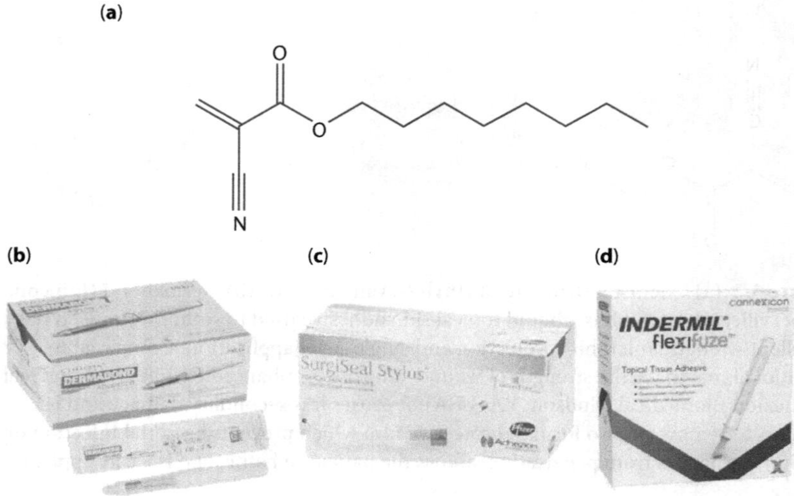

Figure 3.4 (a) Chemical structure of 2-octyl cyanoacrylate. (b) Dermabond (Ethicon, Somerville, NJ, USA): DERMABOND ADVANCED™ is used only to secure local, easily-closed skin edges of surgical incision wounds, including punctures from minimally invasive surgery, and simple, thoroughly-cleaned traumatic lacerations. (c) SurgiSeal (Adhezion Biomedical, Hudson, VA, USA): 2-octyl-cyanoacrylate is a liquid monomer with plasticity and flexibility, which changes into a polymer by exothermic reaction to form strong tissue bonds when exposed to a humid surface, such as the skin. The wound edges are covered with 2-3 tissue adhesive layers. (d) Indermil (Sherwood-Davis & Geck, St Louis, MO, USA): INDERMIL® flexifuze™ Topical Tissue Adhesive is a sterile, liquid topical tissue adhesive, composed of an octyl cyanoacrylate. INDERMIL® flexifuze™ is utilized for the closure of skin wounds, including laparoscopic incisions and trauma-induced lacerations and may be used in conjunction with deep dermal stitches.

3.2.1.1.5 Allyl Cyanoacrylate-Based Adhesives

Allyl cyanoacrylate provides heat resistance up to 250 °C (Figure 3.5). Although its uses are limited, new application areas that utilize the characteristics of the allyl group are being developed.

3.2.1.2 Fibrin Glue-Based Adhesives

When a tissue is injured, fibrinogen exudes from the surrounding capillary, along with the blood coagulation component, to form fibrin, which cross-links around the wound. In Europe, tissue adhesives containing fibrinogen, thrombin, calcium chloride, and blood clotting factor XIII replace or reinforce peripheral nerve and microvascular sutures [3, 4]. Surgical adhesives (Figure 3.6) are used in vascular, cranial nerve, and orthopedic

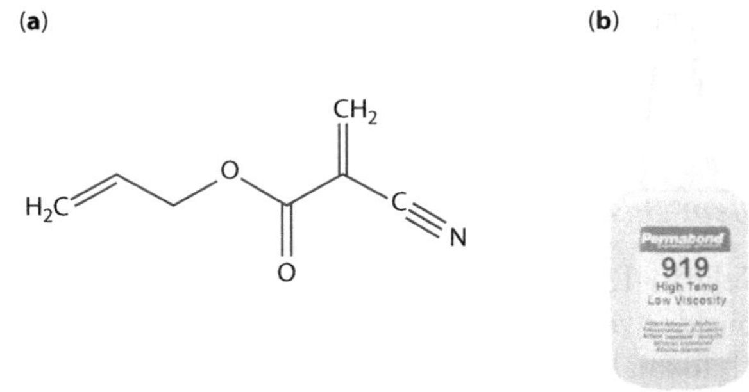

Figure 3.5 (a) Chemical structure of allyl 2-cyanoacrylate. (b) Permabond 919 (low viscosity, highly heat resistant (250 °C), Permabond, USA).

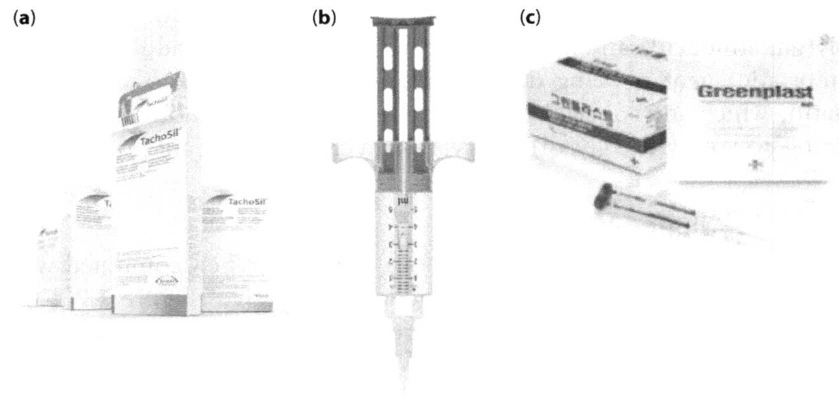

Figure 3.6 (a) Tachosil (Baxter Healthcare Corporation, Deerfield, IL, USA): Surgical, trauma-related hemostatic agent. Treatment of bile, lymph, fluid, and air leaks.
(b) TISSEEL (Baxter Healthcare Corporation, Deerfield, IL, USA): Upon mixing, human sealer protein and thrombin combine into fibrin that adheres to the wound surface and achieves hemostasis and sealing. TISSEEL mimics the final coagulation cascade step by forming a clot. (c) Greenplasty (Green Cross Corporation, Seoul, Korea): Assists in all areas of surgery (general, cardiac, thoracic, pediatric, neuro-, orthopedic, plastic, trauma, obstetric, urology, otolaryngology, dental and oral, ophthalmic, etc.).

surgeries, such as bone adhesion and hemostasis in lacerated patients [3]. Fibrin glue has biological advantages, such as fast adhesion, no unnecessary heat or pressure, suitable coagulation process with platelets, excellent tissue compatibility, and proper absorption; however, it has weak adhesion and high risk for infection [3]. To negate these shortcomings, its design should be ameliorated to slow the decomposition rate of fibrin polymers, while maintaining the adhesion effect, thus inhibiting fibrinase and fibroblast proliferation into stabilized polymers, consequently promoting tissue repair [4].

3.2.1.3 *Protein-Based Adhesives*

3.2.1.3.1 Gelatin-Resorcinol-Formaldehyde (GRF)

Gelatin is a bio-derived adhesive, developed by crosslinking gelatin and resorcinol with formaldehyde (GRF) [3]. It has high tissue adhesion; however, formaldehyde, which is used as a crosslinking agent also crosslinks the *in vivo* proteins, resulting in toxicity. GRF is currently manufactured in France and has not yet been approved; however, its effectiveness as a surgical hemostatic agent is being reviewed. A new adhesive that does not contain formaldehyde was developed using a cross-linking reaction of gelatin, poly(glutamic acid), and water-soluble carbodiimide [3]. In addition, a local hemostatic agent is being developed by synthesizing photo-crosslinked gelatin, which introduces benzophenone groups as a photoreactor, and mixing poly(ethylene glycol) diacrylate as a cross-linking aid [3].

3.2.1.3.2 Albumin Glue

Albumin solidifies under acidic aqueous solution. When combined with glutaraldehyde, albumin can act as a strong adhesive (Figure 3.7). The albumin/glutaraldehyde complex can resist 500-800 mmHg of pressure, more than four times the normal blood pressure, and thus has excellent adhesion strength [21].

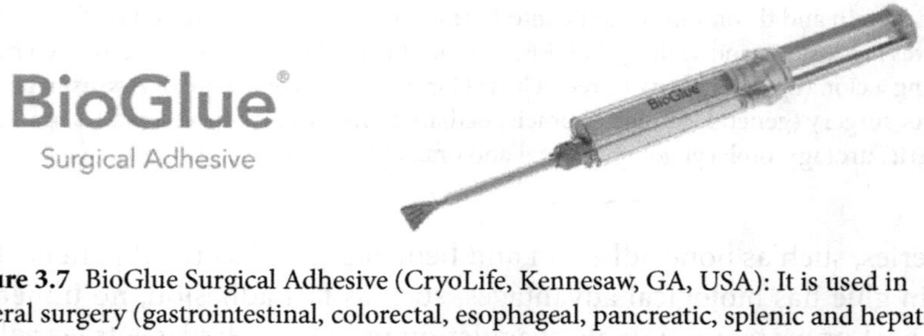

Figure 3.7 BioGlue Surgical Adhesive (CryoLife, Kennesaw, GA, USA): It is used in general surgery (gastrointestinal, colorectal, esophageal, pancreatic, splenic and hepatic repair, and hernia mesh fixation).

3.2.1.4 *Polyurethane-Based Adhesives*

As it is applied to a soft tissue, the urethane-based adhesive should preferably be an elastic adhesive, in which the flexibility of the bonded part is maintained after curing. This adhesive absorbs water present on the surface of the biological tissue to enhance adhesion to the tissue, reacts with water to cure within minutes, exhibits rubber elastic flexibility, is slowly biodegradable, and reduces stress concentration at the soft tissue, thus increasing the aging stability of the adhesive (Figure 3.8). Although the synthetic raw material, aromatic diisocyanate, is biotoxic, it is expected to be useful when using non-toxic aliphatic diisocyanate, even with delayed curing rate [3].

3.2.1.5 *Poly(ethylene glycol) (PEG)-Based Adhesives*

Poly(ethylene glycol) (PEG) hydrogel sealants protect and adhere to tissues. They contain biodegradable lactides and both ends are made of polymerizable acrylate [22]. Depending on the initiator, polymerization with visible and ultraviolet light is facilitated, and the polymeric hydrogel is obtained by crosslinking. The hydrogel thus produced has mechanical properties with a tensile modulus of 50 kPa or more which is enough mechanical performance in various animal experiments [22]. It is mainly used to apply at the suturing area when rebuilding the vein and to prevent leakage of cerebrospinal fluid after brain surgery [23, 24].

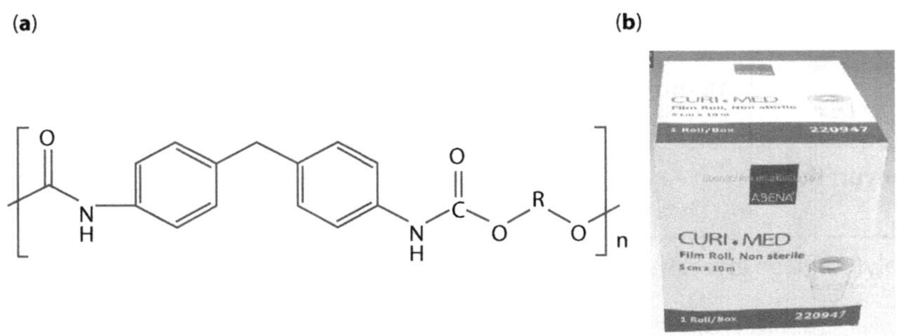

Figure 3.8 (a) Chemical structure of polyurethane. (b) Surgical adhesive film (ABENA): It is used for wound dressing or surgical supplies.

3.2.1.6 *Mussel-Derived Protein-Based Adhesives*

Mussel secretes thread-like adhesive proteins and grows on rocks without disintegrating them in seawater. It is a hybrid biological adhesive that uses mussel's adhesive protein by applying this principle [25, 26]. In previous studies, by introducing a repetitive sequence of fp-1 protein (foot protein type 1) that adheres and coats fp-5 (foot protein type 5) secreted by mussel's feet, the separation and purification process was simplified, and excellent adhesion of the fp-5 protein was maintained [25, 26]. Problems such as low concentration, insolubility, low yield, and separation purification, which were the biggest obstacles to the practical use of fp-5 protein

Table 3.2 Soft tissue adhesives: strengths and weaknesses.

Type of soft tissue adhesive		Strengths	Weaknesses
Cyanoacrylate		Short-term curing with moisture, high adhesion, hemostatic effect.	Poor heat resistance and water resistance.
Fibrin glue		No heat or pressure required, excellent tissue compatibility.	Weak adhesion, risk of infection.
Protein glue	Gelatin-resorcinol-formaldehyde (GRF)	Good handling.	Tissue toxicity of formaldehyde.
	Albumin glue	Pressure-resistant, highly adhesive.	N/A
Polyurethane		Hardens in minutes, flexible.	Diisocyanate is bio-toxic.
Poly(ethylene glycol)		Easy to polymerize with visible light as well as ultraviolet light.	N/A
Mussel-derived adhesive protein		Used in various ways as a bioactive peptide, safe in the human body.	N/A

have been solved [25–27]. When 40 mg is used, this hybrid adhesive can adhere and lift an object of approximately 10 kg over an area of 1 cm^2. Mussel adhesive proteins can be used in a variety of ways, from general products to high-value medical adhesives for drug delivery, and cell culture immobilization by introducing various bioactive peptides [25–27]. Cyanoacrylate-based adhesives, which are widely used as medical adhesives, have cancer-causing potential and toxicity; however, mussel-derived adhesives can be safely used in the human body and can be applied as next-generation medical adhesives. Table 3.2 lists soft tissue adhesives.

3.2.2 Hard Tissue Adhesives

The currently used hard tissue adhesives are categorized as dental cement and bone cement, applicable for bony tissue.

3.2.2.1 Dental Cement

Dental cement functions as an adhesive that attaches to inlays, crowns, bridges, and provides mechanical support to the repair thus protecting from harmful external stimuli [28–30].

3.2.2.1.1 Zinc Phosphate-Based Dental Cement

Zinc phosphate has been the most popular material used in dentistry for a long time. It is mainly used for inlays, crowns, bridges, orthodontic bands,

(a) (b)

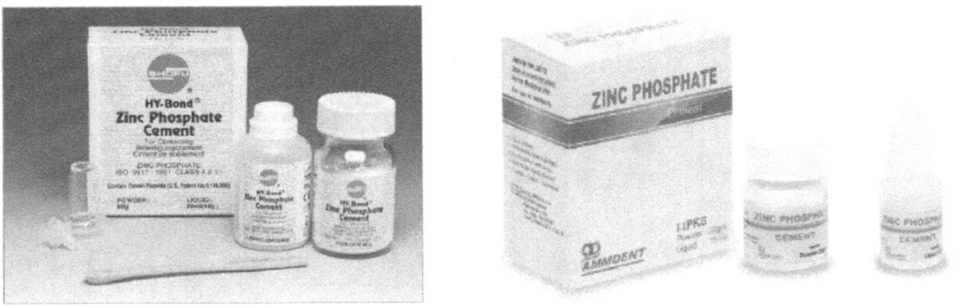

Figure 3.9 (a) Zinc phosphate cement (SHOFU, USA): a superior zinc phosphate cement containing tannin fluoride and offering high compressive strength (seals dentinal tubules and minimizes post-operative sensitivity, is excellent for permanent prostheses, possesses high compressive strength (98 MPa), shows high level of radiopacity, can be applied directly to prepared dentin, neutralizes the harshness of phosphoric acid). (b) Ammdent Zinc phosphate cement: it is used for luting permanent metal restoration.

and other oral devices. It has a long life, can be used as a permanent adhesive, and has a thermal conductivity similar to that of ivory; therefore, it is widely used in clinical practice (Figure 3.9). Compared to other cements, its strength is high, and the solubility is low. Initially, cement is used after applying a liner or varnish to provide dimensional protection; however, it can disturb the dimensions owing to its low pH [28–30].

3.2.2.1.2 Zinc Oxide Eugenol-Based Dental Cement

Zinc oxide Eugenol-based dental cement is biocompatible and relieves pain when sedation is required. Zinc oxide Eugenol cement is mainly used as a temporary cement for temporary filling and is also used as an impression material or periodontal treatment agent (Figure 3.10). Eugenol, when used alone, is allergic and can irritate the tissue after direct contact [28–30].

3.2.2.1.3 Glass Ionomer-Based Dental Cement

Glass ionomer cement is a silicate and polycarbonate complex that releases fluoride and adheres to hemorrhoids. It has various applications, such as cervical and milk tooth repair, and is used as a sealant for assisting with restoration, and cavity lining (Figure 3.11) [28–30].

3.2.2.1.4 Resin-Based Dental Cement

Resin cement has been used for inlays, crowns, bridges, brackets for calibration, and tooth preparation (Figure 3.12). Resin cement is composed of

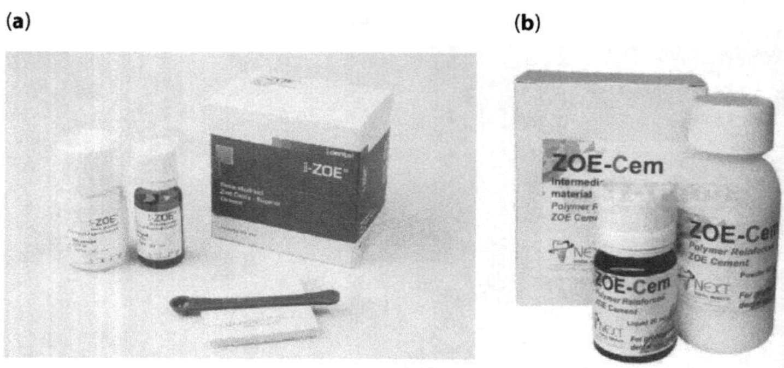

Figure 3.10 (a) I-ZOE N (i-dental, Siauliai, Lithuania): It is used for temporary fillings, fixing crowns and bridges. It has the advantages of high strength, quick setting, and available modification of resin. (b) Zoe-Cem Jumbo pack (NEXT IMPLANTS, Israel): it is an excellent foundation or temporary cement once it cures and protects the pulp from thermal shock or irritation from freshly mixed cement.

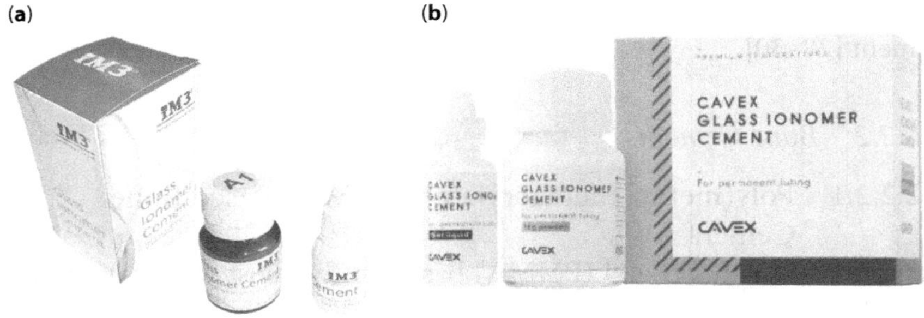

Figure 3.11 (a) IM3 (iM3 Dental Limited, Ireland): It is used for fissure sealing, filling of small noncarious defects of hard dental tissues, temporary fillings, and as cavity fillings. (b) Cavex Glass ionomer cement (Cavex, Nederlands): It is a high-quality dental cement for crown, bridge, and inlay cementation. The cement is prepared by gradually adding small amounts of powder to the liquid while mixing to enhance absorption of both powder and liquid and is optimized to attain a high-quality luting agent with aesthetic properties.

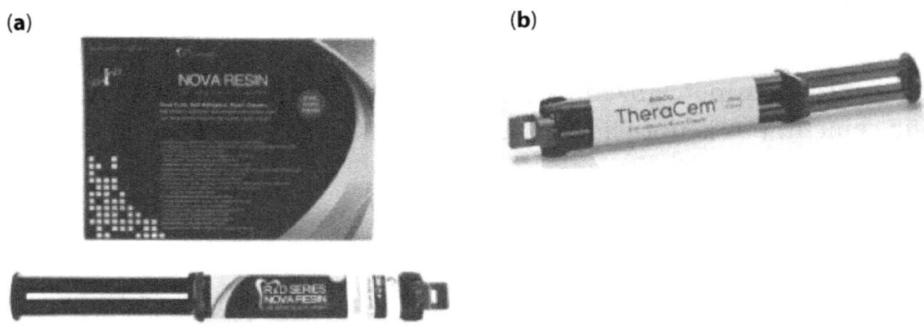

Figure 3.12 (a) Nova Resin Cement (Imicry Dental, Anatolia): self-adhesive dual cure (self-cure and light cure) resin cement for the cementation of indirect restorations made of high strength all ceramic, composite, metal-ceramic, and metal. When cement is used, additional bonding agents are not required. (b) TheraCem Resin Cement (BISCO, USA): It is a dual-cured, calcium and fluoride-releasing, self-adhesive resin cement indicated for luting crowns, bridges, inlays, onlays and posts (prefabricated metal/non-metal/fiber posts). Delivering a strong bond to Zirconia and most substrates, along with easy clean-up and high radiopacity, TheraCem offers the clinician a reliable and durable cementation for indirect restorations.

poly(methyl methacrylate) powder, filler, and liquid methyl methacrylate cement [28–30].

3.2.2.2 *Bone Cements*

3.2.2.2.1 Poly(methyl methacrylate) (PMMA)-Based Bone Cement

Poly(methyl methacrylate) (PMMA) has been the most widely used biological adhesive since it was first presented as an adhesive between the bone and implant by Sir John Charnley in the 1960s [31]. Self-curing of PMMA osteoclasts is easy to perform and can effectively distribute pressure between the bones and implants to reduce pain and facilitate a rapid recovery of patients. However, PMMA bone cement is pointed out as having permanent initial strength and adhesion, aseptic dissociation due to penetration of new tissues between bones and implants, necrosis of surrounding cells due to polymerization heat, increased cell toxicity due to monomer elution after a certain period of time, and inflammatory reactions. PMMA bone cement is used in hip, knee, and other joints surgery to fix the polymer of metallic artificial implants to living bones (Figure 3.13) [32].

3.2.2.2.2 Calcium Phosphate-Based Bone Cement

Calcium phosphate-based bone cement has been studied since 1976 and is currently commercialized as a general product, such as PMMA-based bone cement, but is not widely used at present. There are two main types

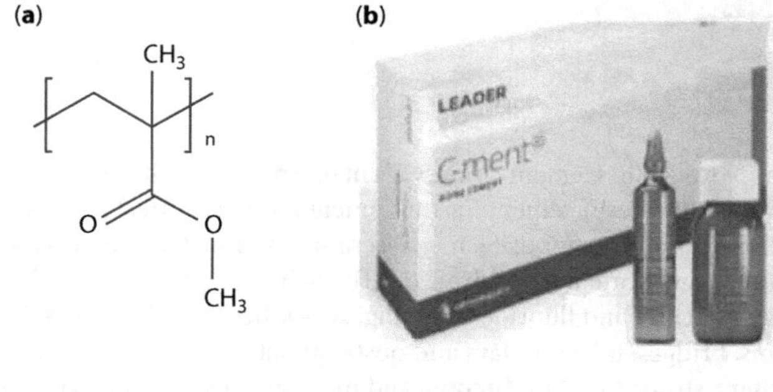

Figure 3.13 (a) Chemical structure of PMMA. (b) Orthopedic poly(methyl methacrylate) bone cement (Teknimed, France): it is radiopaque and self-hardening bone cement indicated for arthroplasty surgeries, standard (high) viscosity bone cement for manual application, and low viscosity bone cement with optimum fluidity to allow an easy application with a syringe or injection gun.

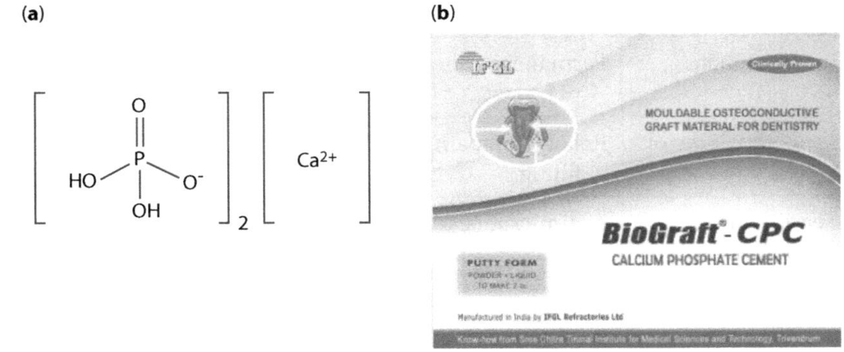

Figure 3.14 (a) Chemical structure of calcium phosphate. (b) BioGraft® Calcium Phosphate Cement (IFGL, Kolkata, India): It is ideal for repair of infra bony defects in dentistry and oral and maxillofacial segments.

of curing method. First, calcium phosphate bone cement has a curing temperature of 80 °C or higher and a long curing time. Second, organic acids have excellent durability in vivo, but their use is limited due to cell toxicity and acid degradation (Figure 3.14) [33].

3.2.2.2.3 Bisphenol-A-Glycidylmethacrylate (Bis-GMA)-Based Bone Cement

Bis-GMA bone cement, which has expanded its application to the orthopedic area (Figure 3.15), was proposed by Bowen in 1963 and is emerging as a material that can compensate for the shortcomings of PMMA bone cement or calcium phosphate bone cement [34]. Bis-GMA bone cement has superior physical and mechanical properties and biocompatibility comparable to PMMA bone cement. The Bis-GMA monomer itself rarely generates high heat when polymerized, and the fact that the monomer itself has low cytotoxicity is also highlighted as an advantage. However, compared to

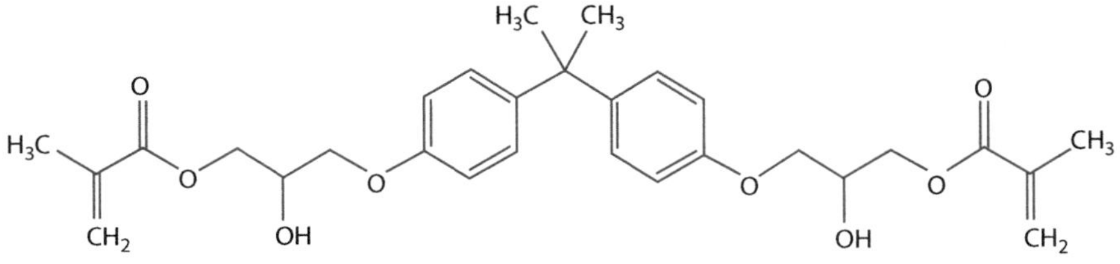

Figure 3.15 Chemical structure of Bis-GMA.

Table 3.3 Advantages and disadvantages of different types of dental cements [36].

Zinc phosphate	Permanent adhesive, high strength, low solubility, low pH.
Zinc oxide Eugenol	Relief from pain and allergic reactions, temporary filling.
Glass ionomer	Antimicrobial activity, tooth repair, cavity lining.
Resin cement	Low solubility, harmful irritation to pulp.

Table 3.4 Advantages and disadvantages of bone cements [36].

PMMA	Reduces pain, increases resilience, aseptic dissociation, necrosis of cells, cytotoxicity, inflammatory reactions.
Calcium Phosphate	High durability, high curing temperature, long-time procedure, cytotoxicity, acid degradation.
Bis-GMA	Excellent biocompatibility, high viscosity, weak physical properties.

MMA monomers, Bis-GMA monomers are not easy to handle because of their high viscosity, and the use of diluents to improve these involves a decrease in strength due to the relative decrease in organic matrix content compared to other bone cements. In addition, the dissolution of monomers on the surface where polymerization is suppressed and the resulting weakening of the physical properties are mentioned as disadvantages [35].

Table 3.3 lists advantages and disadvantages of different types of dental cements.

Table 3.4 lists advantages and disadvantages of bone cements.

3.3 Advances in Adhesives Development for Biomedical Uses

3.3.1 Bioglue in Surgery

Biological adhesives save time during wound repair and provide a flexible waterproof protective coating. They are more effective in scar recovery than suturing. These can also overcome the disadvantages of conventional sutures, such as additional tissue damage, side effects of infection, and body fluids or air leakage.

BioGlue® Surgical Adhesive was developed as an adjunct for achieving hemostasis during cardiovascular surgery, and it was approved for use in the United States by the Food and Drug Administration in 2001. When applied to cardiovascular tissues, glutaraldehyde and bovine serum albumin, which comprise BioGlue, produce chemical crosslinking that bonds tissues and wound area. These features have made BioGlue particularly well-suited for preventing bleeding from fragile cardiovascular anastomoses, such as those inherent in the repair of acute aortic dissection.

BioGlue demonstrates excellent tensile and shear strengths compared to other surgical adhesives and sealants. These features make BioGlue particularly well-suited for preventing bleeding in fragile cardiovascular anastomoses. BioGlue adheres to woven and knitted synthetic vascular grafts by creating mechanical interlocks within the interstices of the graft material. In addition, because BioGlue is stored and used at room temperature (it has a shelf- life of 3 years at 25 °C) and because it requires no preparation before use, this product offers the advantage of immediate availability.

BioGlue requires a bloodless area to adhere; thus, it is not useful for controlling active bleeding. Therefore, BioGlue is generally applied to an anastomose before perfusion is restored (i.e., during the clamping procedure). Before BioGlue is applied to the tissue, the applicator tip is carefully coated to evacuate air and ensure that the components are properly mixed. This adhesive has a very low viscosity, which makes it difficult to fully control during application, particularly if it is rapidly released. Slowly releasing glue during application increases its initial viscosity and improves its control. BioGlue is applied such that a layer approximately 2 mm thickness covers the tissue anastomoses. The adhesive begins to polymerize within 20–30 s and reaches its maximum bonding strength in 2–3 min [37].

3.3.2 Wound Dressing

Healthcare professionals involved in wound care have access to many wound dressings. Many of these require the use of retention bandages or adhesive systems to keep them secure in position. These adhesive systems may be separate entities (e.g., tapes) or integral components of dressings. There are a number of so-called bordered or island dressings that incorporate adhesives, either over their entire surface or around a central absorptive pad.

Tissue trauma caused by the removal of adhesive tapes and dressings increases wound size, exacerbates wound pain, and delays healing [38].

These factors can adversely affect patients' quality of life and have cost implications for health care providers.

Historically, wound dressings have been categorized into three broad groups (adherent, low-adherent, and non-adherent) based on their interaction with the skin and wound bed. The main limitation of this simplistic categorization is that it fails to consider the possibility for trauma to the peri-wound skin and the wound bed caused by the removal of adhesive products [39].

Medical dressings have several functions. These protect wounds to reduce the risk of infectious agents entering the area and becoming colonies. They are also collected as exudate, a substance that bounces off the wound. Dressing can also be used to keep the wound moist and to promote healing.

The dressing may be made of various materials, including cotton and synthetic fibers. Accordingly, the dressing procedure makes the dressing materials being slippery and is generally permeable so that the liquid permeated from the wound can pass through it and be absorbed into the dressing. Also, when the bandage must be removed, the wound dressing area should be considered as easy to peel off without tearing.

3.4 Summary

Commercially available biomedical adhesives are categorized based on the applied region of tissues, such as soft and hard tissues. The corresponding features, strengths, and weaknesses of the main component involved in the adhesives are summarized. For the development of biomedical adhesives, adhesion capability, protecting against infection, inflammatory reactions, and reducing scar possibility should be considered. Biomedical adhesives in bioglues for medical surgery and wound dressing are exemplified by their functions and roles in biomedical applications.

3.5 Acknowledgements

This work was supported by a National Research Foundation of Korea (NRF) grant, funded by the Korean government (MSIP: Ministry of Science, ICT & Future Planning) (NRF-2021R1F1A105923711), and the BK21 FOUR Program of the National Research Foundation of Korea (NRF), funded by the Ministry of Education, Korea.

References

1. Supporting Expenses and Aging Index (OECD), KOSIS, last modified.09.30 (2021), accessed.08.08 (2022), https://kosis.kr/statHtml/statHtml.do?orgId=101&tblId=DT_2KAA202_OECD&conn_path=I2.
2. S. W. Kim and E. S. Son, Trends in Polymer Technology for Biomaterials, Korea Science & Technology Corporation KISTI Technology Trend Analysis Report (2002).
3. H. S. Jeong, M. K. Song and S. H. Yoo Medical Adhesive Manufacturing Technology, Korean Science KISTI Technology Value Assessment Report (2004).
4. D. G. Han Recent Research Trends in Medical Biological Adhesives, Korea Science and Technology Researcher (KIST) (2005).
5. J. W. McBain and D. G. Hopkins, On adhesives and adhesive action, *J. Phys. Chem.* 29, 88 (1925).
6. E. M. Borroff and W. C. Wake, Adhesion of rubber and textiles Part III–Examination of test techniques. *Trans. Inst. Rubber Industry* 25, 199-210 (1949).
7. A. N. Gent and R. P. Petrich, Adhesion of viscoelastic materials to rigid substrates, *Proc. R. Soc., A,* 310, 433-448 (1969).
8. A. N. Gent and J. Schultz, Effect of wetting liquids on the strength of adhesion of viscoelastic material, *J Adhesion*, 3: 281 (1972).
9. W. C. Wake, *Adhesion and the Formulation of Adhesives*, Applied Science Publishers, London (1982).
10. B. V. Deryaguin and N. A. Krotova, Adhesion, *Nauk SSSR*, 61, 843 (1949).
11. B. V. Deryaguin. Problems of adhesion. *Research*, 8, 70 (1955).
12. B. V. Deryaguin, N. A, Krotova, V. V. Karassev, Y. M. Kirillova, and I. N. Aleinikova, *Proceedings of the 2nd International Congress on Surface Activity-III*, p 417, Butterworths, London (1957).
13. B. V. Deryaguin and V. P. Smilga, *Adhesion, Fundamentals and Practice*, pp. 152-156, Maclaren and Sons, London (1969).
14. B. V. Deryaguin and N. A. Krotova, and V. P. Smilga, *Adhesion of Solids, Studies in Soviet Science* (1978).
15. L.H. Sharpe and H. Schonhorn, Surface energetics, adhesion and adhesive joints, in: *Contact Angle, Wettability and Adhesion, Advances in Chemistry* Series No. 43, pp. 189-201, American Chemical Society, Washington D.C. (1964).
16. S. S. Voyutskii, *Autohesion and Adhesion of High Polymers*, Wiley-Interscience, New York (1963).
17. E. P. Plueddemann, *Silane Coupling Agents*, Plenum Press, New York, (1982).
18. A. N. Gent and A. Ahagon, Effect of interfacial bonding on the strength of adhesion, *J. Polym Sci. Polym Phys.* 13, 1285-1300 (1975).
19. P. Delescluse, J. Schultz, and M. E. R. Shanahan, in: *Adhesion* 8, K. W. Allen (Ed.), p 79, *Elsevier Applied Science*, London (1984).

20. J. J. Bikerman, *The Science of Adhesive Joints*, Academic Press, New York (1961).
21. MedMarket Diligence. Worldwide market for surgical sealants, glues, wound closure and anti-adhesion, (2012-2017).
22. D. G. Han and G. D. Ahn. Medical adhesive agent. *Polymer Sci Technol* 11, 436-442 (2000).
23. C. J. Burkett, S. Patel, M. H. Tabor, T. Padhya and F. L. Vale. Polyethylene glycol (PEG) hydrogel dural sealant and collagen dural graft matrix in transsphenoidal pituitary surgery for prevention of postoperative cerebrospinal fluid leaks. *J Clinical Neurosci* 18, 1513-1517 (2011).
24. K. D. Than, C. J. Baird and A. Olivi, Polyethylene glycol hydrogel dural sealant may reduce incisional cerebrospinal fluid leak after posterior fossa surgery, *Neurosurgery,* 63, 182-186 (2008).
25. D. S. Hwang, Y. Gim, H. J. Yoo, and H. J. Cha, Practical recombinant hybrid mussel bioadhesive fp-151, *Biomaterials*, 28, 3560-3568 (2007).
26. D. S. Hwang, S. B. Sim, and H. J. Cha, Cell adhesion biomaterial based on mussel adhesive protein fused with RGD peptide, *Biomaterials*, 28, 4039-4046 (2007).
27. D. S. Hwang, H. J. Yoo, J. H. Jun, W. K. Moon, and H. J. Cha, Expression of functional Recombinant Mussel Adhesive Protein Mgfp-5 in *Escherichia coli*, *Appl. Environ. Microbiol.*, 70, 3352-3359 (2004).
28. M. Gladwin and M. Bagby, *Clinical Aspects of Dental Materials: Theory, Practice, and Cases, third edition*, pp. 88-90, Lippincott Williams & Wilkins, Philadelphia (2011).
29. W. S. Eakle and C. D. Hatrick, *Dental Materials: Clinical Applications for Dental Assistants and Dental Hygienists, third edition*, pp. 220-242, Elsevier Health Sciences, St. Louis (2015).
30. S. K. Kim, Y. K. Kim, and H. K. Kang, *Conservation Dentistry*, pp. 108-128, Komoonsa, Seoul (2017).
31. N. S. Eftekhar, The life and work of John Charnley (August 29, 1911-August 5, 1982), *Clin Orthop Relat Res.* 211, 10-22 (1986).
32. E. J. Harper and W. Bonfield, Biomed. Mechanical properties of hydroxyapatite reinforced poly(ethylmethacrylate) bone cement after immersion in a physiological solution: influence of a silane coupling agent, *J. Biomed. Mater. Res.* 53, 605-609 (2000).
33. J. A. Planell, M. M. Vila, F. J. Gil, and F. C. M. Driessens, *Encyclopedic Handbook of Biomaterials and Bioengineering*, p 879, Taylor & Francis, London (1995).
34. R. L. Bowen, Dental filling material comprising vinyl silane treated fused silica and a binder consisting of the reaction product of bis phenol and glycidyl acrylate, US Patent 3 066,112 (1962).
35. S. Shinzato, T. Nakamura, T. Kokubo, and Y. Kitamura. Bioactive bone cement: effect of filler size on mechanical properties and osteoconductivity, *J. Biomed. Mater. Res.* 54, 491-495 (2001).

36. D. G. Han, Recent Research Trends in Medical Biological Adhesives, *KIST* (2005).
37. C. M. Bhamidipati, J. S. Coselli, and S. A. LeMaire. BioGlue® in: *What Is Its Role in Cardiac Surgery*, 6-12 (2012).
38. H. Hollinworth and R. White, The clinical significance of wound pain, in: *Trauma and Pain Wound Care*, R. White and K. Harding (Eds.) pp. 3-16, Wounds UK, Aberdeen (2006).
39. S. Thomas. *Atraumatic dressings*. World Wide Wounds (2003). http://www.worldwidewounds.com/2003/january/Thomas/Atraumatic-Dressings.html (last accessed 15 October 2007).

4

Bioadhesion: Fundamentals and Mechanisms

Amit Porwal and Kamla Pathak*

Faculty of Pharmacy, Uttar Pradesh University of Medical Sciences, Saifai, Etawah, Uttar Pradesh, India

Abstract
Bioadhesion is defined as an event in which natural or synthetic substances adhere to biological surfaces. An interpretation of the underlying mechanisms that control bioadhesion is of considerable interest for various researchers involved in development of new biomaterials, treatments, and innovative applications. In this chapter, bioadhesion as well as the fundamental mechanisms along with various examples have been discussed. This is followed by discussion regarding mucosal layer, barriers to drug delivery, and factors affecting bioadhesion. Also various theories of bioadhesion have been discussed which distinctly describe the underlying mechanisms of adhesion along with a detailed discussion of the formation and behaviour of bioadhesive bonds. Finally, routes for administration of bio- and mucoadhesive Drug Delivery Systems have been described.

***Keywords*:** Bioadhesion, mucoadhesion, mechanisms, theories, factors affecting mucoadhesion

4.1 Introduction

Bioadhesion has been an area of growing interest, in light of the fact that a critical advancement in therapeutic efficacy might be due to an increased residence time at higher level of activity or absorption [1]. The term 'bioadhesion' refers to broadly diversified occurrences, which all include the adhesion of substances (natural or synthetic) to biological surfaces. As an

**Corresponding author*: kamlapathak5@gmail.com

K.L. Mittal and S. Neogi (eds.) Adhesives in Biomedical Applications, (71–98) © 2023 Scrivener Publishing LLC

interfacial phenomenon, bioadhesion is like ordinary adhesion, apart from the extraordinary characteristics of biological surfaces and structures. So, fundamentally bioadhesion includes the adhesive characteristics of both natural surfaces (e.g. cells) and synthetic substances. The term bioadhesion can also be applied to the utilization of bioadhesives to bond two surfaces together, which is significant in surgical, dental and drug delivery applications. Consequently, because of its indications for the advancement of new technological products, e.g. biosensors, therapies and biomaterials, researchers have shown extensive interest in research related to bioadhesion [2–5]. Bioadhesive systems have the following recognised benefits: (i) due to adhesion, the dosage form remains for a prolonged time at the site of application/action resulting in enhanced bioavailability of the therapeutic moiety; (ii) targeting a specific site or tissue is possible by utilizing specific bioadhesive molecules, for example, in the gastrointestinal tract; and (iii) lower administration frequency can be achieved by combining increased retention time with controlled release of a drug [6].

4.2 Bioadhesion in Biological Systems

Bioadhesion can also be defined as the utilization of adhesive substances in clinical use for biological surfaces. An example of naturally occurring bioadhesion is the attachment of mussels to submerged surfaces, for example ship or coral surfaces. The organism from its shell extends its foot to release a pad of mussel adhesive proteins (MAPs) on the submerged surfaces. It then retreats to the shell interior, leaving a string that ties down the mussel to the pad. This cycle is repeated several times, resulting in a strong bond [7]. These MAPs are being utilized as the blueprint for the development of mucoadhesive substances. Furthermore, bioadhesion is also involved in the deposition of fatty substances on the walls of organism arteries, which is part of the mechanism resulting in atherosclerosis [8–10]. Other examples of bioadhesion between natural biomaterial surfaces include lectins, which are proteins or glycoprotein complexes having binding capacity to polysaccharides (sugars). Lectins are present in both plants and animals. Calcium-dependent binding lectin (also known as "C-type") is one of the common animal lectins. One of these C-type lectins, the asialoglycoprotein lectins, adhere specifically to liver cells and are associated with biological functions in animals [11]. Also, plant lectins are plentiful and are generally found in common plants like kidney beans, soybeans and tomatoes. Plant lectins may cause toxic effects if they bond to animal cells [11]. Ascidians, a group of marine animals, also possess the property of bioadhesion [12].

Bioadhesion in biological systems can be distinguished into four categories:

- Adhesion of a normal cell to another normal cell, for example in platelet aggregation
- Adhesion of a normal cell to a pathological cell, for example in healing of wounds
- Adhesion of a cell to a foreign substance, as in case of adhesion of cells to culture dishes and formation of biofilms on prosthetic inserts and devices
- Adhesion of an adhesive to a biological substance, for instance the ability of synthetic hydrogels to adhere to soft tissues [12, 13]

Bioadhesion between substances can also be disastrous which is known as biofouling. A biofilm is an accumulation of microbes on a solid substrate, which may form during the insertion of biomaterials into the biological systems. For humans and animals, the development of biofilms on both nonliving (such as catheters and stents) and living (cells) surfaces has been recognised as a reason for illness. Diseases, for example, cystic fibrosis, endocarditis and other infections that are generally resistant to antibiotics are known to occur due to bacterial biofilms. Another example of biofilm is dental plaque that can be comprised of more than 500 bacterial strains [5, 14].

4.3 Bioadhesion/Mucoadhesion

Bioadhesion implies the adhesion of substances (bioadhesives) to biological surfaces. Almost all of these biological surfaces which can be utilised for drug delivery are mucosal surfaces, and they are mostly surrounded by a layer of mucus: a complex hydrated gel-like substance consisting of a large glycoprotein mucin. This is especially true in case of lining of the small intestine, the main body surface for drug delivery. So a new term 'mucoadhesion' is introduced to characterize molecules that bind to the mucus layer. It is very challenging to differentiate between adhesion of molecules to mucus layer (mucoadhesion) and to the cell surfaces (bioadhesion). At the same time various bioadhesive molecules bind to both. Due to the presence of mucus layer, it can be concluded that mucosal surfaces seldom, if at all, exhibit 'naked' cell surfaces. A popular example of mucoadhesion is the attachment of mussels to submerged surfaces, for example ship or coral surfaces. One more significant example of mucoadhesion relates to drug delivery systems, in which the active ingredient is incorporated in a bioadhesive system, which then sticks to the mucus gel layer present on the mucosal surface [6].

4.3.1 Specific and Nonspecific Bioadhesion

The term bioadhesion can be considered to include both 'specific' and 'nonspecific' adhesion. The latter refers to materials, generally synthetic polymers, that will stick to the mucus and/or most biological surfaces. In contrast, 'specific' bioadhesion refers to those biological substances that identify and then adhere or bind to specific target chemical compounds on the cell surface or in mucus. Through an experiment, specificity can be illustrated by showing that the target substance in soluble form will inhibit bioadhesion. Lectins are the best substances to exemplify such molecules. These are mostly plant proteins that perceive and adhere to specific arrays of sugar molecules. Also different lectins will identify distinct combinations of sugars. Thus, if the cell surface is covered with a large number of sugar arrays, the lectins will adhere very well to these cell surfaces, and hence are bioadhesive. Bacterial invasins and fimbrins are other examples of 'specific' bioadhesives. Information regarding these specific bioadhesives provides a concept for targeting specific cells of tissues, which might be recognised by variations in their cell-surface sugar arrays [6].

4.4 The Mucosal Layer and Barriers to Drug Delivery

The walls of various body cavities, for example the gastrointestinal and respiratory tracts, are lined with moist surfaces known as mucosae or mucus membranes [15, 16]. These consist of a connective tissue layer (the lamina propria) with an epithelial layer whose surface is maintained moist, generally due to the presence of a mucus layer. The epithelia might be monolayered (as in case of bronchi, stomach, small and large intestines) or stratified/multilayered (for example in the cornea, vagina and oesophagus). The monolayered epithelia consists of goblet cells that release mucus directly onto the epithelial surfaces; the stratified epithelia consists of or are adjoining to tissues containing specialized glands, for example, salivary glands that release mucus onto the epithelial surface [15–17]. Mucus is available either as a gel layer adhering to the surface of mucosa or as a luminal soluble or suspended form. All mucus gels consist mainly of mucin glycoproteins, inorganic salts, lipids and water, the last-mentioned accounts for nearly 95% of their weight, resulting in a highly hydrated system [15–17]. Among these, the most significant constituent of the mucus are the mucin glycoproteins, as they could make intra- and/or intermolecular disulfide bonds with their cysteine-rich subdomains. The mucus is negatively charged due to the presence of sulfate and sialic acid. Though the mucus layer is uninterruptedly secreted, it is also consistently degraded by enzymatic and mechanical action on the luminal surface. The thickness of this mucus layer varies extensively with different mucosal surfaces from under 100 μm in the oral cavity to as much as 450 μm in the stomach [18]. Lubrication and protection are the important functions of mucus. Bioadhesive systems are generally utilized for sustained drug delivery across biological barriers. Nonetheless, based on the site of application, a quick administration is also achievable. Utilization of these systems to the oral, nasal, rectal, and vaginal mucosa or skin bypasses hepatic first-pass metabolism by delivering the drug directly to the systemic circulation to improve bioavailability. Improved delivery of biotherapeutics, for example peptides and proteins, is also achievable by bypassing low pH environment of the stomach and gastrointestinal enzymes and avoiding the low permeability of the small intestine. As mucosal sites contain many immunomodulatory molecules (produced by the goblet cells), vaccination efficiency can be enhanced [19].

4.5 Barriers to Mucosal Drug Delivery

The barriers to mucosal drug delivery primarily include the mucus gel layer and the mucosal tissue. The thick mucus gel formed over an epithelial cell surface poses a significant barrier to drug permeation. The mesh space of mucus is a measure of the size of the pores between the entangled mucin fibres. In a study using human cervicovaginal mucus, a mean mesh space of 340 ± 70 nm was found with 80 % of the mesh spaces being larger than 200 nm [20]. The mesh space of mucus varies greatly with its anatomical origin. Further, the rigidity of the mucus network is equally important for drug delivery. The rigidity of the mucus network depends on the numbers and types of intermolecular interactions involved, such as electrostatic interactions. The hydrophilic nature and the existence of numerous negative charges allowing for interaction of drug with the components of mucus have also been shown to retard diffusion and subsequently result in decreased drug absorption. Further, the mucosal tissue also plays a significant role as an obstacle to drug delivery. The morphology of the epithelial absorption barrier changes depending on the particular physiological site. Consequently, the nasal, lung and intestinal epithelia are made of a single layer of cells interconnected by tight junctions; while in the oral and vaginal cavities, epithelia are stratified and not tightly interconnected by junctions, but by an intercellular lipid matrix, resulting in an obstruction for absorption. As a consequence, the epithelial absorption profile of therapeutic agents may shift with regard to the rate and the lag time depending on the particular mucosal epithelium targeted as well as the physicochemical characteristics of the therapeutic agent. Absorption through the epithelia may take place by passive paracellular or transcellular diffusion. Hydrophilic substances, for example peptides, prefer the paracellular route as its main route for absorption. Substances being delivered will follow transport through paracellular or transcellular route, depending on the physical and chemical properties of the drug. Paracellular transport takes place via only passive diffusion, while transcellular transport can occur by passive, active, or facilitated transport processes. An active transport system can be used for transporting only a limited number of medications. For example, the intestinal absorption of di- and tri-peptides occurs via active transport by the oligopeptide transporter. For transport via the mucus pathways, either paracellular or transcellular transporters are involved; still, in a number of cases, both pathways are favoured in the absorption process. Molecules having a radius more than 15 Å show restricted transport through the tight intersections, representing the restricted area for the paracellular route [15].

4.6 Factors Affecting Mucoadhesion

The factors that may affect mucoadhesion are as follows:

1. ***Polymer (mucoadhesive) related factors:***

   i) Molecular weight
   ii) Concentration of active polymer
   iii) Flexibility of polymer chains
   iv) Spatial conformation
   v) Swelling
   vi) Hydrophilicity

2. ***Environment related factors:***

   i) pH at polymer - mucus interface
   ii) Applied force
   iii) Initial contact time

3. ***Physiological factors:***

    i) Mucin turnover
    ii) Presence of disease

Molecular weight
The interpenetration of polymer molecules is favoured by low molecular weight polymers; on the other hand, high molecular weight polymers adhere by entanglement. The type of polymer determines the optimum molecular weight for achieving highest mucoadhesion. A few investigations concluded that with an increase in the molecular weight of the bioadhesive polymer up to 100,000 Da, the bioadhesion strength increases and thereafter it becomes constant [21]. The adhesive characteristic is directly proportional to the molecular weight in case of a linear polymer, whereas in a non-linear polymer, it may or may not depend on the molecular weight [12].

Concentration of active polymer
An optimum concentration of polymer should be used. After an optimum concentration, the adhesion forces diminish significantly because of the separation of coiled polymeric regions so that the length of the polymer chain available for permeation becomes limited [12].

Flexibility of polymer chains
It is necessary for interpenetration and entanglement. As the degree of cross-linking increases in water-soluble polymers, the flexibility of the individual polymer chain diminishes. With further increase in cross-link density, the effectual length of the chain which can penetrate into the mucus layer further reduces significantly and mucoadhesion strength is decreased [15].

Spatial conformation
Bioadhesion will also be influenced by the spatial conformation of the polymer molecule. The polymers such as poly(ethylene glycol) having linear conformation support bioadhesion unlike dextran polymers having helical conformation which may cover numerous active groups which are responsible for adhesion [15].

Swelling and Cross-linking
The degree of swelling increases with a decrease in the cross-link density and vice versa [22]. The lower the cross-link density, the higher the

flexibility and hydration rate; the larger the surface area of the polymer, the better the mucoadhesion. Thus, a slightly cross-linked polymer is preferred to attain a high degree of swelling. However, the presence of excess moisture and high degree of swelling may result in slippery mucilage leading to an easy removal from the substrate [23]. In addition, the mucoadhesion of cross-linked polymers can be increased by including adhesion promoters, for example free polymer chains and polymers attached onto the preformed network [24].

Hydrophilicity

Numerous hydrophilic functional groups like hydroxyl and carboxyl are present on bioadhesive polymers. These groups form hydrogen bonds with the substrate, swell in aqueous media, thereby leading to maximum exposure of potential anchoring sites. Also, swollen polymers have the maximum distance between their chains resulting in enhanced flexibility of polymer chains and effective penetration into the substrate.

pH

The charge on the mucosal surface and the polymer is affected by the pH due to difference in the degree of dissociation of the functional groups of the amino acid and polymeric backbone. The degree of hydration of a polymer depends on its chemical structure and the pH [15].

Applied force and initial contact time

The initial applied force will also influence mucoadhesion [25]. Higher force results in increased interpenetration and greater bioadhesion strength [2]. Also, the longer the initial contact time between bioadhesive and substrate, higher the swelling and interpenetration of the polymer [13].

Mucin turnover and presence of disease

The residence time of the mucoadhesive systems on the mucus layer is limited by the natural mucin turnover. The rate of mucin turnover is different at various mucosal surfaces (anatomic differences) and in different individuals. The mucin turnover time in humans has been found to be in the range of 12–24 h and depends on many other physiological factors and presence of food. Besides, disorders like common cold, gastric ulcers, inflammation, bacterial or fungal infections can affect the mucosal physicochemical characteristics [13, 15].

4.6.1 Bioadhesive Interactions

The attachment and bonding between bioadhesive systems and mucosal surfaces commonly takes place by means of interpenetration followed by secondary non-covalent bonding which generally occurs by forming hydrogen bonds between them [26]. The bioadhesive polymers contain hydrophilic functional groups like carboxyl (-COOH), hydroxyl (-OH), amino ($-NH_2$) and sulfate groups ($-SO_4H$) that are suitable for targeted delivery. So, it can be concluded that fundamentally hydrogen bonds are responsible for the formation of a strong bond. Therefore, polymers having higher number of hydrogen bonding groups can interact with the glycoproteins with greater bioadhesion strength. A bioadhesive polymer that can easily adhere to the mucosal surface should have the following properties. Firstly, it should have adhesive property when placed in water. Secondly, it should bind non-covalently and non-specifically. Finally, the polymer should bind specifically to the receptor sites on the cells or mucosal surfaces. The preferable basic characteristics of bioadhesive polymers are anionic surface charge, high molecular weight, and low surface tension to spread on mucus layer [12, 27].

4.7 Mechanisms of Bioadhesion

4.7.1 Interface Energetics

In case of bioadhesive systems, the spreading of one material phase over another is influenced by the liquid environment. Consider the example of a polymer adhering to the mucus layer on a mucosal membrane in the presence of a liquid medium (such as drug-delivery systems). Figure 4.1 is a schematic illustration of the components of interfacial free energy that must be considered in assessing the thermodynamic work of adhesion of the system under consideration. During adhesion, a unit interface between the polymer and the liquid and between the mucus and the liquid disappears, whereas an interface between the polymer and mucus forms. Therefore, the thermodynamic work of adhesion (${W_{PM}}^{adh}$) between the polymer and mucus is expressed as follows [28, 29]:

$$ {W_{PM}}^{adh} = \gamma_{PL} + \gamma_{ML} - \gamma_{PM} \tag{4.1} $$

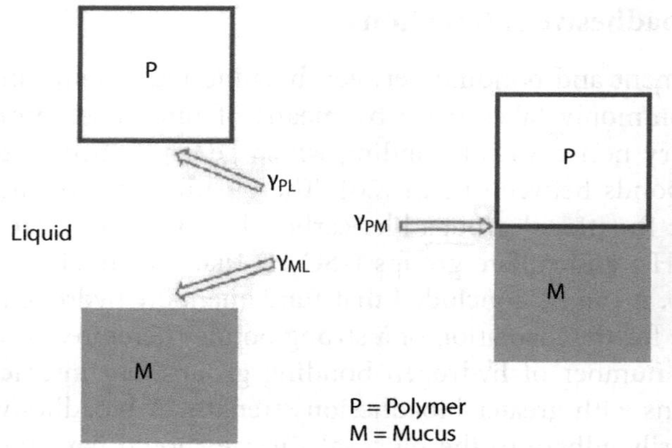

Figure 4.1 Schematic illustration of the applicable interfacial free energies (γ) determining the formation of a bioadhesive bond between a polymer (P) and the mucus layer (M) on the mucosal membrane immersed in a liquid medium (L).

Where γ stands for surface or interfacial free energies and the subscripts P, M, and L are for polymer, mucus, and liquid, respectively.

To attain an effective adhesion between the two surfaces, the value for $W_{PM}{}^{adh}$ should be positive.

4.7.2 Chemical Interactions

The process of bioadhesion can be facilitated by various interactions between two chemically active surfaces (i.e., not inert). In case the two surfaces are capable of creating either ionic, covalent or metallic bonds, strong adhesion can take place. Simultaneously, bonding between the two surfaces is also facilitated by weaker interaction forces, like hydrogen bonding, polar (dipole–dipole) bonding, or van der Waals forces [11, 29]. An example illustrating adhesion due to chemical interaction between the contacting surfaces is the case of mussel adhesion to submerged surfaces. The initial interaction between the mussel and the submerged surface requires the removal of weak boundary layer (mostly water). In case, the submerged surface is non-polar, the water boundary layer deposits via weak dispersion interactions. Because of the larger size of a MAP as compared to a water molecule, the protein encounters more prominent dispersion interactions with the non-polar surface relative to water, leading to the displacement of the water boundary layer and adhesion of the protein. If the submerged surfaces are polar, the water boundary layer cannot be easily displaced. Under these conditions, the MAP utilizes its hydrophilic amino acid side chains, which contain hydroxyalkyl, aminoalkyl and phenolic groups like DOPA (3,4-dihydroxyphenylalanine), all of which have the capability to form strong hydrogen bonds. Thus, the MAP becomes capable of displacing water and the mussel can attach to polar submerged surfaces also [30]. The extraordinary chemical composition and characteristics of the MAP have prompted the evolution of synthetic analogs for possible use as mucoadhesives for drug delivery systems.

4.7.3 Mechanical Effects

Though during the initial phases of bioadhesion, interfacial contact and chemical bonding interactions are required, the interdiffusion or interpenetration between the molecules of the two contacting surfaces will further continue the adhesive bond. In case a polymer is involved as one of the contacting surfaces in the bioadhesion interaction, the interpenetration process involves the mobility of the individual chains and their entanglement in the biological surface. An associated hypothesis is the swelling index of the polymer, which is the ratio of the wet to dry weights. Higher chain mobility and a greater propensity towards interpenetration for a given polymer will be due to its higher swelling index [28]. For instance, bioadhesion in dental prostheses involves mechanical interlocking or interpenetration as the main adhesion mechanism. When bonding to enamel, surface engraving is done to produce micropores on the surface leading to increased penetration of the restorative substance. Additionally, dentin contains tubules that radiate from the pulp. These tubular structures promote diffusion of resin monomers into the dentin and their retention once the resin has been polymerized [31].

4.8 Theories of Bioadhesion

For the purpose of drug delivery, bioadhesion means attachment of a drug delivery system to a precise anatomical/biological location. The biological location can be epithelial tissue or the mucus layer on the surface of a tissue. If the adhesive attachment is intended for mucus coating, it is referred to as *mucoadhesion*. Leung and Robinson described mucoadhesion as the interaction between a mucin surface and a synthetic or natural polymer [32].

For bioadhesion to occur, a series of steps are mandatory. The first step is an intimate contact between a bioadhesive component and a biological surface, either accomplished through wetting of the bioadhesive surface or through swelling of the bioadhesive component. In the second stage, after the intimate contact is established, penetration of the bioadhesive into the crevices of the tissue surface or interpenetration of the chains of the bioadhesive in the substrate proceeds. After this, secondary chemical bonds can then establish themselves [33].

Of the many significant factors, one of the prominent factors for bioadhesion is the roughness of tissue surface. The body tissue surfaces often exhibit undulations and crevices that offer both attachment points and mechanical interlocking for polymeric adhesives. Such surface irregularities have been observed by Merrill [34], who affirmed that the surface topography can be described in terms of an aspect ratio of maximum depth (d) to maximum width (h) as shown in Figure 4.2. The tissue surface that depicts a ratio of $< 1/20$ is considered too smooth for adequate bioadhesion [34]. For higher ratio values, the adhesives of low viscosity can penetrate into the tissue topography to establish mechanical or physical bonds.

Of the various factors, viscosity and wetting are also significant considerations in bioadhesion [35]. Bioadhesion can be considered as a wet-stick adhesion process [36]. During wet-stick, the polymer chains are released from restraining dry lattice forces by hydration. These chains then move and entangle into the substrate. Thereafter, interaction may occur at molecular level between the bioadhesive component and the substrate. The interaction is a result of both attraction and repulsion between the polymeric chains. The attraction is a result of van der Waals forces, hydrogen bonding, electrostatic attraction and hydrophobic interaction, whereas repulsion is a result of electrostatic and steric repulsions. The attractive and/or repulsive interactions may occur between the bioadhesive component and the lipids, glycolipids, proteins, glycoproteins, and polysaccharides found on epithelial cell membranes. Alternatively, the bioadhesive may interact with the glycocalyx on the cell surface. In the body regions, where mucus covers the epithelial surface (the gastrointestinal tract), the interaction may be an outcome of interplay between the bioadhesive and the mucin glycoproteins rather than the epithelial cell itself [11]. Conclusively,

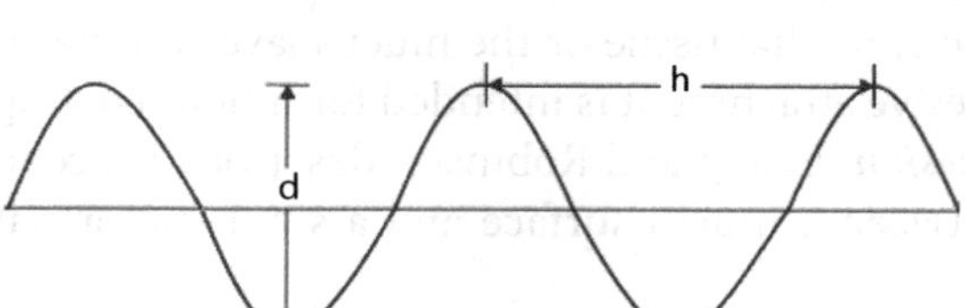

Figure 4.2 Schematic representation of the surface roughness of a soft tissue.

for bioadhesion to occur, the attractive interaction should be larger than non-specific repulsion [37].

Various theories exist to explain the bioadhesion process. However, five major theories can be described. Unfortunately, each theoretical model can explain a limited number of the diverse interactions that constitute the bioadhesive bond [32, 38].

4.8.1 Wetting Theory of Bioadhesion

The wetting theory is best applicable to liquid bioadhesives, describing the interactions between the bioadhesive, its contact angle and the thermodynamic work of adhesion. The work done is related to the surface tension of both the adhesive and the substrate, as described by Dupré equation [39, 40];

$$W_A = \gamma_b + \gamma_s - \gamma_{bs} \tag{4.2}$$

Where W_A is the specific thermodynamic work of adhesion and γ_b, γ_s, and γ_{bs} represent the surface tension(s) of the bioadhesive polymer (b), the substrate (s), and the interfacial tension, respectively. The work of adhesion is a sum of the surface tensions of the two adherend phases, minus the interfacial tension between the two phases [41]. Figure 4.3 represents a drop of liquid bioadhesive spreading over a soft tissue surface. Horizontal resolution of the forces leads to the Young equation (Eq. 4.3),

$$\gamma_{ta} = \gamma_{bt} + \gamma_{ba} \cos\theta \tag{4.3}$$

Where θ is the contact angle, γ_{bt} is the interfacial tension between the soft tissue (t) surface and bioadhesive polymer (b), γ_{ba} is the surface

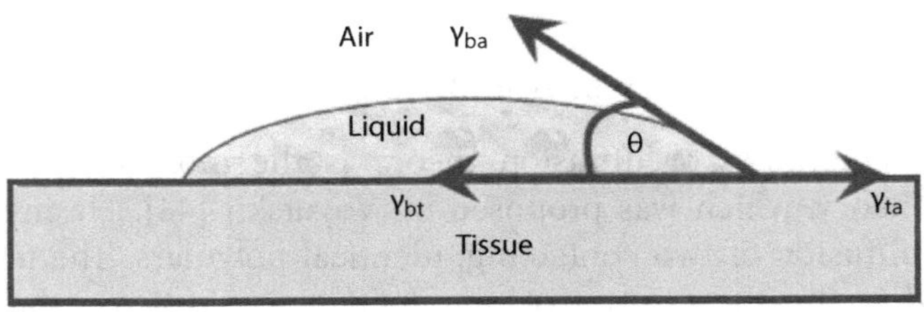

Figure 4.3 Diagrammatic representation of spreading of a bioadhesive liquid (b) over a typical soft tissue surface (t).

tension between bioadhesive polymer and air and γ_{ta} is the surface tension between tissue and air. Equation 4.3 states that if the contact angle θ is greater than zero, the wetting will be incomplete. If γ_{ta} greatly exceeds $\gamma_{bt} + \gamma_{ba}$, that is:

$$\gamma_{ta} \geq \gamma_{bt} + \gamma_{ba} \tag{4.4}$$

then θ will approach zero and wetting will be complete. If a bioadhesive material is to successfully adhere to a biological surface it must first displace barrier layers and then spontaneously spread across the underlying substrate, either tissue or mucus. The spreading coefficient, S_b, is defined as shown in equation 4.5,

$$S_b = \gamma_{ta} - \gamma_{bt} - \gamma_{ba} \tag{4.5}$$

The equation clearly states that bioadhesion is possible if S_b is positive. In other words, bioadhesion is favoured by large values of γ_{ta} or by small values of γ_{bt} and γ_{ba} [41, 42].

4.8.2 Electrostatic Theory of Bioadhesion

Separation of two distinct bodies is achieved by overcoming the electrostatic forces that are a result of formation of an electrical double layer between a substrate and an adhesive. The electrostatic theory of repulsion defines the transfer of electrons across the adherend-adhesive interface, establishing the electrical double layer and a series of attractive forces responsible for maintaining contact between the two layers [43]. The adherend-adhesive interface is considered to be analogous to a parallel-plate condenser, such that the work must be done against any electrical charges before separation is achieved. The work of adhesion can be equated to the energy of the condenser, provided that no work is done in overcoming van der Waals forces.

4.8.3 Diffusion Theory of Bioadhesion

The interpenetration or diffusion theory is the most widely accepted physical theory, which was proposed by Voyutskii [44]. He investigated the auto-diffusion of two contacting, identical polymers. The term *auto-diffusion* describes the diffusion of long chain molecules across the interface from one polymer into another identical one. Many adhesives are applied to the adherends as a solvent or aqueous solution and on evaporation of the vehicle the polymer residues are combined. Failure of the joint is not normally expected along the interface because diffusion results in formation of single layer, with a loss of the interface [41].

In context to bioadhesion, where the polymeric chains from the bioadhesive and the biological substrate are involved, they intermingle and reach a sufficient depth within each other to allow formation of a semi-permanent bond [11]. The process initiates with initial contact, followed by a concentration gradient that drives the polymeric chains of the bioadhesive into the mucus network and the glycoprotein mucin chains into the bioadhesive matrix until an equilibrium penetration depth is achieved as shown in Figure 4.4. Though the precise depth needed for effective bioadhesive bonds is unclear, but is proposed to be in the range of 0.2–0.5 μm [2]. The mean diffusional depth of the bioadhesive polymer segments, *S*, is given by equation 4.6,

$$S = \sqrt{2tD} \tag{4.6}$$

Where *D* is the diffusion coefficient and *t* is the contact time. Duchene *et al.* [33] adapted equation 4.6 to derive Eq. 4.7, which can be used to determine the time to bioadhesion (t) of a particular polymer:

$$t = \frac{x^2}{\mathrm{D_b}} \tag{4.7}$$

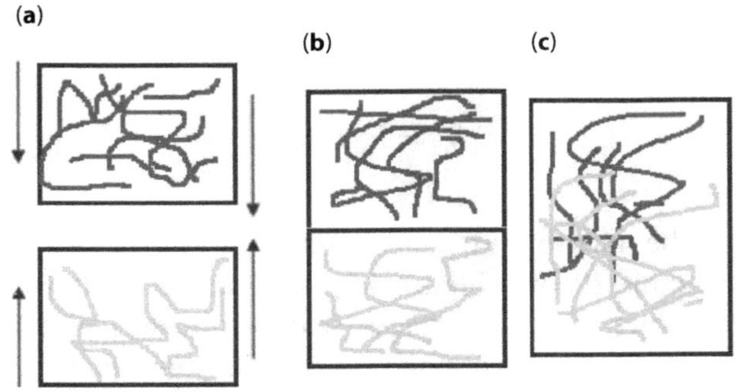

Figure 4.4 Schematic representation of the diffusion theory of bioadhesion. Blue colored polymer layer and pink coloured mucus layer before contact (a); upon contact (b); the interface becomes diffuse after contact for a period of time (c).

in which x represents the interpenetration depth and D_b the diffusion coefficient of the bioadhesive in the substrate. Once intimate contact is achieved, the substrate and adhesive chains move along their respective concentration gradients into the opposite phases.

The depth of diffusion is dependent on the diffusion characteristics of both phases. Reinhart and Peppas [45] reported that the diffusion coefficient is dependent on the molecular weight of the polymeric strand, and that it decreases with increasing cross-link density [32].

4.8.4 Adsorption Theory of Bioadhesion

According to the adsorption theory, after an initial contact between two surfaces, the materials adhere because of surface forces acting between the functional groups at the two surfaces [46]. When polar molecules or groups are present, they re-orient at the interface [41]. Chemisorption may be manifested when adhesion is specifically strong. The theory maintains that adhesion to tissue is due to the net result of one or more secondary forces. The three types of secondary forces are (i) van der Waals forces, (ii) hydrogen bonding, and (iii) hydrophobic interaction. The formation of primary chemical bonds such as ionic or covalent bonds is undesirable in bioadhesion as these may result in permanent adhesion [46].

4.8.5 Fracture Theory of Bioadhesion

This is perhaps the most utilized theory in studies to determine the mechanical strength of mucoadhesion. It assesses the force required to separate two surfaces after adhesion is established. This force, S_m, is frequently calculated in resistance to rupture tests by the ratio of the maximal detachment force, F_m, and the total surface area, A_0, involved in the adhesion interaction.

$$S_m = F_m / A_0 \tag{4.8}$$

Where S_m is the maximum tensile stress.

Since the fracture theory (Figure 4.5) is concerned only with the force required to separate the two surfaces, it does not take into account the interpenetration or diffusion of polymer chains. Consequently, it is appropriate for use in the calculations for rigid or semi-rigid bioadhesive materials, in which the polymer chains do not penetrate into the mucus layer [47, 48].

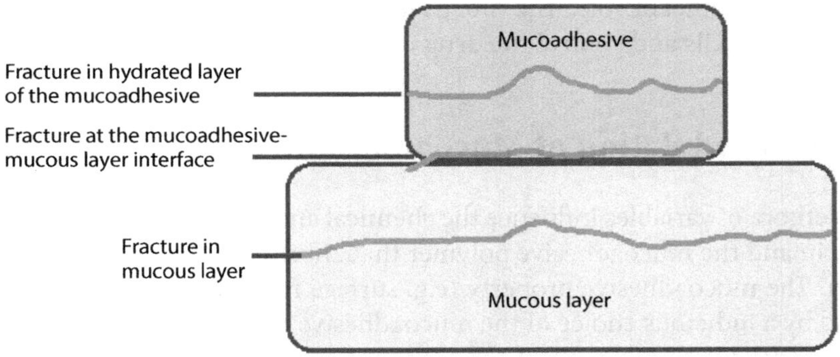

Figure 4.5 Regions that represent rupture of the mucoadhesive bond.

4.9 Stages of Mucoadhesion

For a better understanding of mucoadhesion, the process of mucoadhesion can be differentiated into three stages: wetting, interpenetration, and interaction of mucoadhesive with biological substrate. The mucoadhesive must wet the substrate to develop an intimate contact between the mating partners. The hydrophilic property of the mucoadhesive polymer is an important consideration in mucoadhesion because mucoadhesive interaction occurs in the presence of water (mucus consists of 95% water) [49]. Low contact angle between water and the polymer will encourage the hydration of the mucoadhesive polymer chains and increase the segmental mobility. The spreading of the mucoadhesive polymer over the mucus also promotes intimate contact. So, it is important that the mucoadhesive polymeric chains are able to diffuse into the mucus network so that interdigitation between the interacting materials may occur. The mucoadhesive polymer must have an adequate linear chain length to ensure interpenetration during mucoadhesion [50]. The segmental mobility of the mucoadhesive polymer chain is of great importance in inducing entanglement between the interacting polymeric chains. The proposed mechanism of adhesion of hydrocolloids suggests that upon hydration, the synthetic mucoadhesive polymer molecules become more mobile and are even able to orient themselves at adhesive sites of the substrate. As the level of hydration increases, the adhesion strength reduces since the mucoadhesive bonds become more extended. Based on the rate of diffusion of mucoadhesive polymer in the mucus network and the depth of penetration required for mucoadhesion, interdigitation alone cannot account for the mucoadhesive interaction because it is a time-dependent process. Therefore, secondary bond formation is assumed to play a significant role. The mucus layer acts as a defensive covering to protect the cells and control the drug delivery.

4.10 Modulation of Mucoadhesion

A plethora of variables influence the chemical and physical attributes of the mucin and the mucoadhesive polymer that affect the extent of mucoadhesion. The mucoadhesive property (e.g. surface free energy) can be modulated by a judicious choice of the mucoadhesive polymer [51]. Surface free energy modulation can affect the extent of mucoadhesion by influencing the interaction of mucin with mucosal liquids as listed below [52]:

- Various agents interact with the mucin and are designated as mucous thickening/thinning agents. Accumulation of agent (thickening/thinning) depends on electrostatic charges and composition of the mucin.
- Calcium precipitates mucin, and when mucin is used to vary the tonicity (osmotic pressure gradient) of the medium, it decreases shear stress. A higher value of shear stress shows better bioadhesion between polymer and mucus.
- The level of hydration of mucoadhesive declines with a consequent reduction in tensile stress, and thus expanding the ionic nature of the medium.
- The mucin can be altered by mucolytic agents that decrease the thickness of mucus by adjusting the organization of mucus by breaking the disulfide bonds or by the proteolytic activity of the catalyst. The breaking of disulfide bond leads to splitting of the disulfide bridges.
- Structural breakdown of the mucus polypeptide is affected by sodium deoxycholate and lysophosphatidylcholine.
- Duodenogastric reflux causes alteration in the structure of mucin.
- Some disease states disturb the integrity of the mucin layer. For example, ulceration and irritation of the digestive system cause rupture of the mucin layer, while cystic fibrosis causes thickening of the mucin layer and makes it harder to move air in and out of the lungs [53].

4.11 Molecular Biology in Bioadhesion

The capability of an organism to attach to a specific surface, either temporarily or permanently, is referred to as "bioadhesion". Bioadhesion is observed in many living organisms that have developed numerous ways to adhere to a range of surfaces [54–56]. The knowledge of as to how the organisms manage to adhere in diverse environments can lead to the development of novel bio-inspired adhesives [57] with major application in the areas of surface engineering and biomedicine. Molecular biology can assist in bioadhesion research related to the isolation of genes, their expression and function. The scientific methods in the field have advanced tremendously in recent years, primarily due to advances in DNA and RNA sequencing, and protein analysis.

The advances in molecular biology have provided a wide range of research tools and techniques that are of particular interest to researchers working on bioadhesion of organisms where no reference genome exists. The bioadhesion research is primarily based on histology, biochemistry and mechanics [54, 56] but gradually model systems such as mussels [58], barnacles [59, 60], sandcastle worms [61], starfishes [62], and flatworms [63] are also finding way in molecular biology. Efforts to develop bio-inspired adhesives are most effective when guided by a detailed understanding of the key features and mechanisms of natural adhesives [64]. However, no single protocol can be applied to all organisms [65].

There are four different steps for the transcripts involved in adhesion. These steps include (i) generation of transcriptome and differentially expressed cDNA that are enriched in adhesion-related transcripts; (ii) setting up a basic local alignment search tool (BLAST); (iii) performing an *in-situ* hybridization screening, and (iv) functional analysis of selected genes using RNA interference. These steps are detailed sequentially in the following text. For the generation of a transcriptome that encodes only adhesion related genes of a living organism, it is mandatory to select a specific tissue that contains bioadhesive organs. Selection of appropriate bioadhesive organ is necessary to minimize the complexity of the transcriptome while expressing genes and minimizing cost. The steps for the generation of a transcriptome include isolation of total RNA and selection of polyRNA. After selection, the next step is fragmentation of RNA into 200-300 base pairs followed by reverse transcription into complementary DNA [66]. Thereafter, the sequencing adapters bind chemically by a standard procedure. On the other hand, strand-specific sequencing can be executed. The selection of particular size range base pairs (200) followed by polymerase chain reaction based amplification leads to the generation of DNA [67]. Finally, bioinformatic data analysis is carried out that includes error correction, de-multiplexing and artifact removal. Finally, the data is assembled into a hypothetical transcript that results in the transcriptome of the selected organism [68].

Next is setting up a basic local alignment search tool (BLAST) which is basically a software for the analysis of the sequence databases. By use of this software, adhesion-related transcripts of the organism can be analyzed as well as mass spectrometry results for peptide sequences can be compared with the predetermined transcriptome database [69].

In situ hybridization can be used to detect the spatial and temporal genes expression within a tissue for investigating different kinds of nucleic acids. The principle of *in situ* hybridization includes pretreatment (by paraformaldehyde and proteinase K) of the tissue or the organism with an adhesive organ that consists of staining the organisms [70]. For assessing whether a transcript is expressed into an adhesive organ of the organism or not, functional analysis of both protein and the gene is essential. Of the various techniques used to determine the role of a gene, RNAi gives fast and direct results [12].

4.12 Administration of Bio- and Mucoadhesive Drug Delivery Systems

Mucosa is the moist tissue that lines the organs and body cavities such as mouth, nose, eyes, gut, rectum and genital area. The anatomical differences in mucus membranes are outlined in Table 4.1. The presence of mucous membrane offers the possibility of drug delivery via bio-/mucodhesive drug delivery systems. Examples of mucoadhesive drug delivery systems include powders, compacts, sprays, semisolids, and films. The dosage forms include compacts for drug delivery into the oral cavity [73], powders and nanoparticles for delivery to nasal mucosa [74, 75], and oral strips for the tongue and buccal cavity [76]. Furthermore, the buccal films provide a greater degree of flexibility and comfort than adhesive tablets. They may also prolong the residence time compared to oral gels [77]. The most commonly used polymers for such formulations are cellulose derivatives, poly (acrylic acid) namely Carbopol®, and Gantrez® copolymers, like poly(methylvinylether/maleic anhydride) [36].

Table 4.1 Anatomical differences in mucus membranes [16].

Mucus membrane	Characteristic anatomical features
Buccal [17]	Surface area of the buccal mucosa: approximately 30 cm^2 Layers include: epithelium, basement membrane, and connective tissues Non-keratinized cells constitute buccal mucosa, sublingual, and soft palate Keratinized cells constitute gingival tissue and hard palate Thickness of buccal epithelium is between 500–800 μm, 40–50 cells thick Mucus secreted by salivary glands forms a 0.1–0.7 mm thick layer Mucus turnover time: 5–6 days Permeability barrier property of oral mucosa is attributable to intercellular materials
Nasal [18]	Surface area of the nasal cavity: approximately 160 cm^2 Length of nasal cavity: approximately 60 mm Volume of nasal cavity: approximately 20 mL Mucous membrane lining is comprised of goblet cells, columnar cells, and basal cells Columnar cells are lined with cilia, the anterior part of the nasal cavity is also covered with cilia Presence of both keratinized and non-keratinized epithelial cells within nasal cavity at various locations Mucociliary clearance is affected by ciliary movement Submucosal glands and goblet cells secrete mucus to form a layer 5–20 μm thick Turnover time of mucus is usually ∼10–15 min

(*Continued*)

Table 4.1 Anatomical differences in mucus membranes [16]. (*Continued*)

Mucus membrane	Characteristic anatomical features
Ocular [71]	Made of epithelium, Bowman's layer, stroma, Descemet's membrane, and endothelium. Epithelium is made up of 5–6 layers of cells (basal layer being columnar, and the outermost cells are flattened polygonal cells) Tight junction between the basal cells of the corneal epithelium. Conjunctiva is structurally continuous with the corneal epithelium. Conjunctival tissue is permeable to molecules with molecular weight up to 20,000 Da, whereas the cornea is impermeable to molecules with molecular weight > 5000 Da Conjunctiva is made of around~5 million goblet cells, which synthesize mucins and peptides Volume of mucus secreted /day~2–3 μL Mucus turnover time = 15–20 h Exposed part of the eye is covered by percorneal tear film Tear film thickness:~3–10 μm
Vaginal [45]	Length of vaginal cavity: 6–10 cm Epithelial layer consists of lamina propia and stratified squamous epithelium Cell turnover of about 10–15 layers: about 7 days The surface is usually covered with vaginal fluid Vaginal fluid is made up of cervical mucus Volume, viscosity, and pH of the cervical mucus vary with age and during the menstrual cycle
Rectal [72]	Length:~15–20 cm Surface area: about 300cm^2 Epithelium consists of a single layer of cylindrical cells and goblet cells secreting mucus The surface is relatively flat, without villi and with only three major folds, the rectal valves. Approximately 3 mL of mucus with a neutral pH spread over the surface.

4.13 Prospects

Bioadhesion is one of the primary reasons for enhanced drug delivery to various mucosal sites, and thus has garnered considerable interest in formulators for developing bioadhesive formulations. Bioadhesive dosage forms offer sufficiently prolonged contact at the site of administration, low enzymatic activity, and enhanced drug permeability. Many bioadhesive systems have been investigated and many more are being investigated. The bio/mucoadhesive formulations in the future can be expected to be bioadhesive polymeric coatings or highly specific 'biological' ligands.

The formulations of bio-/mucoadhesive drug delivery systems depend on the judicious selection of bioadhesive polymers with excellent mucoadhesive property and biocompatibility. Researchers are now focusing on next-generation mucoadhesive polymers such as lectins and thiols. These offer stronger bioadhesion and retention of dosage forms. However, these novel mucoadhesive formulations will require considerable research work before these are approved for clinical use.

4.14 Summary

Drug administration to the various mucosal cavities can be improved via bioadhesion. In the near future, various bioadhesive delivery systems are likely to be used for treating a plethora of diseases. Thus it becomes extremely important to understand the fundamental mechanisms, theories and factors that govern bioadhesion so that a viable bioadhesive dosage form can be developed. This is further driven by the fact that dosage forms to satisfy patient's demand for convenient dosage forms are needed. The overview presented in the chapter about bioadhesion might be a useful tool for the efficient design of novel mucoadhesive drug delivery systems. With the influx of a large number of new drug molecules, bioadhesion can serve an important role in delivering these molecules. However, the novel mucoadhesive formulations require considerable research to deliver these clinically for the treatment of both topical and systemic diseases.

References

1. D. Duchene and G. Ponchel, Principle and investigation of the bioadhesion mechanism of solid dosage forms. *Biomaterials* 13, 709-714 (1992).

2. N.A. Peppas and P.A. Buri, Surface, interfacial and molecular aspects of polymer bioadhesion on soft tissues. *J. Controlled Release* 2, 257–275 (1985).
3. S.R. Mobley, J. Hilinski, and D.M. Toriumi, Surgical tissue adhesives. *Facial Plastic Surgery Clinic North America* 10, 147–154 (2002).
4. T.K. Vaidyanathan and J. Vaidyanathan, Recent advances in the theory and mechanism of adhesive resin bonding to dentin: a critical review. *J. Biomed. Mater. Res. B* 88, 558–578 (2009).
5. M.L. Palacio and B. Bhushan, Bioadhesion: A review of concepts and applications. *Philos. Trans. Royal Soc. A* 370, 2321-2347 (2012).
6. J. Woodley, Bioadhesion: new possibilities for drug administration. *Clinical Pharmacokinetics*. 40, 77–84 (2001).
7. J.R. Burkett, J.L. Wojtas, J.L. Cloud and J.J. Wilker, A method for measuring the adhesion strength of marine mussels. *J. Adhesion* 85, 601–615 (2009).
8. Anonymous. Hardening of the arteries. A.D.A.M. Medical Encyclopedia. http://www.ncbi.nlm.nih.gov/pubmedhealth/PMH0001224 (2011).
9. W.D. Spotnitz and S. Burks, Hemostats, sealants, and adhesives: components of the surgical toolbox. *Transfusion* 48, 1502–1516 (2008)
10. R. Noort, *Introduction to Dental Materials*, 3rd edn. Elsevier, London (2007).
11. X. Yang and J.R. Robinson, Bioadhesion in mucosal drug delivery. In: *Biorelated Polymers and Gels*, T. Okano (Ed.). pp. 135-192, Academic Press, San Diego (1994).
12. K. Pathak and R Malviya, Introduction, theories and mechanisms of bioadhesion, In: *Bioadhesives in Drug Delivery*, K.L. Mittal, I. S. Bakshi and J. K. Narang (Eds.), pp.3–28, Wiley-Scrivener, Beverly, MA (2020).
13. S. Asati, S. Jain and A Choubey, Bioadhesive or mucoadhesive drug delivery system: A potential alternative to conventional therapy. *J. Drug Deliv. Therapy*. 9, 858-867 (2019).
14. I. Kroes, P.W. Lepp, and D.A. Relman, Bacterial diversity within the human subgingival crevice. *Proc. Natl Acad. Sci. USA* 96, 547–552 (1999).
15. S. Pattnaik, K, Pathak and K. Swain, *Novel Drug Delivery Systems*, pp. 4.1-4.26, Doonville Publishers (P) Ltd, Dehradun, India (2020).
16. R. Shaikh, T.R. RajSingh, M.J. Garland, A.D. Woolfson and R.F. Donnelly, Mucoadhesive drug delivery systems. *J. Pharm. Bioallied Sci.* 3, 89–100 (2011).
17. N. Salamat-Miller, M. Chittchang, and T.P. Johnston, The use of mucoadhesive polymers in buccal drug delivery. *Adv. Drug Deliv. Rev.* 57, 1666–1691 (2005).
18. M.I. Ugwoke, R.U. Agu, N. Verbeke, and R. Kinget, Nasal mucoadhesive drug delivery: background, applications, trends and future perspectives. *Adv. Drug Deliv. Rev.* 57, 1640–1665 (2005).
19. E. Larrañeta and R. F. Donnelly, Bioadhesive polymers for drug delivery, in: *Polymers for Biomedicine: Synthesis, Characterization, and Applications*, C. Scholz (Ed.), pp. 559-601, John Wiley & Sons (2017).

20. S.K. Lai, Y.Y. Wang, K. Hida, R. Cone, and J. Hanes, Nanoparticles reveal that human cervicovaginal mucus is riddled with pores larger than viruses. *Proc. Natl. Acad. Sci. USA*, 107, 598-603 (2010).
21. A. Lohani and G.P. Chaudhary, Mucoadhesive microspheres: A novel approach to increase gastroretention. *Chronicles of Young Scientists* 2, 121-128 (2012).
22. L. Gudeman and N.A. Peppas, Preparation and characterisation of pH-sensitive, interpenetrating networks of poly (vinyl alcohol) and poly(acrylic acid). *J Appl. Polym. Sci.* 55, 919-928 (1995).
23. P.A. McCarron, A.D. Woolfson, and R.F. Donnelly, Influence of plasticiser type and storage conditions on the properties of poly(methyl vinyl ether-co-maleic anhydride) bioadhesive films. *J Appl. Polym. Sci.* 91,1576-1589 (2004).
24. N.A. Peppas, M.D. Little, and Y. Huang, Bioadhesive controlled release systems. in: *Handbook of Pharmaceutical Controlled Release Technology*, D.L. Wise (Ed.), pp. 255-269, Marcel Dekker, New York (2000).
25. J.D. Smart, An *in vitro* assessment of some mucoadhesive dosage forms. *Int J Pharm* 73, 69-74 (1991).
26. S. Punitha and Y. Girish, Polymers in buccal mucoadhesive drug delivery system - A review. *Intl. J. Res. Pharm. Sci.* 2, 170-186 (2010).
27. S.U. Zate, P.I. Kothawade, G.H. Mahale, K.P. Kapse and S.P. Anantwar, Gastro retentive bioadhesive drug delivery system: A review. *Intl. J. Pharm Technol. Res.* 5, 1227-1235 (2010).
28. G. Buckton, *Interfacial Phenomena in Drug Delivery and Targeting*, pp. 237-281, Harwood Academic Publishers, Chur, Switzerland (1995).
29. B. Bhushan, *Introduction to Tribology*, pp. 164-170, Wiley, New York (2002).
30. J.H. Waite, Nature's underwater adhesive specialist. *Int. J. Adhesion Adhesives.* 7, 9–14 (1987).
31. E. Asmussen, E.K. Hansen, and A. Peutzfeldt, Influence of the solubility parameter of intermediary resin on the effectiveness of the Gluma bonding system. *J. Dental Res.* 70, 1290–1293 (1991).
32. S.H.S. Leung and J.R. Robinson, The contribution of anionic polymer structural features related to mucoadhesion. *J. Controlled Release* 5, 223–231 (1988).
33. D. Duchene, F. Touchard, and N.A. Peppas, Pharmaceutical and medical aspects of bioadhesive systems for drug administration. *Drug Dev. Ind. Pharm.* 14, 283–318 (1988).
34. E. W. Merrill, Properties of materials affecting the behaviour of blood at their surfaces. *Annals N. Y. Acad. Sci.* 283, 6–16 (1977).
35. P.B. Van Wachem, T. Beugeling, J. Feijen, A. Bantjes, J.P. Detmers, and W.G. Van Aken, Interaction of cultured human cells with polymeric surfaces of different wettabilities. *Biomaterials* 6, 403–408 (1985).

36. A.D. Woolfson, D.F. McCafferty, and G.P. Moss, Development and characterisation of a moisture-activated bioadhesive drug delivery system for percutaneous local anaesthesia. *Int. J. Pharm.* 169, 83–94 (1998).
37. K.R. Kamath and K. Park, Mucosal adhesive preparations. In; *Encyclopedia of Pharmaceutical Technology*, J. Swarbrick and J.C. Boylan, (Eds.), pp. 133-152, Marcel Dekker, New York (1992).
38. K.L. Mittal (Ed.), *Acid-Base Interactions: Relevance* to *Adhesion Science and Technology*, Vol. 2, CRC Press, Boca Raton, FL (2000).
39. W.H. Pritchard, The role of hydrogen bonding in adhesion. In; *Aspects of Adhesion*, Vol. 6, K.W. Allen (Ed.), pp. 11–23, Wiley, New York (1971).
40. K.L. Mittal, The role of the interface in adhesion phenomena, *Polym. Eng. Sci.*, 17, 467-473 (1977).
41. W.C. Wake, *Adhesion and the Formulation of Adhesives*. pp. 53-55, Applied Science Publishers, London, UK (1982).
42. D.E. Packham, The mechanical theory of adhesion, In; *Handbook of Adhesive Technology*, second edition, A. Pizzi and K.L. Mittal (Eds.), pp. 69-93, CRC Press, Boca Raton, FL (2003).
43. B.V. Deraguin and V.P. Smilga, Mechanisms of adhesion. In; *Adhesion: Fundamentals and Practice*. pp. 56-100, McLaren, London, UK (1969).
44. S.S. Voyutskii, *Autoadhesion and Adhesion of High Polymers*. John Wiley & Sons, New York (1963).
45. C.P. Reinhart and N.A. Peppas, Solute diffusion in swollen membranes. Part II: Influence of crosslinking on diffusion properties. *J. Membrane Sci.* 18, 227–239 (1984).
46. A. Ahuja, R.K. Khar, and J. Ali, Mucoadhesive drug delivery systems. *Drug Dev. Ind. Pharm.* 23, 489–515 (1997).
47. J.D. Smart, The basics and underlying mechanisms of mucoadhesion. *Adv Drug Deliv Rev.* 57, 1556-1568 (2005).
48. H. Hägerström, K. Edsman, and M. Strømme, Low-frequency dielectric spectroscopy as a tool for studying the compatibility between pharmaceutical gels and mucus tissue. *J Pharm Sci* 92, 1869-81 (2003).
49. S. Rossi, M.C. Bonferoni, F. D'Autilia, G. Sandri, F. Ferrari, C. Caramella, E. Mortara, V. Giannini and F. Gasparri, Associations of natural polymers to modulate mucoadhesion of vaginal rinse-off and leave-on formulations. *J. Drug Deliv. Sci. Technol.* 5, 496-502 (2014).
50. J. Schultz and M. Nardin, Theories and mechanisms of adhesion, in: *Handbook of Adhesive Technology*, A. Pizzi and K.L. Mittal (Eds.) pp. 19-33, Marcel Dekker, New York (1994).
51. J. das Neves, M.F. Bahia, M.M. Amiji and B. Sarmento, Mucoadhesive nanomedicines: Characterization and modulation of mucoadhesion at the nanoscale. *Expert Opinion Drug Delivery* 8, 1085-1104 (2011).
52. L. Bacakova, E. Filova, M. Parizek, T. Ruml and V. Svorcik, Modulation of cell adhesion, proliferation and differentiation on materials designed for body implants. *Biotechnology Advances* 6, 739-767 (2011).

53. F. Brochard-Wyart, P.G. de Gennes, L. Leger, Y. Marciano and E. Raphael, Adhesion promoters. *J. Phys. Chem.* 98, 9405-9410 (1994).
54. A.Chiovitti, T.M. Dugdale and R. Wetherbee, Diatom adhesives: Molecular and mechanical properties. In; *Biological Adhesives*, A.M. Smith and J.A. Callow, (Eds.), pp. 79-103, Springer-Verlag: Berlin (2007).
55. S. Gorb (Ed.) *Adhesion and Friction in Biological Systems.* Springer (2011).
56. J. Von Byern and I. Grunwald, *Biological Adhesive Systems: From Nature to Technical and Medical Application*. Springer, Vienna, Austria (2011).
57. S.Y. Yang, E.D. O'Cearbhaill, G.C. Sisk, K.M. Park, W.K. Cho, and J.M. Karp, A bio-inspired swellable microneedle adhesive for mechanical interlocking with tissue. *Nature Commun.* 4, 1-23 (2013).
58. Z. Bai, H. Zheng, J. Lin, G. Wang, and J. Li, Comparative analysis of the Transcriptome in tissues secreting purple and white Nacre in the pearl mussel *Hyriopsis cumingii. PLoS One 8,* e53617 (2013).
59. Z.F. Chen, K. Matsumura, H. Wang, S.M. Arellano, X. Yan, I. Alam, J.A.C. Archer, V.B. Bajic, and P.Y. Qian, Toward an understanding of the molecular mechanisms of barnacle larval settlement: A comparative transcriptomic approach. *PLoS One 6,* e22913 (2011).
60. H.C. Lin, Y.H. Wong, L.M. Tsang, K.H. Chu, P.Y. Qian, and B.K.K. Chan, First study on gene expression of cement proteins and potential adhesion-related genes of a membranous-based barnacle as revealed from Next-Generation Sequencing technology. *Biofouling 30,* 169–181 (2014).
61. B.J. Endrizzi, and R.J. Stewart, Glueomics: An expression survey of the adhesive gland of the sandcastle worm. *J. Adhesion. 85,* 546–559 (2009).
62. E. Hennebert, R. Wattiez, J.H. Waite, and P. Flammang, Characterization of the protein fraction of the temporary adhesive secreted by the tube feet of the sea star *Asterias rubens. Biofouling* 28, 289–303 (2012).
63. B. Lengerer, R. Pjeta, J. Wunderer, M. Rodrigues, R. Arbore, and P. Ladurner, Biological adhesion of the flatworm *Macrostomum lignano* relies on a duo-gland system and is mediated by a cell type-specific intermediate filament protein. *Frontiers Zoology,* 11, 1-15 (2014).
64. R.J. Stewart, Protein-based underwater adhesives and the prospects for their biotechnological production. *Appl. Microbiol. Biotechnol.* 1, 27-33 (2011).
65. M. Rodrigues, B. Lengerer, T. Ostermann and P. Ladurner, Molecular biology approaches in bioadhesion research. *Beilstein J. Nanotechnol.* 5, 983-993 (2014).
66. B.T. Wilhelm and J.R. Landry, RNA-Seq-quantitative measurement of expression through massively parallel RNA-sequencing. *Methods* 3, 249-257 (2009).
67. Z. Wang, M. Gerstein and M. Snyder, RNA-Seq: A revolutionary tool for transcriptomics. *Nature Rev Genetics* 1, 57-63 (2009).
68. C. Camacho, G. Coulouris, V. Avagyan, N. Ma, J. Papadopoulos, K. Bealer and T.L. Madden, BLAST: Architecture and applications. *BMC Bioinformatics* 10, 1-9 (2009).

69. A.Y. Chen and A. Chen, Fluorescence *in situ* hybridization. *Investigative Dermatology* 5, 1-4 (2013).
70. R. Pennati and U. Rothbacher, Bioadhesion in ascidians: A developmental and functional genomics perspective. *Interface Focus* 1, 1-10 (2015).
71. A. Ludwig, The use of mucoadhesive polymers in ocular drug delivery. *Adv. Drug Deliv. Rev.* 57, 1595–1639 (2005).
72. K. Edsman and H. Hägerström, Pharmaceutical applications of mucoadhesion for the non-oral routes. *J. Pharm. Pharmacol.* 57, 3–22 (2005).
73. G. Ponchel, F. Touchard, D. Duchene, and N.A. Peppas, Bioadhesive analysis of controlled-release systems. I. Fracture and interpenetration analysis in poly (acrylic acid)-containing systems. *J. Controlled Release* 5, 129–141 (1987).
74.. T. Nagai and R. Konishi, Buccal/gingival drug delivery systems. *J. Controlled Release* 6, 353–360 (1987).
75. B. Sayin, S. Somavarapu, X.W. Li, M. Thanou, D. Sesardic, and H.O. Alpar, Mono-N-carboxymethyl chitosan (MCC) and N-trimethyl chitosan (TMC) nanoparticles for non-invasive vaccine delivery. *Int. J. Pharm.* 363, 139–148 (2008).
76. R. Ilango, *In-vitro* studies on buccal strips of glibenclamide using chitosan. *Indian J. Pharm. Sci.* 59, 232-235 (1997).
77. R. Anders and H.P. Merkle, Evaluation of laminated muco-adhesive patches for buccal drug delivery. *Int. J. Pharm.* 49, 231–240 (1989).

Part 2

SPECIFIC ADHESIVES, CHARACTERISTICS AND APPLICATIONS

5

Fibrin Glue: Sources, Characteristics and Applications

Anindya Karmaker[1,2] and Shoeb Ahmed[1]*

[1]Department of Chemical Engineering, Bangladesh University of Engineering and Technology, Palashi, Dhaka, Bangladesh
[2]Department of Chemical Engineering, University of California Davis, Davis, CA, USA

Abstract

Fibrin glue is one of the most useful bioadhesives that mimics the last stage of wound healing by sealing the affected area and stopping the blood flow. In various surgical processes such as cardiovascular surgeries, skin grafting, it plays a crucial role. In this chapter, the evolution and history of fibrin glue have been discussed and various types of adhesives and their working mechanisms have also been reviewed. With the advent of modern purification methods and research work by prominent researchers in this field, the process for making the glue has changed drastically, however, the working principle of the glue has remained essentially the same. The production process for fibrin glue has been discussed in the latter part of the chapter which highlights the typical production stages, which are fibrin and thrombin separation, purification, quantification, and storage. Furthermore, a brief comparison of some commercial fibrin adhesives like CryoSeal®, Vivostat®, Tissucol®, and Evicel® has also been presented. The formulations of the commercial fibrin adhesives are mostly patented and these vary from one company to another based on the source and type of components, however, the properties are quite similar. The chapter concludes with a discussion on the recent developments and future trends of fibrin glue highlighting the research significance and diverse application areas of fibrin adhesives.

***Keywords*:** Fibrin glue, fibrin, thrombin, hemostasis, wound healing, factor XIII

**Corresponding author*: shoebahmed@che.buet.ac.bd

K.L. Mittal and S. Neogi (eds.) Adhesives in Biomedical Applications, (101–120) © 2023 Scrivener Publishing LLC

5.1 Introduction

Adhesives play a critical role in different surgical operations. Other than fixing man-made objects, adhesives are also extensively used for biological applications such as sealing or attaching tissues, skin, and even organs. Bioadhesives are also known to enhance the healing process significantly. When we get a cut and bleeding starts, a fibrin clot forms spontaneously to stop bleeding, initiating the healing process. This course of stoppage of bleeding or hemorrhage is coined as hemostasis. This natural healing mechanism has inspired the creation and development of different kinds of bioadhesives. The three most prevalent forms of bioadhesives utilized in surgical applications are hemostatic agents, tissue adhesives, and sealants [1]. Even though the adhesives have various names, they all operate in a similar manner and are categorized based on the application stage and purpose. Hemostatic drugs improve primary hemostasis, induce fibrin production, or prevent fibrinolysis [2]. Tissue adhesives are a class of natural and synthetic chemicals that are utilized in a range of local applications for hemostasis, wound closure, and fistula healing. The most commonly used tissue adhesives include cyanoacrylates, fibrin glues, and thrombin [3, 4]. A sealant is a substance that is used to prevent fluids from passing through a surface, joints, or holes in materials. Fibrin glue, which is produced from blood plasma, is one of the most frequently used commercial bioadhesives. It can be used as a biodegradable tissue adhesive or sealer to help with hemostasis or to keep air and fluid out. Human plasma and bovine thrombin were utilized to produce fibrin glue in the early phases of research and application experiments. However, bovine thrombin had many negative side effects. Separation and purification technologies have advanced to the point where pure thrombin can now be separated. This fibrin glue was later commercially developed and sold under various names by varying the composition and production methodology. Some of the commonly known brands are Biocol™ (L & R, USA), Tissucol®/Tisseel® (Baxter, USA), Beriplast® (CSL Behring, USA), Vivostat® (Vivostat A/S, Denmark), etc. All fibrin adhesives, however, have one thing in common: they all are relatively expensive. Despite the increased demand, their prices have not decreased substantially. Furthermore, the creation of this type of glue takes a long time from plasma separation to final adhesive form, and despite rigorous purification efforts, still some of them induce serious allergic responses in a small percentage of patients [5, 6].

The commercial fibrin glues are being used in major surgical operations over autologous fibrin glue due to the ease of application and effective adhesion performance. These developments did not occur in a day and took several years of research to reach this state. Hence, it is imperative to study the evolution of fibrin glue to understand its impact on society and the health sector in general.

5.2 Evolution of Fibrin Glue

Suture was first employed to stop bleeding in the early 1100 BC when grass and other materials were used to seal wounds [7, 8]. Suture has been used for wound closure and bleeding control for a long time because of its high tensile strength and low chances of wound opening. Unfortunately, suturing has a high disease transfer rate, particularly for internal surgical operations or long stitch lengths, is inconvenient to handle, and raises concerns about the possibility of transmitting blood-borne diseases by using the same needle [9, 10]. Various hemostasis agents, clips, staples, tapes, and tissue adhesives have all been created to assist with faster and better bleeding control and wound closure which overcomes these problems [8]. These techniques have grown in popularity because of their high effectiveness rate, and they are being continually explored and refined, leading to the development of new types of tissue adhesives with improved characteristics.

Bergel was the first to employ dry plasma as a source of fibrinogen and fibrin fleece to produce hemostasis [11]. Fibrin was first used as a scaffold for tissue regeneration and as a local hemostatic agent in the 1910s. During World War I, fibrin patches were widely used to control bleeding from parenchymatous organs. Young and Medawar [12] discovered the ability of fibrin to function as an adhesive during microsurgery for peripheral nerve sutures in an animal model. Its efficacy in humans was proved later by Seddon and Medawar [13]. Fibrinogen and thrombin were used as biological adhesives for human skin grafting in the 1940s. Tidrick and Warner [14] combined fibrinogen with bovine thrombin to speed up the production of fibrin clots and used it to attach human skin grafts in 1944 [14]. This paved the way for the creation of effective fibrin sealants. Early products, however, had inadequate adhesive property due to the lack of fractionation technologies for delivering a concentrated source of fibrinogen. As a result, these investigations focused on the direct administration of dry fibrin or fibrin tampons to hemorrhagic surfaces.

It was feasible to create concentrated fibrinogen due to the development of industrial plasma fractionation technologies, which resulted in a significant increase in the fibrin sealant's rheological characteristics (elasticity, tensile strength, and adhesiveness) [15]. Furthermore, improvements in microsurgical methods have enhanced the product's utilization. The first commercial multi-donor fibrin sealant solutions were introduced in Europe in the late 1970s. Most of today's freeze-dried standardized products include significant levels of clottable fibrinogen and factor XIII. They are dissolved in an antifibrinolytic substance (typically aprotinin), combined with thrombin in the presence of $CaCl_2$, and injected into the operation site using separate syringes. These products offer enhanced and reproducible features, increasing the likelihood of predicted surgical performance [16].

In several European countries, Canada, and Japan, only three commercial fibrin adhesives were registered up to 1997 and were developed by Immuno (Austria), Behring (Germany), and CRTS-Lille (France) [16]. Due to the history of large-pool fibrinogen concentrates for intravenous use being connected to hepatitis B virus transmission prior to the introduction of efficient viral inactivation methods, the US Food and Drug Administration (FDA) did not authorize any commercial fibrin sealants in the United States until recently. In 1978, the FDA revoked the license for fibrinogen concentrates. Furthermore, because fibrin sealant is a mixed product, each component must be proved to have a demonstrable benefit to the claimed effect in the mixture if it is not naturally copurified, making the registration process with FDA more difficult. The difficulty of performing double-blind, sufficient, and well-controlled clinical studies, as well as establishing relevant clinical outcomes hindered the production and registration of commercial fibrin sealants in the United States in numerous surgical applications. Because of surgeons' interest piqued by the clinical success of fibrin sealants in Europe, US hospitals began using single-donor or small pool cryoprecipitates, stored human plasma, and autologous fibrinogen preparations as a source of human fibrinogen, usually after mixing with commercial bovine thrombin preparations [17]. The ability to manufacture highly concentrated fibrinogen led to the production of fibrin glue with stronger adhesion property and, therefore, accelerated the use of fibrin glue [18].

The FDA approved several commercial fibrin adhesives in the United States in the year 1998 [19]. Fibrin glues mimic the last stage of blood clotting, in which fibrinogen is transformed to fibrin clot through a complex coagulation cascade. The two major components of fibrin glue are concentrated human-derived fibrinogen (along with factor XIII and other blood plasma proteins) and human or bovine thrombin in combination with calcium chloride solution [18, 20]. When the two components are mixed, thrombin converts fibrinogen into fibrin monomers, which are subsequently cross-linked to create a polymer. Meanwhile, in the presence of calcium chloride, thrombin transforms factor XIII to factor XIIIa, which stabilizes the network by creating amide bonds between fibrin molecules, stopping the blood flow. Antifibrinolytic agents (such as aprotinin) are used in some formulations to prevent fibrinolysis (the disassociation of fibrin clots caused by the plasmin enzyme). Before being used in some formulations, the fibrinogen and thrombin components are usually freeze-dried and then solubilized using aprotinin and calcium chloride solutions.

A family of fibrin glues has been widely employed in clinical applications. The majority of documented applications were in surgical procedures to control bleeding and leaking during and after operations [21]. Numerous research investigations on the use of fibrin glue as a hemostatic agent and sealant in cardiovascular surgery have been published, with the goal of preventing bleeding from the suture line and graft area, which is a common issue in these procedures [21–26]. Fibrin sealants are also used in neurosurgery to keep the cerebrospinal fluid out during operations on the central nervous system, as well as for peripheral nerve repair and grafting [21, 27, 28]. The use of fibrin glues in the treatment of gastrointestinal diseases such as bleeding peptic ulcers has also been investigated to replace surgical treatments with noninvasive endoscopic fibrin sealant injections. Sealants with accurate flow and crosslinking characteristics, as well as ample time to handle and inject them through lumen catheters, are required for this application [21]. Numerous studies on the use and efficiency of fibrin sealants in many medical fields, including plastic surgery and skin grafts [29], ENT (ear, nose, and throat), and head and neck surgery [30] trauma surgery [31], urology [21, 32], and ophthalmology [33] are also available.

Commercial fibrin glues, despite having numerous advantages like short cure time, biocompatibility, and biodegradability, can pose some risks and safety concerns. Thrombin produced from bovine sources might cause allergic responses in certain people. Furthermore, certain antibodies are produced in reaction to bovine factor V or thrombin, which can cross-react with human clotting factors, resulting in severe bleeding. There is a risk of infectious pathogens being transferred to humans when bovine-source thrombin is utilized. Human-derived thrombin might be utilized to address these concerns [16]. Another issue with utilizing autologous fibrin sealant is the risk of blood-borne diseases transmission, such as HIV and hepatitis A, B, and C, as a result of extracting fibrinogen and thrombin from pooled human plasma [34, 35]. To solve this problem, several fibrin sealants employ fibrinogen and/or thrombin taken from the patient's blood plasma (such as Vitagel™ (Orthovita Inc., USA) and CryoSeal® (ThermoGenesis Corp., USA) System). The adverse effects of antifibrinolytic agents, which are intended to prevent premature fibrinolysis (the dissolution of a fibrin clot), are another possible cause of worry with fibrin glue [35]. Despite several drawbacks, fibrin glue is still the most used surgical adhesive. New research and development in improving fibrin glues have already overcome some of the major limitations. Moreover, various types of fibrin adhesives have been developed for specialized surgical operations. The next section covers the various types of fibrin adhesives and their basic working mechanisms.

5.3 Types of Fibrin Adhesives and their Working Mechanisms

Thrombin and fibrin are the two major components of fibrin adhesives. However, various polymers and compounds are added to enhance the adhesion and binding properties [36–38]. Figure 5.1 shows a typical dual syringe fibrin adhesive applicator used for common surgical operations. Based on the type of additive, fibrin-based adhesives can be categorized into few groups. Table 5.1 highlights the type, composition, and functioning mechanism of various types of fibrin adhesives [39].

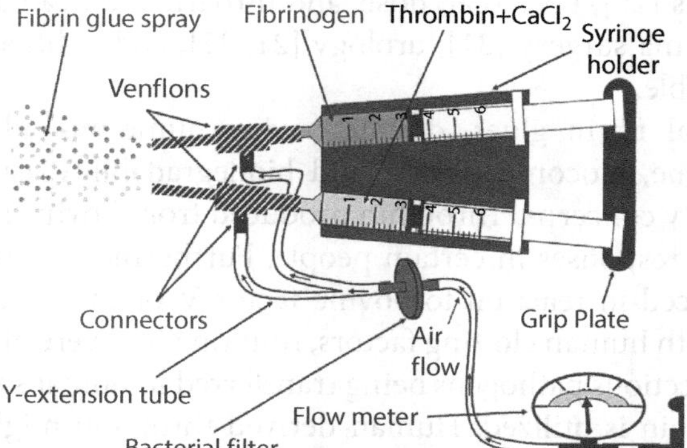

Figure 5.1 Typical spray applicator for use with various types of fibrin adhesives [40].

Table 5.1 List of major hemostats, sealants, and adhesives available for clinical use.

Type	Major components	Functioning mechanism
Mechanical hemostats	Bovine collagen Porcine gelatin Oxidized regenerated cellulose Polysaccharide spheres	Available in various forms of powder, sponge, microfibrillar, sheets, and flowable products. These hemostats swell on contact with blood and mechanically impede bleeding
Active hemostats	Human pooled thrombin Bovine thrombin Recombinant thrombin Bovine gelatin with human thrombin. Porcine gelatin with thrombin	Comprise thrombin as an active ingredient, which converts fibrinogen available in patient's blood to fibrin clot, accelerating cessation of bleeding.
Sealants	Fibrin sealant and hemostat Poly(ethylene glycol) (PEG) based sealants	Typically consist of two or more components that undergo chemical reaction upon mixing, forming a solid body or hydrogel that seals the wound area.
Adhesives	Albumin and glutaraldehyde	Create covalent/secondary bonds with biological surfaces, which result in adhesion to tissue and/or approximating wound edges to close wounds and control bleeding.

Both natural and artificial additives are often added to the commercial fibrin glues as well. Lundblad in his study summarized the effects of various additives on the fibrin polymerization and gel structure which are shown in Table 5.2 [36].

Table 5.2 Effects of various additives/components on fibrin polymerization and gel structure.

Additive/ component	Effect
Fibronectin	Some positive impact on polymerization; cross-linking mediated by factor XIIIa allows it to be integrated into the gel structure which is critical for wound retraction
Calcium ions	Enhance polymerization by accelerating protofibril lateral association into fibers; it is necessary for thrombin activation of factor XIII
Factor XIII/XIIIa	Fiber size is increased, and the final fibrin gel structure is stabilized, as demonstrated by an increase in tensile strength. Most of the evidence suggests that it plays a role in gel cohesive force, but not in adhesion forces
Heparin	Unfractionated heparin promotes fibrin protofibril lateral polymerization, resulting in fibrin fibers; low-molecular-weight heparin prevents lateral polymerization, resulting in thin fibers. The serum amyloid P component enhances the effect of unfractionated heparin
Collagen	The result is determined by the type of collagen used. Collagen type IV promotes lateral polymerization and protofibril development. Collagen incorporated in a clot improves resistance to fibrinolysis, which may indicate reorganization into thin fibers or, more likely a result of collagen incorporation into the clot through fibronectin cross-linking. The elasticity of collagen monomer appears to be increased, but the adhesion strength appears to be decreased
Platelet factor 4	Facilitates the development of fibrin fibers by improving the lateral association of fibrin protofibrils
Hyaluronan	Binds to fibrinogen and speeds up the polymerization of fibrin, resulting in thicker fibrin fibers
IgG	Lateral polymerization is inhibited, resulting in thinner fibers. The resulting gel structure though forms slowly is more rigid

5.4 Production Methods of Commercial Fibrin Adhesives

Commercial fibrin sealants are prepared from fibrinogen sources such as plasma cryoprecipitate or Cohn fraction I [41]. Cryoprecipitate is produced by centrifuging plasma after it has been thawed at 2–4°C. Fibrinogen,

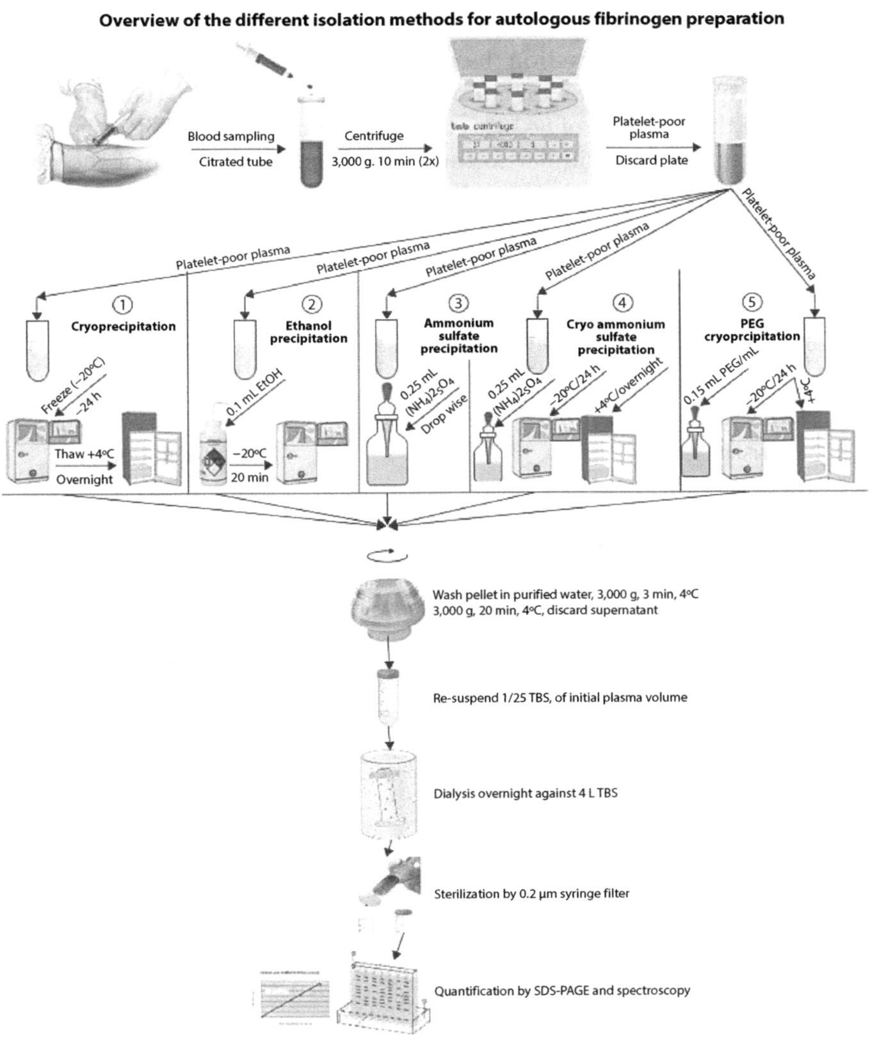

Figure 5.2 Methods for autologous fibrinogen preparation. Adapted from [42].

fibronectin, von Willebrand factor, factor VIII, and other plasma components (albumin, IgG, IgA, IgM, vitamin K-dependent clotting factors) captured during the centrifugation process make up cryoprecipitate. However, cryoprecipitate may be low in biologically active factor XIII. Figure 5.2 highlights some of the common ways of separating and purifying the fibrinogen.

Ferreira *et al.* outlined the action mechanism of the fibrin sealant in their paper which is summarized in Figure 5.3 [43]. It summarizes the interaction between the various components and the formation of stable fibrin clot.

To generate Cohn fraction I or equivalent fractions, whole human plasma is exposed to a precipitation step at low temperature (usually –3°C to –5°C) in the presence of 8–10% ethanol at neutral pH [41]. The proteins in Cohn fraction I are similar to those in cryoprecipitate, albeit the ratio of fibrinogen to fibronectin, as well as the quantity of factor XIII, may be higher. To enhance the purity of fibrinogen, several precipitation processes

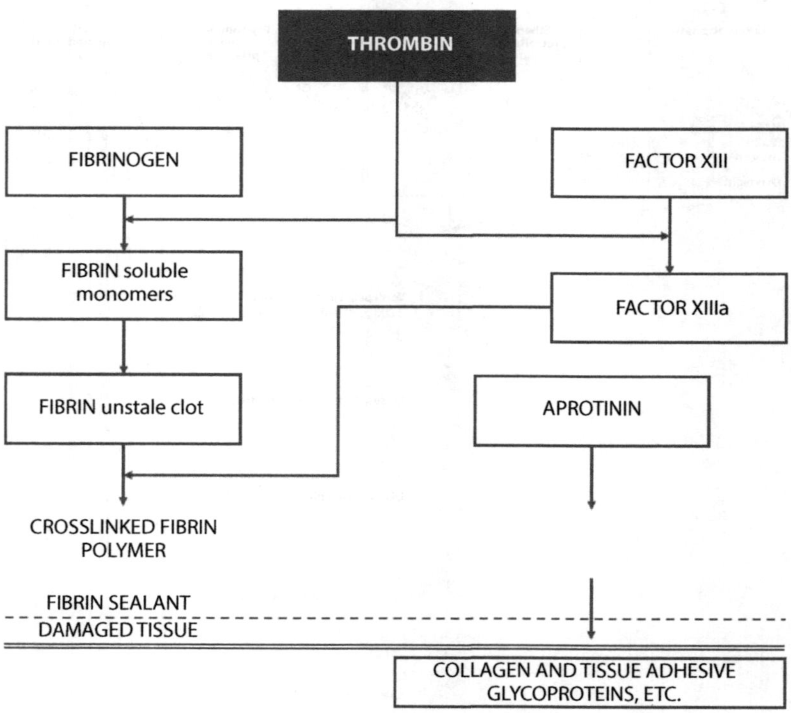

Figure 5.3 Simplified representation of the action mechanism of the fibrin sealant. Adapted from [43].

are generally employed to reduce the relative quantities of fibronectin, albumin, and immunoglobulins. The coagulability of fibrinogen is not affected by the purifying procedure. The fibrinogen concentration is subsequently extracted from the solution by centrifugation. Before being precipitated with 10% cold ethanol, the entire plasma is thawed at 37°C for at least 6 hours. After solubilization in a Tris/citrate/lysine buffer at a near to neutral pH following the viral inactivation stage by solvent/detergent (SD) process, the protein solution is cooled to 4°C and exposed to a second 10% ethanol precipitation phase for 10–12 hours. This allows for the purification of fibrinogen and the removal of virus-sterilizing chemicals. The precipitate is solubilized in a neutral pH Tris/NaCl buffer and ultrafiltered to reach the required protein content after centrifugation. After sterile filtration, this product is aseptically dispensed into vials and freeze-dried. Some of the typical procedures used to inactivate viruses in commercial fibrin sealant human plasma components include solvent/detergent process (Biocol™), two-step vapor heating (Tissucol®/Tisseel®), and pasteurization (Beriplast®).

5.5 Comparison of Some Commercial Fibrin Adhesives

Different methods for developing fibrin adhesives have been discovered to date and several of them have been employed in producing commercial fibrin adhesives. Some fibrin adhesives require external components like blood or plasma and others are available in ready-to-use dual syringes. The composition of proteins and stability of clots could be used to determine the efficacy of fibrin adhesives. CryoSeal® (ThermoGenesis Corp., USA) and Vivostat® are two commercial systems used to produce autologous fibrin glue. The CryoSeal® system produces cryoprecipitates which are used for producing fibrin glue [44]; however, the Vivostat® system uses batroxobin, a snake venom protease enzyme, to convert fibrinogen to fibrin [45]. On the other hand, Tissucol/Tisseel®, a homologous fibrin sealant, is based on virus inactivated pooled donor plasma [46]. Evicel® (Johnson & Johnson, USA), a novel formulation of a previously existing fibrin sealant, Quixil® (Omrix Biopharmaceuticals S.A., Belgium) or Crosseal™ (Johnson & Johnson, USA), is one such fibrin sealant made entirely of human products. Evicel® differs from Quixil®/Crosseal™ in that its fibrinogen component (made up of clottable proteins, mostly fibrinogen) does not contain tranexamic acid, an antifibrinolytic drug that prevents fibrinogen breakdown [47].

In the study reported by Buchta *et al.* [46], the protein composition of fibrin sealants was analyzed by U/SDS-PAGE and the concentrations of the various protein components were determined by densitometry. In both the fibrinogen and thrombin components of CryoSeal®, albumin was the most common protein. The fibrinogen component had a low quantity of fibrinogen, although fibronectin was present. Many other proteins were also determined but they could not be identified. Fibrin 1 and albumin were the most common proteins found in the Vivostat® fibrin 1 component; neither fibronectin nor significant quantities of other proteins were found. Tissucol/Tisseel® fibrinogen component is comprised of three main proteins: fibrinogen, fibronectin, and albumin. Albumin is the most prevalent protein in Tissucol/Tisseel® thrombin component. The protein composition for various commercial fibrin adhesives are summarized in Table 5.3.

Though the direct protein concentration in the individual components of Evicel® and Quixil® were not found, in a later study it was shown that they contain approximately double the amount of thrombin present in Tisseel® and Beriplast® [47]. However, Beriplast® P contains bovine aprotinin, which may cause hypersensitivity reactions, whereas Tisseel® contains synthetic aprotinin [48]. Buchta *et al.* [46] also determined the clot stability using U/SDS-PAGE obtained 24 hours after clot formation. After one day, CryoSeal® clots were partly dissolved, and after two days, they were fluid. Vivostat® clots were entirely dissolved after 1 day in their tranexamic acid-free formulation but remained stable for 4 days and showed partial dissolution after 5 days in their tranexamic acid-containing formulation. Over the course of 5 days, there was no change in the appearance of Tissucol®/Tisseel® clots.

Although there is no direct comparison of tensile strength values of all commercial fibrin glues, some data on a few commercial fibrin adhesives are available. In terms of fibrin clot tensile strength, *in vitro* research revealed that clots generated with Evicel® exhibited stronger resistance to stretching and higher tensile strength than clots formed with Tisseel®, implying that Evicel® clots were better able to preserve their structure against applied force [49]. The presence of factor XIII in the fibrinogen component of the sealant (9 IU/mL in Evicel® versus undetectable in Tisseel®) may account for the increased strength and durability of Evicel® clots. In a rat pancreaticojejunal attachment model, however, results from an *in vivo* study comparing fibrin sealants (Quixil® and Tisseel®), a semisynthetic sealant (Bioglue®), and cyanoacrylates (Histoacryl®, Dermabond®, and Glubran®) suggested that the tensile strength (as measured by pull test) of Quixil® was low compared to other bioadhesives [50].

Table 5.3 Composition of proteins in Vivostat®, CryoSeal® and Tissucol/Tisseel® fibrin adhesives [46].

	Vivostat®		CryoSeal®			Tissucol/Tisseel®[e]		
	Fibrin 1 component	**Clot (7:1)[a]**	**Fibrinogen component**	**Thrombin component**	**Clot**	**Fibrinogen component**	**Thrombin component**	**Clot**
Total protein (mg/ml)	37.3	32.64	70.00	40.4	55.20	115	50	83
Fibrinogen/ Fibrin (mg/ml)	18.69	16.35	19.88	0.00	9.94	90	0	45
Fibronectin (mg/ml)	0.00	0.00	4.20	0.00	2.10	6	0	3
Albumin (mg/ml)	14.17	12.40	37.31	40.4	38.86	15	50	33
Plasminogen (μg/ml)	41[b,c], 59[b,d]		107[b]			40-120		

[a]Vivostat® product is mixed with buffer in 7:1 ratio for clot formation
[b]Determined by ELISA
[c]With Tranexamic acid (tAMCHA)
[d]Without Tranexamic acid (tAMCHA)
[e]Tissucol/Tisseel® protein composition is based on official product specification

5.6 Recent Developments and Future Trend of Fibrin Adhesives

Fibrin glue has been prevalent for over 100 years [51]. Various types of fibrin glues have been developed with specific applications and better biocompatibility. Recent research studies focus on improving tissue bonding, controlled degradation, drug release, wound healing, and other specific properties [52–54]. In a recent research study, poly(glycolic acid) (PGA) sheets were used in conjunction with autologous fibrin glue in a rat model with introduced epithelial defects. It was observed that wound healing and epithelial regeneration were superior compared to the responses due to the commercial homologous fibrin glues based on the evaluation of pathological measurements and immunohistochemical findings [53]. This research shows the impact of various additives in enhancing the properties of autologous fibrin glue.

In another study, autologous fibrin glue was used to co-deliver antibiotics for topical treatment after sinus surgery. The slow release of antibiotics for more than 7 days cured *Pseudomonas aeruginosa* in eight out of nine patients by the final visit [54]. These results showcase another practical application of autologous fibrin glue in controlled drug release applications. A new research paper highlights the potential application of fibrin glue for delivering stem cells to targeted zones [55]. In this review paper [55], the clinical trial and evaluation of Mesenchymal Stem Cells (MSCs) delivery via fibrin glue was discussed. Based on this paper, it is certain that MSCs, alone or in combination with bone substitute, are a potential treatment option for lesions in bone or cartilaginous tissue when delivered with fibrin glue to the targeted areas.

Based on the newest research studies on fibrin adhesives, it is certain that the true potential of fibrin adhesives is beginning to be realized. The newest research trends mostly focus on imparting additional properties to the fibrin glue. Most of the newer studies are based on the drug delivery and wound healing aspects of fibrin adhesives. A few other studies are centered around controlled degradation and tissue binding properties of fibrin adhesives. Considering the scientific interest, it is apparent that unique biological and medical applications of fibrin adhesives are being discovered every day. Commercial fibrin adhesives have also changed their formulations and synthesis processes to improve the properties and impart newer ones. Based on the recent studies and considering the fact that fibrin adhesives are extremely useful, versatile, and cost-effective adhesives, these possess enormous potential in the future for diverse biological applications.

5.7 Summary

Fibrin adhesives are one of the most used surgical adhesives to date. Though the discovery of the fibrin adhesives dates back to the 19th century, the developed product back then and the ones that exist now are very similar in their working principle. However, the adhesion performance and biocompatibility have increased tremendously due to advancements in purification of fibrinogen and addition of various compounds to stabilize the crosslinked polymer. Different types of commercial fibrin adhesives are available in the market depending on the specific surgical requirements and wound characterisitics. Though the demand for fibrin adhesives is quite high, yet the cost of the adhesives has not decreased significantly. Various researchers are working on improving the performance of fibrin adhesives and decreasing the production cost. Some are also working on alternative adhesives which are much cheaper and offer better performance than fibrin adhesives. However, these adhesives are not certified by the regulatory authorities due to biocompatibility concerns and are undergoing animal testing and require extensive clinical trials before commercialization. Even with the discovery of newer types of adhesives, it is certain that fibrin adhesive will hold their dominant role in the market for a long time since it is still the most studied and extensively used adhesive by surgeons all around the world.

References

1. W. D. Spotnitz and S. Burks, Hemostats, sealants, and adhesives: components of the surgical toolbox, *Transfusion* 48, 1502–1516 (2008).
2. M. M. Levi, R. Vink, and E. de Jonge, Management of bleeding disorders by prohemostatic therapy, *Int. J. Hematol.* 76, 139–144 (2002).
3. M. Ryou and C. C. Thompson, Tissue adhesives: A review, *Techniques Gastrointestine Endoscscopy,* 8, 33–37 (2006).
4. E. M. Petrie, Cyanoacrylate adhesives in surgical applications, *Rev. Adhesion Adhesives,* 2, 253–310 (2014).
5. M. Saffarzadeh, A. Mulpuri, and J. S. Arneja, Recalcitrant anaphylaxis associated with fibrin sealant: treatment with "TISSEEL-ectomy," *Plast. Reconstr. Surgery - Global Open* 9, e3382 (2021).

6. R. Kanazawa, S. Sato, N. Iwamoto, and A. Teramoto, Allergic reaction following Arachnoid Plasty with a fibrin sealant—Case report—, *Neurologia Medico-Chirurgica (Tokyo)*, 50, 608–610 (2010).
7. G. Majno, *The Healing Hand: Man and Wound in the Ancient World*, Harvard University Press, Cambridge, MA (1975).
8. R. D. Forrest, Early history of wound treatment, *J. Royal Soc. Medicine*, 75, 198–205 (1982).
9. L. Mahesh, V. R. Kumar, A. Jain, S. Shukla, J. M. Aragoneses, J. M. M. González, M. Fernández-Domínguez, and J. L. Calvo-Guirado, Bacterial adherence around sutures of different material at grafted site: a microbiological analysis, *Materials (Basel)*. 12, 2848 (2019).
10. D. Millbourn, Y. Cengiz, and L. A. Israelsson, Effect of stitch length on wound complications after closure of midline incisions: a randomized controlled trial, *Archives Surgery*, 144, 1056–1059 (2009).
11. S. Bergel, Ueber wirkungen des fibrins, *DMW-Deutsche Medizinische Wochenschrift* 35, 663–665 (1909).
12. J. Z. Young and P. B. Medawar, Fibrin suture of peripheral nerves: measurement of the rate of regeneration, *Lancet* 236, 126–128 (1940).
13. H. J. Seddon and P. B. Medawar, Fibrin suture of human nerves, *Lancet* 240, 87–88 (1942).
14. R. T. Tidrick and E. D. Warner, Fibrin fixation of skin transplants, *Surgery* 15, 90–95 (1944).
15. H. Matras, F. Braun, H. Lassmann, H. P. Ammerer, and B. Mamoli, Plasma clot welding of nerves (experimental report), *J. Maxillofacial Surgery*, 1, 236–247 (1973).
16. M. Radosevich, H. I. Goubran, and T. Burnouf, Fibrin sealant: scientific rationale, production methods, properties, and current clinical use, *Vox Sanguinis*, 72, 133–143 (1997).
17. W. D. Spotnitz, P. D. Mintz, N. Avery, T. C. Bithell, S. Kaul, and S. P. Nolan, Fibrin glue from stored human plasma. An inexpensive and efficient method for local blood bank preparation, *Amer. Surgeon*, 53, 460–462 (1987).
18. M. Brennan, Fibrin glue, *Blood Rev.* 5, 240–244 (1991).
19. W. D. Spotnitz, Fibrin sealant: the only approved hemostat, sealant, and adhesive—a laboratory and clinical perspective, *Intl. Scholarly Res. Notices*, 1–28 (2014).
20. D. H. Sierra, Fibrin sealant adhesive systems: a review of their chemistry, material properties and clinical applications, *J. Biomater. Applications*, 7, 309–352 (1993).
21. D. M. Albala, Fibrin sealants in clinical practice, *Cardiovasc. Surgery*, 11, 5–11 (2003).
22. W. D. Spotnitz, Fibrin sealant in the United States: Clinical use at the University of Virginia, *Thrombosis Haemostasis* 74, 482–485 (1995).

23. G. Köveker, Clinical application of fibrin glue in cardiovascular surgery, *Thorac. Cardiovasc. Surgeon,* 30, 228–229 (1982).
24. J. Rousou, S. Levitsky, L. Gonzalez-Lavin, D. Cosgrove, D. Magilligan, C. Weldon, C. Hiebert, P. Hess, L. Joyce, and J. Bergsland, Randomized clinical trial of fibrin sealant in patients undergoing resternotomy or reoperation after cardiac operations. A multicenter study, *J. Thorac. Cardiovasc. Surgery,* 97, 194–203 (1989).
25. J. Stark and M. de Leval, Experience with fibrin seal (Tisseel) in operations for congenital heart defects, *Annals Thorac. Surgery,* 38, 411–413 (1984).
26. W. G. Schenk, C. A. Goldthwaite, S. Burks, and W. D. Spotnitz, Fibrin sealant facilitates hemostasis in arteriovenous polytetrafluoroethylene grafts for renal dialysis access, *Amer. Surgeon,* 68, 728-732 (2002).
27. C. I. Shaffrey, W. D. Spotnitz, M. E. Shaffrey, J. A. Jane, E. R. Laws, L. N. Sekhar, and O. Al-Mefty, Neurosurgical applications of fibrin glue: Augmentation of dural closure in 134 patients, *Neurosurgery* 26, 207–210 (1990).
28. M. R. Patel, W. Louie, and J. Rachlin, Postoperative cerebrospinal fluid leaks of the lumbosacral spine: Management with percutaneous fibrin glue, *Amer. J. Neuroradiol.* 17, 495–500 (1996).
29. L. J. Currie, J. R. Sharpe, and R. Martin, The use of fibrin glue in skin grafts and tissue-engineered skin replacements: A review, *Plastic Reconstructive Surgery,* 108, 1713–1726 (2001).
30. O. Staindl, Tissue adhesion with highly concentrated human fibrinogen in otolaryngology, *Annals Otol. Rhinol. Laryngol.* 88, 413–418 (1979).
31. M. G. Ochsner, Fibrin solutions to control hemorrhage in the trauma patient, *J. Long. Term. Effects Medical Implants,* 8, 161–173 (1998).
32. J. C. Wheat and J. S. Wolf, Advances in bioadhesives, tissue sealants, and hemostatic agents, *Urologic Clinics North Amer.* 36, 265–275 (2009).
33. L. Gauthier and F. Lagoutte, Use of a fibrin glue (Tissucol) for treating perforated or pre-perforated corneal ulcer, *J. Fr. Ophtalmol.* 12, 469–476 (1989).
34. C. Joch, The safety of fibrin sealants, *Cardiovasc. Surg.* 11, 23–28 (2003).
35. J.V. Quinn, *Tissue Adhesives in Clinical Medicine,* second edition, BC Decker Inc., Hamilton, Ontario, Canada (2005).
36. R. L. Lundblad, *Application of Solution Protein Chemistry to Biotechnology,* CRC Press, Boca Raton, FL (2009).
37. A. Karmaker, M. Hasan, M. Sharifuzzaman, and S. Ahmed, Development of natural fiber-supported high-strength autologous fibrin glue from human plasma, *J. Adhesion Sci. Technol.* 34, 1898–1911 (2020).
38. A. Karmaker, M. Hasan, S. Ali, K. M. Asif, and S. Ahmed, A novel high-strength autologous fibrin glue augmented with biocompatible polymers, *J. Adhesion,* 98, 1–16 (2022).
39. M. Mehdizadeh and J. Yang, Design strategies and applications of tissue bioadhesives, *Macromol. Biosci.* 13, 271–288 (2013).
40. T. A. E. Ahmed, E. V Dare, and M. Hincke, Fibrin: a versatile scaffold for tissue engineering applications, *Tissue Eng. Part B Reviews,* 14, 199–215 (2008).

41. M. Burnouf-Radosevich, T. Burnouf, and J. J. Huart, Biochemical and physical properties of a solvent-detergent-treated fibrin glue, *Vox Sanguinis,* 58, 77–84 (1990).
42. A. Noori, S. J. Ashrafi, R. Vaez-Ghaemi, A. Hatamian-Zaremi, and T. J. Webster, A review of fibrin and fibrin composites for bone tissue engineering, *Intl. J. Nanomedicine* 12, 4937–4961 (2017).
43. P. Ferreira, H. Gil, and P. Alves, An overview in surgical adhesives, in: *Recent Advances in Adhesion Research*, A. McFarland and M. Akins (Eds.) pp. 59-85, Nova Science Publishers, Hauppauge, NY (2013).
44. C. Buchta, M. Dettke, P. T. Funovics, P. Höcker, P. Knöbl, M. Macher, P. Quehenberger, C. Treitl, and N. Worel, Fibrin sealant produced by the CryoSeal® FS system: product chemistry, material properties and possible preparation in the autologous preoperative setting, *Vox Sanguinis,* 86, 257–262 (2004).
45. K. Stocker, H. Fischer, and J. Meier, Thrombin-like snake venom proteinases, *Toxicon,* 20, 265–273 (1982).
46. C. Buchta, H. C. Hedrich, M. Macher, P. Höcker, and H. Redl, Biochemical characterization of autologous fibrin sealants produced by CryoSeal® and Vivostat® in comparison to the homologous fibrin sealant product Tissucol/Tisseel®, *Biomaterials* 26, 6233–6241 (2005).
47. S. Dhillon, Fibrin sealant (Evicel® [Quixil®/Crosseal™]), *Drugs* 71, 1893–1915 (2011).
48. J. H. Levy, Efficacy and safety of aprotinin in cardiac surgery, *Orthopedics* 27, S659–S662 (2004).
49. W. L. Hickerson, I. Nur, and R. Meidler, A comparison of the mechanical, kinetic, and biochemical properties of fibrin clots formed with two different fibrin sealants, *Blood Coagulation Fibrinolysis,* 22, 19–23 (2011).
50. T. Lämsä, H.-T. Jin, J. Sand, and I. Nordback, Tissue adhesives and the pancreas: biocompatibility and adhesive properties of 6 preparations, *Pancreas* 36, 261–266 (2008).
51. I. Flemming, Fibrin glue in face lifts, *Facial Plastic Surgery,* 8, 79–88 (1992).
52. V. R. Lepsky, S. Natan, O. Tchaicheeyan, A. Kolel, M. Zussman, M. Zilberman, and A. Lesman, FITC-dextran release from cell-embedded fibrin hydrogels, *Biomolecules,* 11, 337 (2021).
53. A. Kouketsu, Y. Shimizu, S. Nogami, M. Yamada-Fujiwara, H. Nagai, K. Yamauchi, H. Miyashita, H. Saito, K. Odashima, Y. Yanagisawa, and T. Takahashia, Wound healing effect of autologous fibrin glue and polyglycolic acid sheets in a rat back skin defect model, *Transfusion Apheresis Sci.* 60, 103144 (2021).

54. K. Aanaes, K. G. Nielsen, E. Arndal, C. von Buchwald, T. Pressler, and N. Høiby, Autologous fibrin sealant co-delivered with antibiotics is a robust method for topical antibiotic treatment after sinus surgery, *Acta Oto-Laryngologica,* 141, 181–186 (2021).
55. A. de C. Ortiz, S. O. M. Fideles, K. T. Pomini, C. H. B. Reis, C. R. de S. Bueno, E. de S. B. M. Pereira, J. de O. Rossi, P. C. Novais, J. P. G. Pilon, and G. M. Rosa Jr., Effects of therapy with fibrin glue combined with Mesenchymal Stem Cells (MSCs) on bone regeneration: A systematic review, *Cells* 10, 2323 (2021).

6

Herbal Bioactives-Based Mucoadhesive Drug Delivery Systems

Shristhi Sohan Rawat, Arya Rai, Deepika Raina and Inderbir Singh*

Chitkara College of Pharmacy, Chitkara University, Punjab, India

Abstract
Mucoadhesive drug delivery systems are strategically designed and developed for adhering to the mucosal surfaces of gastrointestinal, ocular, nasal, vaginal, and rectal areas. Increased residence time provides effective drug targeting with pronounced therapeutic efficacy and reduced toxic effects at the site of action. The concepts of mucoadhesion, mucoadhesive polymers, and mucoadhesive drug delivery systems for ocular, nasal, oral, pulmonary, gastrointestinal, vaginal, and rectal routes have been elaborated. The applications of herbal bioactive-based mucoadhesive drug delivery systems have been systematically presented in tabular form. Various clinical studies and patents filed under mucoadhesive formulations for different herbal bioactives have also been included. In conclusion, mucoadhesive formulations are an effective, safe and preferred system for the delivery of herbal bioactives. To investigate the economic potential of herbal bioactive-based mucoadhesive drug delivery systems, regulatory and ethical considerations must be addressed.

***Keywords*:** Bioactives, bioadhesion, theories of bioadhesion, factors affecting mucoadhesion, 3D printing

6.1 Introduction

Bioadhesion is defined as the state in which two materials at least one of which is biological are held together by interfacial forces for an extended

**Corresponding author*: inderbir.singh@chitkara.edu.in

K.L. Mittal and S. Neogi (eds.) Adhesives in Biomedical Applications, (121–150) © 2023 Scrivener Publishing LLC

period. It is an approach to drug delivery that aims to exploit the natural adhesion between soft tissues such as the gastrointestinal mucosa and certain polymers. This allows the delivery of drugs and other molecules which would otherwise be unable to cross the mucosal barrier and access the bloodstream. Bioadhesive drug delivery systems (bioadhesives) can be designed to be reversible, allowing for drug delivery and then removed from the body once the desired therapeutic effect has been achieved [1].

As more innovative medicinal substances have become available for drug delivery, herbal bio-adhesive drug delivery has become more important for delivering novel molecules. The application of herbal medicine has shown promising effects in prevention/treatment of various diseases. The lack of proper scientific explanation has been a drawback of herbal medicine. Problems with herbal preparations have been observed such as standardization, extraction, and identification of active ingredients from the polyherbal system. Storage-related problems have been encountered such as stability issues, physical and chemical instability, drug interaction, deterioration, and decomposition. The problems associated with herbal bioactives delivery to the target site can be overcome by novel drug delivery approaches such as use of nanoparticles, matrix systems, liposomes, mucoadhesion drug delivery systems, etc. Bioadhesion phenomenon or mucoadhesive drug delivery system can be a good candidate for delivery of herbal bioactives to the target site by accumulating the required amount of drug to the site to give the expected therapeutic effect [2, 3].

6.2 Mucous Membrane

The moist lining walls of various body cavities, such as the gastrointestinal and respiratory systems, are known as mucous membranes (mucosae). They are made of a connective tissue layer, a layer of epithelial cells, and a layer of fluid. The purpose of the mucous membranes is to protect the underlying tissues and organs and to regulate fluid flow. The layers of epithelial cells vary according to the site, as at the buccal site it is about 40 – 50 cell layers thick, while at the sublingual site the epithelial cell layers are fewer in number. Mucus is a gelatinous fluid secreted by mucous cells that cover the tissue layers at respiratory, digestive, reproductive, and urinary tracts. Mucus plays a critical role in the protection, health maintenance, and lubrication of the body's mucosal surfaces. The two major forms of mucus are mucin-containing mucus sheets (also known as mucous membranes) and mucus-containing secretions that are freely mobile (also known as mucus). The mucus sheets are composed of mucin, fats, water, and inorganic salts. Mucous sheets form a continuous layer on different types of mucosal surfaces, ranging in thickness from 50 to 450 μm in the stomach to less than 1 μm in the oral cavity. The most essential structure-forming component of mucus is mucin, which is constituted of glycoproteins and glycolipids. The main functions of mucin are to provide a protective layer, to maintain the structural integrity of the mucus layer, and to provide a lubricating layer. Mucus also contains lysosomal enzymes that help to break down large molecules and to clear cellular debris [4]. The structure and arrangement of cells in the stomach are represented in Figure 6.1.

The mucoadhesion phenomenon has not yet completely been understood but the proposed theoretical consideration to understand the mechanism is discussed here. The mechanism has been bifurcated into two stages: contact stage, and consolidation stage (Figure 6.2). In the contact stage, the receptors on the mucosal surface bind to the mucin, and the receptor-mucin contact area is increased, which increases mucoadhesion. In the consolidation stage, the mucin is transported to the deeper parts of the mucus layer which increases mucoadhesion. In general, several hypotheses have been proposed to explain the fundamental mechanism of adhesion. One or more theories can explain the creation of bioadhesive bonds in a particular system. Bioadhesive bonds are of two types of bonds: chemical and mechanical bonds. The chemical bonds include strong bonds (covalent bonds) and weaker bonds (van der Waals and hydrogen bonds). Mechanical bond suggests the interlocking of mucin to the adhesive polymer. The flexibility of the adhesive polymer determines its penetration into the mucus layers [2].

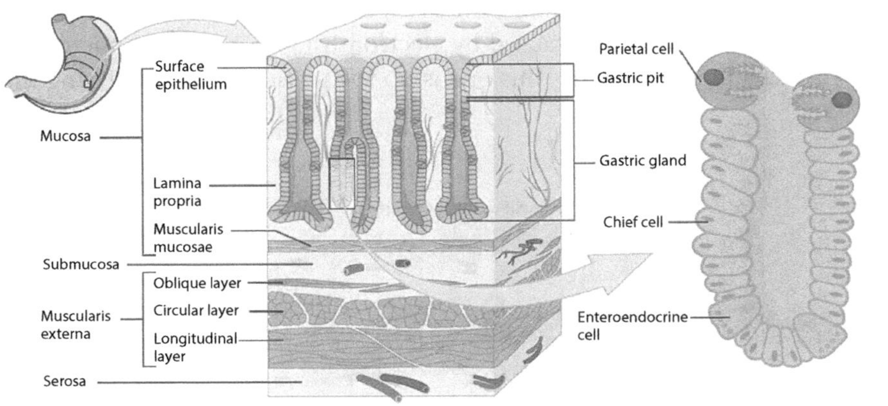

Figure 6.1 Cellular structure of gastrointestinal mucosa.

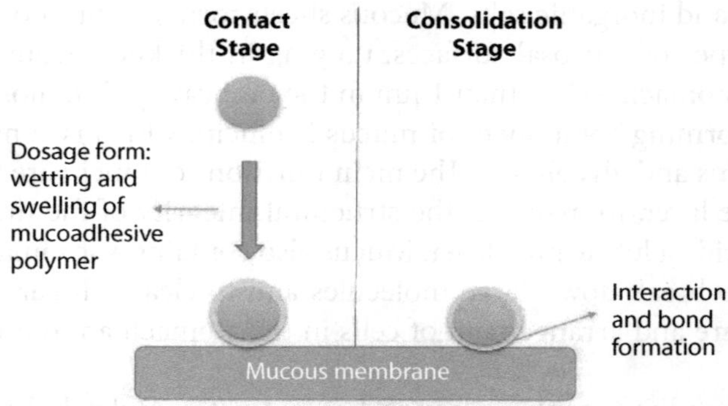

Figure 6.2 Contact and consolidation stages for understanding the mechanism of mucoadhesion.

6.3 Theories of Adhesion

Mucoadhesion is a complex phenomenon involving different theories to explain the interaction between the adhesive polymer and the mucosal membrane [5].

6.3.1 Electronic Theory

This theory is based on electron transfer and subsequent creation of an electrical double layer between the adhesive and the mucin network (Figure 6.3). One of the most critical criteria in evaluating the success of a mucoadhesive system is the adhesion of biological material to mucus. Differences in the electrical structures of the mucus and the mucoadhesive system promote adhesion, which is caused by electron transfer between mucus and the mucoadhesive system. This opens up the possibility of modifying the

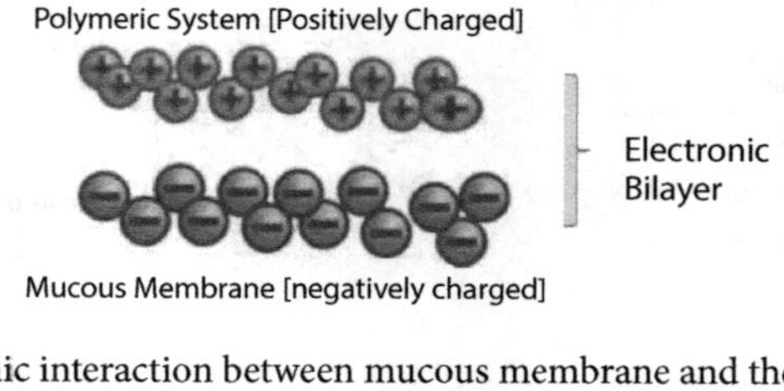

Figure 6.3 Electronic interaction between mucous membrane and the adhesive polymeric system.

system's adhesion by adjusting the electrical structure of the mucus and the mucoadhesive system. Biomimetic systems have been shown to have better adhesion than conventional systems, suggesting that tailoring the electrical structures of the mucus and the mucoadhesive system is a viable strategy for improving adhesion. The development of forces of attraction within this double layer is the end outcome of such a process [6].

6.3.2 Adsorption Theory

This theory suggests that the interaction of the adhesive polymer with the mucous membrane is a surface phenomenon. The adhesion mechanism begins when the mucoadhesive polymer comes in contact with the surface of the mucous membrane and gets adsorbed by bonding forces (Figure 6.4). In this scenario, the bonding force is bifurcated into two types of chemical interaction forces, including primary bonds and secondary bonds. The primary bond (covalent bond) is a strong bond. The secondary bond has weak bonding such as hydrogen bonds and van der Waals forces, which play a vital role at the interface [7].

6.3.3 Wetting Theory

The wetting theory refers to liquid systems that have a strong attraction to a surface and spread across it (Figure 6.5). The contact angle is used to determine this affinity. The rule of thumb is that the smaller the contact

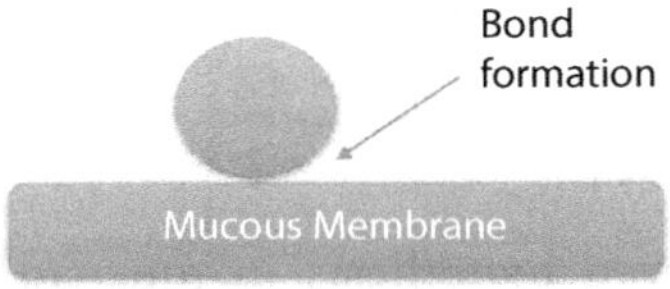

Figure 6.4 Adsorption mechanism of mucoadhesion.

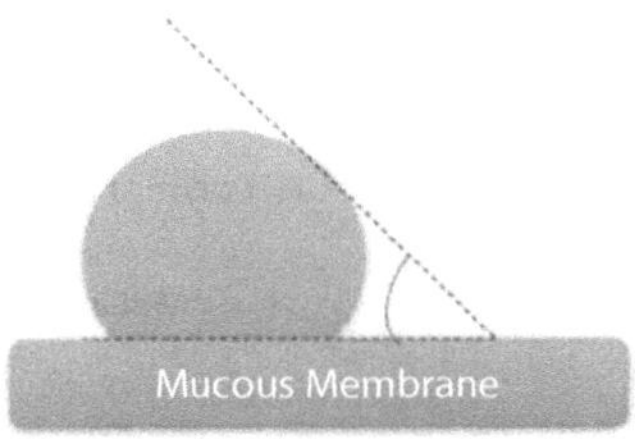

Figure 6.5 Wetting mechanism of mucoadhesion.

angle, the stronger the bond. To ensure proper spreadability, the contact angle should be equal to or near zero. The spreadability can be determined using the equation:

$$S_{AB} = \gamma_B - \gamma_A - \gamma_{AB}$$

Where, γ_A and γ_B denote the surface free energies of bioadhesive material (A) and biological membrane (B) respectively and γ_{AB} represents interfacial free energy. S_{AB} denotes spreading coefficient. This hypothesis explains that lowering the contact angle, and lowering the surface free energy of bioadhesive material as well as the interfacial free energy leads to improved adhesion [8, 9].

6.3.4 Diffusion Theory

Diffusion theory (Figure 6.6) depicts how polymer and mucin chains penetrate to a depth sufficient to produce a semi-permanent adhesive connection. The adhesion strength is assumed to be increased by the degree of penetration of the polymer chains. The penetration rate is influenced by the diffusion coefficient, the flexibility and composition of the mucoadhesive chains, motility, and contact time. The penetration depth necessary to produce an efficient bioadhesive binding, according to literature, is between 0.2 and 0.5 μm. The adhesion strength of a polymer is obtained when the penetration depth of the polymer is almost equal to the polymer chain size. Diffusion mechanism requires sufficient solubility of the bioadhesive material to interact with the mucous membrane [5, 10].

6.3.5 Fracture Theory

It is most widely utilized hypothesis in mucoadhesion which includes mechanical strength studies. It examines the amount of force required to separate two surfaces once adhesion has been established (Figure 6.7). It describes the separation of two surfaces (mucous layer and adhesive

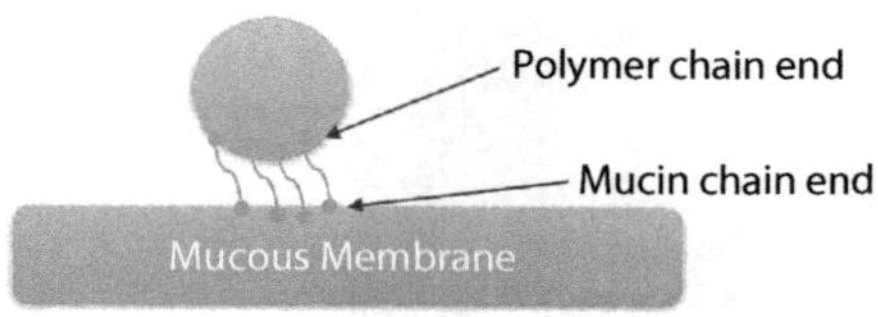

Figure 6.6 Diffusion mechanism of mucoadhesion.

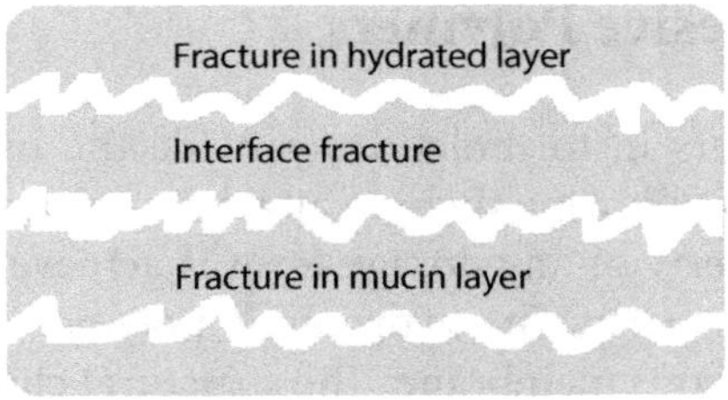

Figure 6.7 Fracture mechanism of mucoadhesion.

polymer) after adhesion. The fracture strength is directly proportional to the adhesion strength.

$$S_m = (E\varepsilon / L)^{1/2}.$$

Where S_m represents fracture strength or force required to break the bond between the two surfaces, E represents Young's modulus of elasticity of mucoadhesive polymer film, ε represents fracture energy to break the bond, and L is length of fractured mucous surface [11, 12].

The theories are useful to understand the adhesion behavior of the mucoadhesive polymer with the biological membrane. A brief summary is shown in Table 6.1.

Table 6.1 Different theories of mucoadhesion.

Theory	Comments	Reference(s)
Electronic Theory	Attractive electrostatic forces between mucoadhesive and biological component	[13]
Adsorption Theory	Bioadhesive bond formed between an adhesive polymer and mucosa is ascribable to surface forces (covalent bond, ionic bond, hydrogen bond and van der Waals forces)	[8]
Wetting theory	Ability of bioadhesive polymer to spread and develop intimate contact with the mucous membrane	[9]
Diffusion Theory	Entanglement of mucin strands and flexible polymer chain	[5, 10]
Fracture Theory	Analyses the force required to detach two surfaces after adhesion	[11, 12]

6.4 Mucoadhesive Polymers

Hydrophilic polymers in formulations provide the mucoadhesive property of dosage forms, allowing them to attach to mucosal membranes. The mucoadhesive property of any dosage form is achieved by polymers that have hydrophilic property. The hydrophilic polymers have the potential to adhere to the mucous membrane. The structural characteristics for any polymer in mucoadhesive formulations are as follows:

- Requires to possess strong hydrogen bonding groups, such as carboxyl, hydroxyl, amino and sulfate groups,
- Requires to have strong anionic or cationic charges,
- Requires to have high molecular weight,
- Requires to have long-chain flexibility,
- Requires having good spreadability onto mucus.

Some examples of adhesive polymers are mentioned in Table 6.2. The adhesive polymers are classified according to the origin and chemical structure, aqueous solubility, and type of charge on adhesive polymer.

Ideal characteristics of a mucoadhesive polymer

1. The adhesive polymer should adhere to the desired surface and preferably have some site-specificity.
2. Should be compatible with the drug and have no interference with the release of the drug.
3. Must be non-irritant to the mucus surface.
4. Its by-products must be harmless and easily absorbed through the mucosal membrane.
5. It must be in a stable state during storage and for the duration of the product's shelf-life [14].

MDDS have gained a lot of interest in deploying bioadhesive drug delivery systems because of their significant potential for controlled release drug delivery. Mucoadhesive drug delivery has since been extended to numerous routes.

Table 6.2 Classification of adhesive polymers [15].

Source	Class	Adhesive polymers
Based on Source: Natural	Protein-based polymers	Collagen, Albumin, Gelatin
	Polysaccharides	Alginates, Cyclodextrins, Chitosan, Dextran, Agarose, Hyaluronic acid, Starch, Cellulose
Based on Source: Synthetic	Biodegradable polymers	
	Polyesters:	Poly(lactic acid), Poly(glycolic acid), Poly(hydroxyl butyrate), Polycaprolactone
	Polyanhydride:	Poly(adipic acid), Poly(terephthalic acid), Poly(sebacic acid) and various copolymers
	Phosphorous based polymers	Polyphosphate, Polyphosphonate, Polyphosphazene
	Others:	Polycyanoacrylates
	Non-biodegradable polymers	
	Cellulose derivatives:	Carboxymethyl cellulose, Ethyl cellulose, Cellulose acetate, Hydroxypropyl methyl cellulose
	Silicones:	Poly(dimethylsiloxane)
	Others:	Polyvinylpyrrolidone, Ethylene-vinyl acetate, Poloxamines.
Based on aqueous solubility of the polymer	Water soluble polymers	Carbopol, Sodium carboxymethyl cellulose, Cellulose, Sodium alginate
	Water insoluble polymers	Chitosan, Ethyl cellulose, Polycarbophil

(Continued)

Table 6.2 Classification of adhesive polymers [15]. (*Continued*)

Source	Class	Adhesive polymer
Based on type of charge on the polymer	Cationic	Chitosan
	Anionic	Chitosan, Carbopol, Pectin, Sodium alginate and Xanthan gum
	Non-Ionic	Hydroxy ethyl starch, Poly(vinyl alcohol), Polyvinylpyrrolidone, Hydroxypropyl methyl cellulose, Methyl cellulose

6.5 Mucoadhesive-Based Drug Delivery Systems (DDS): Administration Routes

Drug delivery through mucoadhesion can be accomplished through different administration routes. The types of administrations are mentioned below:

6.5.1 Ocular Mucoadhesive DDS

Ocular drug delivery is a complicated approach in the drug delivery system, due to its strong defense mechanism and continuous clearing of the eye. The most common causes for failure of traditional ocular drug delivery systems can be loss of medication via drainage, short residence duration, tear flow, and protein binding. For drug absorption, the duration of drug residence must be prolonged at the precorneal area which can result in absorption of the required amount of drug. Corneal epithelium and conjunctiva serve as absorption platforms for several drugs to absorb. For efficient absorption, the duration of drug residence at the absorption sites i.e., at the cornea and conjunctiva, must be prolonged. The corneal membrane and conjunctiva membrane are covered with a mucus layer, this region is targeted for adhesion of adhesive polymer excipient [16, 17].

Common ocular diseases such as glaucoma and cataract cause vision loss if left undiagnosed or untreated for a long period. In this modern era, cataract has become one of the most common causes of vision loss. If left untreated for a long time, then the last option left to correct it is surgery. Dry eye disease is another example of ocular disease related problem in which the production of tear is reduced [18].

The issues with ocular drug delivery devices are inefficient distribution in certain regions of the eye, nasolacrimal drainage, and short drug contact time on the ocular surface. In such instances, poor drug bioavailability in the retina and systemic absorption has been observed. As a side effect, cataracts, endophthalmitis, retinal detachment, and bleeding might develop. Sometimes the existence of retinal pigment epithelium makes medication delivery to the retina difficult. The effect provided by this method is minimal since the medication may penetrate the globe of the eye, it is difficult to distribute medication to the posterior section, and subretinal administration causes retinal detachment [19]. Some ocular mucoadhesive drug delivery systems containing herbal bioactives are mentioned in Table 6.3.

Table 6.3 Ocular mucoadhesive drug delivery systems containing herbal bioactives.

Scientific name and bioactive	Formulation form	Action	Remarks	Ref.
Pilocarpus microphyllus- Pilocarpine	Liposome	Draining of excess fluid from eye	Liposomal formulations coated with Carbopol 1342 increase the bioavailability of pilocarpine in rabbit eyes compared to an uncoated preparation.	[20]
Coleus forskohlii - Forskolin	Ophthalmic inserts	Reduces intraocular pressure in glaucoma	Poly(vinyl alcohol) 14000 based ophthalmic inserts containing forskolin were fabricated. Showed sustained drug release over a period of 6 hours.	[21]
Camellia sinensis- epigallo-catechingallate	Thermogels	Shows anti -oxidant and anti – inflammatory properties in Dry Eye Disease	A one-time topical administration of an epigallocatechingallate-loaded carrier onto the target site could effectively repair the defective corneal epithelium by relieving cellular inflammation, oxidative stress, and cell apoptosis over a sustained 14 day period in dry eye.	[22]

(*Continued*)

Table 6.3 Ocular mucoadhesive drug delivery systems containing herbal bioactives. (*Continued*)

Scientific name and bioactive	Formulation form	Action	Remarks	Ref.
Brassica oleracea - Quercetin	Thermosensitive gel	Free radical scavenging effect and reducing reactive oxygen species	In order to develop a thermosensitive gel, quercetin and epigallocatechin gallate were used. The resulting formulation demonstrated that the reactive oxygen species inhibition efficacy was improved when poly (D,L-lactic-co-glycolic acid) nanoparticles loaded with quercetin were combined with epigallocatechin gallate.	[23]
Curcuma longa - Curcumin	Polymeric *in-situ* gelling inserts	Shows antioxidant action	*In-situ* gelling polymer inserts loaded with curcumin were prepared, using Hydroxypropyl methyl cellulose, carboxymethyl cellulose sodium, and poly(vinyl alcohol). Acceptable ocular tolerability, improved corneal permeability, and sustained release of curcumin were the reported advantages.	[24]

6.5.2 Nasal DDS

Intranasal route is employed for delivery of low dose drugs like drugs targeting central nervous system, hormones, local and systemic delivery, etc. via nasal route, as it has a much less mucosal surface area for absorption of drugs. The advantages of this delivery route are: rapid action, less dose of Active Pharmaceutical Ingredient (API) is required, increased patient compliance, and can be administered by patients themselves.

Two major challenges faced by any formulator while developing a nasal delivery system are mucociliary clearance and enzymatic degradation. The mucociliary clearance mechanism, which allows for rapid drainage of the supplied formulation from the nasal cavity, is responsible for poor membrane transport. This is especially true when the medicine cannot be absorbed easily through the nasal mucosa. Using bioadhesive excipients in formulations is one way to circumvent this issue. The enzymatic degradation of peptides and proteins causes limited bioavailability across the nasal mucosa [25, 26].

The nasal route consists of vascularized epithelium and a large surface area for drug absorption due to several microvilli. This leads to a rapid uptake of drugs, a higher bioavailability, and a direct transport of drugs into the systemic circulation [27].

Colombo *et al.* developed Kaempferol-loaded mucoadhesive nanoemulsions that reduced the survival of glioma cells by increasing apoptosis, this can be helpful in cancer therapy [28].

Mandal *et al.* developed a curcumin-loaded mucoadhesive microemulsion to improve curcumin absorption in the brain via the intranasal route. The study shows that by using a curcumin-loaded mucoadhesive microemulsion, the curcumin uptake is significantly higher in comparison to intravenous administration [29].

Tiwari and Bajaj formulated eucalyptus oil microemulsions by combining hydrophilic and lipophilic surfactants for calming, stimulating, and exerting antidepressant effects in migraine [30].

6.5.3 Oromucosal DDS

Oral mucosa is the mucous membrane that covers the oral cavity. The buccal mucosa has permeability 44,000 times higher than for the skin epidermis but 44,000 times lower than for the intestinal mucosa [31].

One of the major issues while developing any drug delivery system is bioavailability of the drug at the target site. Another issue while developing oral drug delivery systems is taste masking. It may be difficult to achieve dosage uniformity, and only minimal doses may be used. Another issue to cope with is the possibility of an overdose if not monitored well [32].

The mucoadhesion property of the formulation is critical in buccal drug delivery systems. Bioadhesive polymers are currently utilized in making tablets, patches, tapes, films, semi-solids, and powders, among other dosage forms. Several studies have shown that the addition of various mucoadhesive polymers to drug delivery devices prolongs duration of adhesion and the effectiveness of the drug. The mucoadhesive polymers should have some common physicochemical properties, such as anionic nature, hydrophilicity and a large number of hydrogen bonds forming groups. Natural polymers, as well as their breakdown products, should not be toxic, irritating, or contaminating [33].

The most common oral problems seen with nonsurgical cancer therapy are mucositis and xerostomia. In cancer patients, mucositis is a common adverse effect of chemoradiotherapy [34]. The major cause of mouth cancer is tobacco cancer, which is caused by smoking [35]. These oral problems are better treated using several herbal-based formulations and have been found to have better performance than the conventional dosage forms (Table 6.4).

Table 6.4 Oromucosal drug delivery systems containing herbal bioactives.

Scientific name and bioactive	Formulation	Action	Remarks	Ref.
Chamomilla recutita - Chamazulene, bisabolol oxides, alpha-bisabolol, spiroethers	Oral gel	Relieves pain in chemotherapy-induced oral mucositis	Compared to the conventional preparation, the topical preparation reduces the severity of the mucositis while also lowering pain levels.	[36]
Curcuma longa - Curcumin	Buccal formulation	Oral mucositis (OM)	Curcuminoid has an anti-inflammatory effect by controlling angiogenesis and transforming growth factor beta 1 (TGF-β1) expression. It also has a therapeutic effect on oral mucositis, improves wound healing by re-epithelialization, and has anti-inflammatory property by modulating angiogenesis.	[37]

(*Continued*)

Table 6.4 Oromucosal drug delivery systems containing herbal bioactives. (*Continued*)

Scientific name and bioactive	Formulation	Action	Remarks	Ref.
Garcinia mangostana L.-α-Mangostin	Mucoadhesive Film	Potent anti-inflammatory effect and promotes wound healing without cytotoxicity at a therapeutic dose in oral squamous cell carcinoma	The Mangostin (α-MG) film also shows a potent anti-inflammatory activity and enhances wound healing.	[38]
Curacao Aloe - (Aloin)	Matrix Tablet	Prevention of mucositis in patients undergoing head and neck radiation therapy	Matrix tablet of aloin enhances the accumulation of aloin at buccal mucosa, results in increase its bioavailability and it is used in chemotherapy.	[39]
Camellia sinensis - Epigallocatechin-3-gallate (EGCG)	Nanoparticles	Inhibits prostate cancer cell growth	Characterization and efficacy assessment of a nanotechnology-based oral formulation of chitosan nanoparticles encapsulating epigallocatechin-3-gallate for the treatment of prostate cancer in a preclinical setting.	[40]

(*Continued*)

Table 6.4 Oromucosal drug delivery systems containing herbal bioactives. (*Continued*)

Scientific name and bioactive	**Formulation**	**Action**	**Remarks**	**Ref.**
Nicotiana tabacum - Nicotine	Buccal tablets	Buccal tablets for smoking cessation	Constantly delivering nicotine to the buccal membranes within the mouth, and these are used in nicotine replacement treatment for smoking cessation.	[41]
Vitis vinfera – Ellagic acid	Nanoparticles	Free radical scavenging effect	Ellagic acid encapsulated in poly(D,L-lactide-co-glycolide) nanoparticles was formulated using different stabilizers. The study suggests that the formulation can be a good option for delivery of ellagic acid.	[42]
Syzygium aromaticum - Eugenol	Mucoadhesive Tablets	Relieves pain and has local anesthetic, antibacterial and anti-inflammatory effects.	Administration of eugenol in a mucoadhesive tablet results in controlled release pattern for 8 hours.	[43]
Allium cepa - (Quercetin)	Microparticles	Produces a significant hypoglycemic response after oral administration in diabetics	Safety, efficiency, compatibility and biodegradability of polymers (sodium alginate and succinyl chitosan) with quercetin in diabetes were investigated.	[44]

6.5.4 Pulmonary Mucoadhesive DDS

The main function of the lungs is gas exchange, inhalation of oxygen and exhalation of carbon dioxide. Once air makes its way into the larynx, it continues down the trachea or the windpipe, which splits into two bronchi. Then it enters the lungs and continues to the bronchioles followed by its final destination alveolar sacks. At the sites of alveoli, the gaseous exchange takes place. A mucoadhesive pulmonary drug delivery system can be a good candidate for site-specific delivery [45].

Large surface area, zero first-pass metabolism, considerable drug absorption, minimal dose requirements, and applicability for local and systemic delivery are the advantages of respiratory system. Mucoadhesion is a useful approach for enhancing residence time and, as a result, the delivery system's performance and effectiveness in the treatment of lung disorders [46].

Drug passage for appropriate absorption is hampered by physiological hurdles present in pulmonary system. Mucociliary clearance of the lungs is a natural lung defense process that removes accumulated debris from the conducting airways. Chemicals such as proteolytic enzymes (proteases) and surfactants, which generally disintegrate the inhaled particle, come into contact with the drug as well. Immunological phagocytes, which are located in the alveolar sac and target both potentially dangerous and possibly useful molecules, are another mechanism that destroys the inhaled molecules [47].

The short duration of the effect due to quick clearance of medicines deposited in the peripheral airways, immunogenicity, difficulty of accurate dosing, and chemical instability of the protein all have been noted as drawbacks for protein and peptide delivery [48].

Rao *et al.* [49] and Shahab *et al.* [50] have published research on nicotine replacement therapy in the treatment of smoking cessation by nicotine in a mouth spray formulation.

El-Sherbiny and Smyth [51] have developed and tested curcumin microparticles for application in chronic obstructive pulmonary disease. They observed that curcumin microparticles had a low tendency for inducing tumor necrosis factor production, from the macrophage absorption being significantly delayed and suppressed.

6.5.5 Gastrointestinal Tract Mucoadhesive DDS

The histology of the gastrointestinal system is consistent throughout its length, with minor changes in some sections which alter the function.

Around the lumen, the GI tract is separated into four concentric layers. The mucosa, which surrounds the lumen of the GI tract, is the deepest layer. Each GI tract organ's mucosal layer is highly specialized according to various roles and chemical conditions [52].

Many medications have a terrible taste or emit nasty odors when taken orally, which must be addressed for patients to accept them. Several hurdles and difficulties prevent oral medication absorption. The acidic stomach environment, digesting enzymes, mucus layer barrier diffusion, and epithelial cell tight junctions all inhibit paracellular transfer. Furthermore, food or diseased conditions may hinder proper medicine delivery throughout the GI tract. Also, for medications that are rapidly absorbed, the absorption following oral administration might be quite rapid. This might result in a quick increase or decrease in blood levels, which is not ideal for treatments that require stable plasma concentration over time to work properly [53].

Epidemiological evidence suggests that curcumin use may lower the incidence of some malignancies and provide other biological benefits to people. Curcumin has many medicinal effects, including reducing inflammation, inhibiting tumor development, and promoting wound healing, and having anti-neoplastic activity [53–55].

Suwannateep *et al.* [56] prepared two formulations i.e., curcumin loaded ethylcellulose combined with methylcellulose and curcumin loaded ethylcellulose particles. In tissue culture, both displayed *in vitro* free radical scavenging action as well as a dose-dependent *in vitro* cytotoxic impact on MCF-7 human breast adenocarcinoma and HepG2 hepatoblastoma cells.

Curcumin mucoadhesive microspheres were produced and tested by Ali *et al.* [57] for the treatment of stomach cancer, and gastric ulcers caused by *Helicobacter pylori*. Curcumin mucoadhesive microspheres were made with ethyl cellulose and Carbopol 934P. The modified formulation's drug release demonstrated a controlled-release pattern, indicating that it might include both local and systemic actions for the successful treatment of *H. pylori* infection.

Duan *et al.* [58] produced thiolated chitosan/alginate composite microparticles (CMPs) coated with Eudragit S-100 for colon-specific administration of 5-aminosalicylic acid (5-ASA) and curcumin (CUR), as well as for use as a multi-drug delivery system for the treatment of colitis. Very little medication was released in the upper gastrointestinal tract when the pH was less than 7, and the drug was released in the colon for a long time. This colon site-specific delivery system was able to transport 5-ASA and curcumin to the colon site and effectively decreased inflammation in colitis rats, due to its pH-sensitive, mucoadhesion, and controlled drug delivery capabilities.

Gavini *et al.* [59] demonstrated treatment of hepatic disorders by Silymarin (derived from *Silybummarimum*). This indicates that a mucoadhesive microsphere encapsulated with silymarin is an effective treatment for hepatic disease. Because of its prolonged residence time at the site of pharmacological action, the mucoadhesive formulation may be preferable.

6.5.6 Vaginal Mucosal DDS

The vagina is a female reproductive organ. Its length begins from vulva till the cervix of uterus. The vaginal canal has an outer fibrous adventitia, a middle layer of smooth muscle cells, and an inner layer of mucosa. Throughout the lifespan of a woman, the hormones change the microbial flora of the vaginal region. This change in microbial flora is responsible for alteration in pH of vaginal region [60, 61].

The vaginal route presents a potential venue for local action as well as systemic drug administration because of its large surface area, sufficient blood supply, and mitigation of the first-pass effect [62].

A limitation with vaginal administration is the residence time of active substance. This limitation results in stability and permeability issues. Another setback is the tight junction barrier and sometimes it limits the entry of high molecular weight charged drug molecules. Fluid is not retained due to intravaginal displacement and leakage. Also it is not possible to deliver drugs to a specific cell type that may be implicated in the etiology or transfer of a disease (e.g., cancer cells, immune cells related to HIV transmission) [63].

Candida albicans causes vulvovaginal candidiasis, a widespread disease that affects 75 percent of all women [64]. Salmazi *et al.* [65] designed and evaluated a curcumin-loaded liquid crystal precursor mucoadhesive system for antifungal treatment. The *in vitro* activity of curcumin-loaded formulation against *Candida albicans* was examined. It was found that in comparison to a curcumin solution the curcumin-loaded formulation was more potent against standard and clinical strains. In conclusion, vaginal administration of curcumin-loaded mucoadhesive system for vaginal candidiasis might be a potential treatment strategy.

6.5.7 Rectal Mucoadhesive DDS

Rectal drug delivery has been effectively used to treat local anorectal infections as well as systemic drug delivery as a substitute for oral administration [66].

The anal canal is the last part of the digestive tract. It starts at the anorectal junction, runs for about 4 cm, and ends at the anus region. On different parts, the lining of the anal canal is made of different kinds of epithelia. The epithelium is formed by a single layer of columnar cells on the mucosal surface of the rectum. Goblet cells secrete mucus, which covers the rectal epithelia and aids in the lubrication of feces as these pass through the rectum [67, 68].

Zhao *et al.* [69] developed a colorectal cancer targeted delivery system. The delivery system is microspheres, in which α-lactalbumin peptosomes (PSs) co-loaded with a microRNA (miR)-31 inhibitor (miR-31i) and curcumin are encapsulated in thiolated oxidized Konjac glucomannan (sOKGM) microspheres. The resultant microspheres were found to be more stable and mucoadhesive, resulting in longer retention at colorectal location and better drug absorption and demonstrating colorectal cancer therapeutic activity.

6.6 Clinical Studies

A nanoparticle drug delivery system is a revolutionary drug delivery device that overcomes the limitations of traditional medication delivery systems. India has a significant Ayurvedic knowledge base that has just lately been realized to its full potential. Herbal medicines have long been disregarded for development as new formulations due to a lack of scientific basis and processing difficulties, such as purification, extraction, and identification of particular medicinal components [70].

Numerous studies have been conducted in this phytopharmaceutical field, and many herbal products have been developed using a novel drug delivery mechanism. Several medicines and formulations have already been introduced to the market, and clinical research continues to this day.

dos Santos Filho *et al.* [71] have conducted a randomized phase I clinical trial on a mucoadhesive preparation containing *Curcuma Longa* (Zingiberaceae) and *Bidens Pilosa* Linn (Asteraceae) extract. The resulting formulation can be used for the cure and management of oral mucositis. Patients who are suffering from head and neck cancers undergo chemotherapy for treatment. During chemotherapy, they develop oral mucositis. The purpose of this study was to establish the safety of the formulated mucoadhesive preparation. The report also provides data for phase 2 trial recommendations.

A randomized double-blind clinical trial was conducted by Akbari *et al.* [72]. The goal of the trial was to study the effect of diphenhydramine solution and diphenhydramine-containing *Glycyrrhiza Glabra* for treatment of reappearing aphthous stomatitis. The trial has been filed for registration to the Iranian Registry of Clinical Trials.

Allison *et al.* [73] created a mucoadhesive hydrogel, a commercially available pharmaceutical product, to examine how it would respond in a multi-institutional, double-blind, randomized, placebo-controlled trial. The goal of the trial was to see how effective mucoadhesive hydrogel was at reducing oral mucositis symptoms in patients receiving chemoradiation therapy for head and neck malignancies.

Ghalwash and coworkers [74] developed oromuco-adhesive films for buccal distribution of Propolis extract encapsulated in niosomes to treat recurring aphthous ulcers (RAUs) in the mouth. This research examined the onset of ulcer shrinkage, overall ability to heal, and pain relief in 24 RAU patients who were allocated randomly to one of two groups: medicated or placebo [74].

6.7 Patents on Herbal Bioactive–Based Mucoadhesive Drug Delivery Systems

Table 6.5 highlights patents relating to mucoadhesive drug delivery systems containing herbal bioactives.

Table 6.5 List of patents on herbal bioactives containing mucoadhesive drug delivery systems.

Patent no.	Title	Remarks	Ref.
US20110301213A1	Use of pilocarpine for hypoptyalism treatment	Proposed a method to treat patients suffering from hypoptyalism. The method includes administration of pilocarpine using a bioadhesive polymer.	[75]

(*Continued*)

Table 6.5 List of patent of herbal bioactives containing mucoadhesive drug delivery system. (*Continued*)

Patent no.	Title	Remarks	Ref.
WO 2018/113916 Al	A mucoadhesive oromucosal formulation comprising a nicotine complex	Invented a mucoadhesive oromucosal formulation containing a nicotine complex. The nicotine complex consists of ionic nicotine form with at least one mucoadhesive water-soluble anionic polymer.	[76]
WO2004047813A1	Solid mucoadhesive composition	A mixture of herbal extracts [*Sambucus Nigra, Centella Asiatica* and *Echinacea Purpuria*] is delivered as an adhesive solid dosage form for the treatment of mucosal lesion. Tablet is used as dosage form for mucoadhesive drug delivery.	[77]
US6210699B1	Oral transmucosal delivery of drugs or any other ingredients via the inner buccal cavity	Established a device and a method for the oral transmucosal administration of active substances into the oral cavity using non-plasticized polyvinylpyrrolidone.	[78]

(*Continued*)

Table 6.5 List of patent of herbal bioactives containing mucoadhesive drug delivery system. (*Continued*)

Patent no.	Title	Remarks	Ref.
		The device comprises of two layers: mucoahesive polymer as primary layer and Active Pharmaceutical Ingredient as secondary layer.	
EP 2 749 552 B1	Curcumin-Resveratrol hybrid	Small doses of curcumin can be administered via the nasal route which helps to enhance the effectiveness of curcumin in the brain for prevention and cure of Alzheimer's Disease.	[79]

6.8 Summary

Various inner surfaces of the body are constituted from epithelial tissue/mucous membrane which constitute potential sites for bonding, creating advancement in the drug delivery approach. The concept of bioadhesion opens new areas of research in the biomedical sciences, as MDDS can be used for better adhesion property and for pharmaceutical industries to develop new formulations for improved therapeutic effects.

3D bio-printing is a rapidly developing discipline merging diverse applications of biomedicine, tissue engineering, and materials science for developing customized drug delivery systems for reconstructive and various implantable devices. The aim to develop this delivery system is to reduce toxic levels and obtain high efficacy.

References

1. V.K. Devi, N. Jain and K.S. Valli, Importance of novel drug delivery systems in herbal medicines. *Pharmacognosy Rev.* 4, 27-31 (2010).
2. S.E. Harding, S.B. Davis, M.P. Deacon and I. Fiebrig, Biopolymer mucoadhesives. *Biotechnol. Genetic Eng. Rev.* 16, 41-86 (1999).
3. S. Roy and B. Prabhakar, Bioadhesive polymeric platforms for transmucosal drug delivery systems–a review. *Tropical J. Pharm. Res.* 9, 91-104 (2010).
4. J.D. Smart, The basics and underlying mechanisms of mucoadhesion. *Adv. Drug Del. Rev.* 57, 1556-1568 (2005).
5. B. Chatterjee, N. Amalina, P. Sengupta and U.K. Mandal, Mucoadhesive polymers and their mode of action: A recent update. *J. Appl. Pharm. Sci.* 7, 195-203 (2017).
6. F.L. Leite, C.C. Bueno, A.L. Da Róz, E.C. Ziemath and O.N. Oliveira, Theoretical models for surface forces and adhesion and their measurement using atomic force microscopy. *Int. J. Mol. Sci.* 13, 12773-12856 (2012).
7. M.R. Jimenez-Castellanos, H. Zia and C.T. Rhodes, Mucoadhesive drug delivery systems. *Drug Devel. Ind. Pharm.* 19, 143-194 (1993).
8. N.A. Peppas and P.A. Buri, Surface interfacial and molecular aspects of polymer bioadhesion on soft tissues. *J. Controlled Release.* 2, 257-275(1985).
9. K. L. Mittal, The role of the interface in adhesion phenomenon, *Polym. Eng. Sci.,* 17, 467-473 (1977)
10. A. Ahuja, R.K. Khar and J. Ali, Mucoadhesive drug delivery systems. *Drug Devel. Ind. Pharm.* 23, 489-515 (1997).
11. R. Shaikh, T.R.R. Singh, M.J. Garland, A.D. Woolfson and R.F. Donnelly. Mucoadhesive drug delivery systems. *J Pharm Bioallied Sciences*, *3*(1), 89-100. (2011)
12. G. Ponchel, F. Touchard, D. Duchêne and N.A. Peppas, Bioadhesive analysis of controlled-release systems. I. Fracture and interpenetration analysis in poly (acrylic acid)-containing systems. *J. Controlled Release.* 5, 129-141 (1987).
13. D. Dodou, P. Breedveld and P.A. Wieringa, Mucoadhesives in the gastrointestinal tract: revisiting the literature for novel applications. *Eur. J. Pharm. Biopharm.* 60, 1-16 (2005).
14. V.V. Khutoryanskiy, Advances in mucoadhesion and mucoadhesive polymers. *Macromolecular Biosci.* 11, 748-764 (2011).
15. I. Singh, P. Pawar, E.A. Sanusi, and O.A. Odeku. Mucoadhesive polymers for drug delivery systems, in: *Adhesion in Pharmaceutical, Biomedical and Dental Fields*, K. L. Mittal and F. M. Etzler (Eds.). pp.89-114. Wiley-Scrivener, Beverly, MA (2017)
16. A. Presland, Applied ocular physiology and anatomy. *Anaesthesia Intensive Care Medicine.* 8, 379-382 (2007).
17. J.L. Greaves and C.G. Wilson, Treatment of diseases of the eye with mucoadhesive delivery systems. *Adv. Drug Del. Rev.* 11, 349-383 (1993).

18. P.D. O'Brien and L.M. Collum, Dry eye: diagnosis and current treatment strategies. *Curr. Allergy Asthma Reports.* 4,314-331 (2004).
19. S. Wadhwa, R. Paliwal, S.R. Paliwal and S.P. Vyas, Nanocarriers in ocular drug delivery: an update review. *Curr. Pharm. Design* 15, 2724-2750 (2009).
20. A.M. Durrani, N.M. Davies, M. Thomas and I.W. Kellaway. Pilocarpine bioavailability from a mucoadhesive liposomal ophthalmic drug delivery system. *Int J. Pharm.* 88, 409-415 (1992).
21. V.D. Wagh, K.V. Wagh, P.P. Patil and M.K. Samanta, Formulation and evaluation of ophthalmic insert drug delivery system of forskolin. *Asian J. Pharm.* 2, 221-224 (2014).
22. L.J. Luo, D.D. Nguyen and J.Y. Lai, Long-acting mucoadhesive thermogels for improving topical treatments of dry eye disease. *Mater Sci. Eng. C* 115, 111095 (2020).
23. C. Chittasupho, T. Junmahasathien, J. Chalermmongkol, R. Wongjirasakul, P. Leesawat, and S. Okonogi, Suppression of intracellular reactive oxygen species in human corneal epithelial cells via the combination of quercetin nanoparticles and epigallocatechin gallate and in situ thermosensitive gel formulation for ocular drug delivery. *Pharmaceuticals, 14*, 679-696 (2021).
24. H. Abdelkader, D. Wertheim, B. Pierscionek and R.G. Alany, Curcumin *in situ* gelling polymeric insert with enhanced ocular performance. *Pharmaceutics*, 12, 1158 (2020).
25. D. Khunt and M. Mishra, An overview on anatomical and physiological aspects of the nose and the brain, in: *Direct Nose-to-Brain Drug Delivery*, C. V. Pardeshi and E. B. Souto (Eds.), pp. 3-13, Academic Press (2021).
26. K.R. Jadhav, M.N. Gambhire, I.M. Shaikh, V.J. Kadam and S.S. Pisal, Nasal drug delivery system-factors affecting and applications. *Curr. Drug Therapy* 2, 27-38 (2007).
27. P. Arora, S. Sharma and S. Garg, Permeability issues in nasal drug delivery. *Drug Discovery Today* 7, 967-975 (2002).
28. M. Colombo, F. Figueiró, A. de Fraga Dias, H.F. Teixeira, A.M.O. Battastini and L.S. Koester, Kaempferol-loaded mucoadhesive nanoemulsion for intranasal administration reduces glioma growth in vitro. *Int. J. Pharm.* 543, 214-223 (2018).
29. S.D. Mandal, S. Mandal and J. Patel. Brain targeting efficiency of curcumin loaded mucoadhesive microemulsion through intranasal route. *J. Pharmaceutical Investigation, 46*, 179-188 (2016).
30. N.G. Tiwari and A.N. Bajaj. Formulation development of eucalyptus oil microemulsion for intranasal delivery. *Indian J. Pharm. Sci.* 69, 731-733 (2007).
31. T.J. Canaan and S.C. Meehan, Variations of structure and appearance of the oral mucosa. *Dental Clinics* 49, 1-14 (2005).
32. F.L. Lopez, T.B. Ernest, C. Tuleu and M.O. Gul, Formulation approaches to pediatric oral drug delivery: benefits and limitations of current platforms. *Expert Opin. Drug Del.* 12, 1727-1740 (2015).

33. A. Khare, K. Grover, P. Pawar and I. Singh, Mucoadhesive polymers for enhancing retention in ocular drug delivery: a critical review. *Rev. Adhesion Adhesives* 2, 467-502 (2014).
34. C. Scully, S. Sonis and P.D. Diz. Oral mucositis. *Oral Diseases*, *12*, 229-241 (2006).
35. L.F. Stassen and O. Hammarfjord, Smoking cessation and the role of the dental practitioner. *J. Irish Dental Assoc.* 61, 90-92 (2015).
36. M.A. Elhadad, E. El-Negoumy, M.R. Taalab, R.S. Ibrahim and R.O. Elsaka, The effect of topical chamomile in the prevention of chemotherapy-induced oral mucositis: A randomized clinical trial. *Oral Diseases* 28, 164-172 (2022).
37. T.R. Schmidt, M. Curra, V.P. Wagner, M.A.T. Martins, A.C. de Oliveira, A.C. Batista, M.C. Valadares, R.N. Marreto and M.D. Martins, Mucoadhesive formulation containing *Curcuma longa L.* reduces oral mucositis induced by 5-fluorouracil in hamsters. *Phytotherapy Res.* 33, 881-890 (2019).
38. P. Tangsuksan, J. Chuerduangphui, C. Takahashi Yupanqui, T. Srichana, E. Hitakomate, C. Pientong, T. Ekalaksananan and W. Nittayananta, Mucoadhesive film containing α-mangostin shows potential role in oral cancer treatment. *BMC Oral Health* 21, 1-10 (2021).
39. V. De Caro, A.L. Scaturro, G. Di Prima, G. Avellone, F.M. Sutera, O. Di Fede, G. Campisi and L.I. Giannola. Aloin delivery on buccal mucosa: ex vivo studies and design of a new locoregional dosing system. *Drug Devel. Ind. Pharm.* 41, 1541-1547 (2015).
40. N. Khan, D.J. Bharali, V.M. Adhami, I.A. Siddiqui, H. Cui, S.M. Shabana, S.A. Mousa and H. Mukhtar, Oral administration of naturally occurring chitosan-based nanoformulated green tea polyphenol EGCG effectively inhibits prostate cancer cell growth in a xenograft model. *Carcinogenesis* 35, 415-423 (2014).
41. S. Lewis, G. Subramanian, S. Pandey and N. Udupa, Design, evaluation and pharmacokinetic study of buccal mucoadhesive tablets of nicotine for smoking cessation. *Indian J. Pharm. Sci.* 68, 829-831 (2006).
42. I. Bala, V. Bhardwaj, S. Hariharan, S.V. Kharade, N. Roy, and M.N.V. Ravi Kumar, Sustained release nanoparticulate formulation containing antioxidant-ellagic acid as potential prophylaxis system for oral administration. *Journal of drug targeting*, *14*(1), pp. 27-34. (2006)
43. B.K. Jadhav, K.R. Khandelwal, A.R. Ketkar, and S.S. Pisal, Formulation and evaluation of mucoadhesive tablets containing eugenol for the treatment of periodontal diseases. *Drug Dev Ind. Pharm.* 30, 195-203 (2004).
44. P. Mukhopadhyay, S. Maity, S. Chakraborty, R. Rudra, H. Ghodadara, M. Solanki, A.S. Chakraborti, A.K. Prajapati, and P.P. Kundu, Oral delivery of quercetin to diabetic animals using novel pH responsive carboxypropionylated chitosan/alginate microparticles. *RSC Advances*, *6*, 73210-73221 (2016).
45. H.H. Aung, A. Shivakumar, S.K. Gholami, and S.P. Venkateshwar, An overview of the anatomy and physiology of the lung, in: *Nanotechnology-Based*

Target Drug Delivery System for Lung Cancer, P. Kesharwani (Eds.), pp. 2-21 Academic Press (2019).

46. A.K. Thakur, D.K. Chellappan, K. Dua, M. Mehta, S. Satija and I. Singh, Patented therapeutic drug delivery strategies for targeting pulmonary diseases. *Expert Opinion Therapeutic Patents* 30, 375-387 (2020).
47. S.P. Newman. Drug delivery to the lungs: challenges and opportunities. *Therapeutic Delivery.* 8, 647-661 (2017).
48. S.A. Shoyele and A. Slowey. Prospects of formulating proteins/peptides as aerosols for pulmonary drug delivery. *Int. J.Pharm.* 314, 1-8 (2006).
49. S. Rao, Y. Song, F. Peddie and A.M. Evans. A novel tri-layered buccal mucoadhesive patch for drug delivery: assessment of nicotine delivery. *J. Pharm. Pharmacol.* 63, 794-799 (2011).
50. L. Shahab, L.S. Brose and R.West. Novel delivery systems for nicotine replacement therapy as an aid to smoking cessation and for harm reduction: rationale, and evidence for advantages over existing systems. *CNS Drugs* 27, 1007-1019 (2013).
51. I.M. El-Sherbiny and H.D. Smyth, Controlled release pulmonary administration of curcumin using swellable biocompatible microparticles. *Molecular Pharmacy.* 9, 269-280 (2012).
52. F. Nogueira, I.C. Gonçalves, and M.C.L. Martins, Effect of gastric environment on *Helicobacter pylori* adhesion to a mucoadhesive polymer. *Acta Biomaterialia* 9, 5208-5215 (2013).
53. D.J. Mastropietro, H. Omidian and K. Park, Drug delivery applications for superporous hydrogels. *Expert Opin. Drug Del.* 9, 71-89 (2012).
54. R.K. Maheshwari, A.K. Singh, J. Gaddipati and R.C.Srimal, Multiple biological activities of curcumin: a short review. *Life Sci.* 78, 2081-2087 (2006).
55. G.R. Irving, A. Karmokar, D.P. Berry, K. Brown, and W.P. Steward, Curcumin: the potential for efficacy in gastrointestinal diseases. *Best Practice Res. Clinical Gastroenterology* 25, 519-534 (2011).
56. N. Suwannateep, W. Banlunara, S.P. Wanichwecharungruang, K. Chiablaem, K. Lirdprapamongkol, and J. Svasti. Mucoadhesive curcumin nanospheres: biological activity, adhesion to stomach mucosa and release of curcumin into the circulation. *J. Controlled Release.* 151, 176-182 (2011).
57. M.S. Ali, V. Pandit, M. Jain and K.L. Dhar, Mucoadhesive microparticulate drug delivery system of curcumin against *Helicobacter pylori* infection: Design, development and optimization. *J. Advanced Pharm. Technol. Res.* 5, 48-56 (2014).
58. H. Duan, S. Lü, C. Gao, X. Bai, H. Qin, Y. Wei, X.A. Wu, and M. Liu, Mucoadhesive microparticulates based on polysaccharide for target dual drug delivery of 5-aminosalicylic acid and curcumin to inflamed colon. *Colloids Surfaces B,* 145, 510-519 (2016).
59. V. Gavini, R. Batta, K. Kuchi, and M.S. Murthy. Formulation and characterization of mucoadhesive microparticulate drug delivery system encapsulated

with silymarin. *Intl Pharmaceutical Sciences and Drug Research*, 7, 244-250 (2015).
60. G. D' Amati, C.R.T. di Gioia, L. Proietti Pannunzi, D. Pistilli, E. Carosa, A. Lenzi, and E.A. Jannini, Functional anatomy of the human vagina. *J. Endocrinological Invest.* 26, 92-96 (2003).
61. F. Brako, B. Raimi-Abraham, S. Mahalingam, D.Q. Craig and M. Edirisinghe, Making nanofibres of mucoadhesive polymer blends for vaginal therapies. *Eur. Polym. J.* 70, 186-196 (2015).
62. C. Valenta, The use of mucoadhesive polymers in vaginal delivery. *Adv. Drug Del.Rev.* 57, 1692-1712 (2005).
63. J. das Neves, R. Nunes, A. Machado and B. Sarmento, Polymer-based nanocarriers for vaginal drug delivery. *Adv. Drug Del.Rev.* 92, 53-70 (2015).
64. P. L. Fidel Jr., History and update on host defense against vaginal candidiasis. *Amer. J. Reproductive Immunology* 57, 2-12 (2007).
65. R. Salmazi, G. Calixto, J. Bernegossi, M.A. dos Santos Ramos, T.M. Bauab, and M. Chorilli. A curcumin-loaded liquid crystal precursor mucoadhesive system for the treatment of vaginal candidiasis. *Int. J. Nanomedicine.* 10, 4815-4824 (2015).
66. J. Xu, M. Tam, S. Samaei, S. Lerouge, J. Barralet, M.M. Stevenson and M. Cerruti, Mucoadhesive chitosan hydrogels as rectal drug delivery vessels to treat ulcerative colitis. *Acta Biomaterialia.* 48, 247-257 (2017).
67. E.V. Carrington, S.M. Scott, A. Bharucha, F. Mion, J.M. Remes-Troche, A. Malcolm, H. Heinrich, M. Fox and S.S. Rao, Advances in the evaluation of anorectal function. *Nature Rev. Gastroenterology Hepatology.* 15, 309-323 (2018).
68. J.M. Lee, and N.K. Kim, Essential anatomy of the anorectum for colorectal surgeons focused on the gross anatomy and histologic findings. *Annals Coloproctology*, *34*(2), 59-71 (2018)
69. R. Zhao, S. Du, Y. Liu, Lv, C., Song, Y., X. Chen, B. Zhang, D. Li, S. Gao, W. Cui and M.V. Plikus. Mucoadhesive-to-penetrating controllable peptosomes-in-microspheres co-loaded with anti-miR-31 oligonucleotide and curcumin for targeted colorectal cancer therapy. *Theranostics*, 10, 3594-3611 (2020).
70. A. Goyal, S. Kumar, M. Nagpal, I. Singh and S. Arora, Potential of novel drug delivery systems for herbal drugs. *Indian J. Pharm. Edu. Res.* 45, 225-235 (2011).
71. E.X. dos Santos Filho, D.A.C. Arantes, A.F.O. Leite, A.C. Batista, E.F. de Mendonca, R.N. Marreto, L.N. Naves, E.M. Lima, and M.C. Valadares. Randomized clinical trial of a mucoadhesive formulation containing curcuminoids (Zingiberaceae) and *Bidens pilosa Linn* (Asteraceae) extract (FITOPROT) for prevention and treatment of oral mucositis-phase I study. *Chemico-Biol. Interactions*, 291, 228-236 (2018).
72. N. Akbari, N. Asadimehr, and Z. Kiani. The effects of licorice containing diphenhydramine solution on recurrent aphthous stomatitis: A double-blind,

randomized clinical trial. *Complementary Therapies in Medicine* 50, 102401 (2020).
73. R.R. Allison, A.A. Ambrad, Y. Arshoun, R.J. Carmel, D.F. Ciuba, E. Feldman, S.E. Finkelstein, R. Gandhavadi, D.E. Heron, S.C. Lane, and J.M. Longo, Multi-institutional, randomized, double-blind, placebo-controlled trial to assess the efficacy of a mucoadhesive hydrogel (MuGard) in mitigating oral mucositis symptoms in patients being treated with chemoradiation therapy for cancers of the head and neck. *Cancer* 120, 1433-1440 (2014).
74. M.G. Arafa, D. Ghalwash, D.M. El-Kersh and M.M. Elmazar. Propolis-based niosomes as oromuco-adhesive films: a randomized clinical trial of a therapeutic drug delivery platform for the treatment of oral recurrent aphthous ulcers. *Scientific Reports* 8, 1-14 (2018).
75. P. Perovitch, M. Maury and J. Deymes. Use of pilocarpine for hypoptyalism treatment. United States Patent 9,119,775 (2018)
76. K.S. Nielsen. A mucoadhesive oromucosal formulation comprising a nicotine complex. United States Patent Application 16/471,847 (2020).
77. W. Levine and A. Satter. Solid mucoadhesive composition. United States Patent 7,985,430 (2011).
78. R.N. Acharya and J.L. Baker. Oral transmucosal delivery of drugs or any other ingredients via the inner buccal cavity. United States Patent 6,210,699 (2001).
79. T.M. Dimauro and M.A. Southboro, Curcumin-resveratrol hybrid. European Patent EP 2 749 552 B1 (2015).

7
Adhesive Hydrogels

Proma Bhattacharya[1] **and Sudarsan Neogi**[2*]

[1]*Department of Mechanical Engineering, University of Maryland, College Park, Maryland, USA*
[2]*Department of Chemical Engineering, Indian Institute of Technology, Kharagpur, India*

Abstract
Hydrogels are a special kind of polymeric materials having a three-dimensional interconnected network, exhibiting specialised structure-derived properties which include flexibility, high water content, high surface area, tunable mechanical strength, biodegradability, and excellent biocompatibility. Adhesive hydrogels are promising candidates for biomedical research because a strong and robust adhesion between any biomaterial and living tissues is very desirable for their applicability in biomedicine. This chapter is structured to give a comprehensive outlook of adhesive hydrogels, starting with the types of hydrogels, the recent research on the adhesion mechanism of hydrogels, the different types of polymers currently being researched for adhesive hydrogels, and finally the applications of adhesive hydrogels in cosmetics, textile, biomedicine, and food packaging industries are introduced, and how adhesive hydrogels are useful for the improvement of human life quality is described.

Keywords: Hydrogels, adhesion mechanism, adhesion strength, biocompatible, biomedicine

7.1 Introduction

7.1.1 Hydrogels

Hydrogels are a three-dimensional network of polymer(s) having high water content. According to convention, a substance should have at least 10% water content to be called a hydrogel. These were first reported by

Corresponding author: sneogi@che.iitkgp.ac.in

K.L. Mittal and S. Neogi (eds.) Adhesives in Biomedical Applications, (151–176) © 2023 Scrivener Publishing LLC

Table 7.1 Classification of hydrogels.

Classification based on	Type	Examples (Hydrogels made from)
Origin	Natural	Natural polymers such as alginate, chitosan, dextran.
	Synthetic	Man-made polymers such as poly(vinyl alcohol), poly(ethylene glycol), poly(acrylic acid)
	Hybrid	Combinations of natural and synthetic polymers such as alginate-polyacrylamide, chitosan-poly(vinyl alcohol)-glutaraldehyde
Composition	Homopolymer	Single polymer such as poly(vinyl alcohol), collagen, poly(acrylic acid), poly(ethylene glycol) methacrylate
	Copolymer	Two polymers such as poly(vinyl alcohol)-poly(ethylene glycol), chitosan-poly(vinyl alcohol), cellulose-poly(ethylene glycol).
	Multipolymer	More than two polymers such as chitosan-poly(vinyl alcohol)-glutaraldehyde, cellulose-gelatin-poly(ethylene glycol), heparin-poly(vinyl alcohol)-poly (ethylene glycol)
	Interpenetrating Network	Intertwined between a linear polymer chain penetrating into another cross-linked polymer network such as gelatin-poly(acrylamide), poly(acrylamide) and poly (itaconic acid), poly (vinyl alcohol)-poly (acrylic acid – co – hydroxyl-ethyl-methacrylate).

(*Continued*)

Table 7.1 Classification of hydrogels. (*Continued*)

Classification based on	**Type**	**Examples (Hydrogels made from)**
Ionic Charge	Non-ionic	Poly (N-isopropyl acrylamide), 2-methyl-2-oxazoline.
	Ionic	Having charge leading to pH sensitivity such as cellulose, acrylic acid-2 hydroxyethyl methacrylate, agar-polyacrylamide.
	Ampholytic	Having both charges which balance themselves at iso-electric point such as methylene-bis-acrylamide.
Pore size	Nonporous	Having very few pores
	Microporous	Having pores of micrometre size
	Superporous	Having interconnected microscopic pores leading to high swelling.
Configuration	Amorphous	With no detectable crystalline structure such as carboxy-methylcellulose, alginate, chitosan.
	Semi-crystalline	Mechanically stronger such as poly(vinyl alcohol).
Crosslinking	Physical	Crosslinking achieved without external cross-linker using methods such as freeze-thaw, gamma radiation, ageing.
	Chemical	Crosslinkers such as glutaraldehyde, isocyanates, epoxy compounds to produce stable strong hydrogels.

(*Continued*)

Table 7.1 Classification of hydrogels. (*Continued*)

Classification based on	Type	Examples (Hydrogels made from)
Response to stimulus	Thermo-responsive	Temperature affects the physical and chemical properties of certain hydrogels such as poly(N-iso-propyl-acrylamide), poly(ethylene glycol)-diacrylate.
	pH-responsive	Hydrogels such as poly(allylamine), glycidyl methacrylate with modulation of properties by effect of pH.
	Light and chemical responsive	Exhibit response to light and certain chemicals such as poly(ethylene oxide), poly(2-hydroxylethyl methacrylate-co-acrylic acid).

Wichterle and Lim [1] in 1960. Hydrogels are found to mimic living tissues due to their high flexibility and significant water content. Hydrogels have been classified based on various parameters [2–9] as described in Table 7.1.

7.1.2 Conventional Adhesives in Biomedicine

Wound closure, post-surgery tissue repair and other medical procedures often require re-joining of living tissue surfaces or securing biomedical implants on the substrate [10]. Conventionally, sutures or staples have been commonly used for these purposes. But these sutures can cause more pain, infection, and can leave behind bad scar. Nowadays, modern medical adhesives are synthetic in nature, such as cyanoacrylates, silicone, or polyurethanes [11]. But the main drawbacks of these synthetic adhesives are: (a) their poor efficiency in presence of blood, as they quickly set before bonding the tissues; (b) the adhesive layer is stiff and constrains the movements of the soft tissues; and (c) they can degrade and form toxic intermediates [12]. Also, they may sometimes be required to withstand sterilization via variety of methods, which can lead to their degradation. Certain substrates may necessarily need the usage of primers or other activators to achieve satisfactory performance, which limits their usage and functionality [13]. Table 7.2 presents some of the commercially available medical adhesives and their compositions.

Table 7.2 Commercially available medical adhesives and their manufacturers and compositions.

Commercial name of adhesive	Manufacturer	Composition
Histoacryl Blue®	Braun	Cyanoacrylates
Dermabond®	Closure Medical Corporation	
Indermil®	Vygon	
Granuflex™	ConvaTec	Hydrocolloids
Comfeel™	Coloplast	
Tegasorb™	3M Health Care Ltd	
FocalSeal	Ethicon	Poly(ethylene glycol)
DuraSeal	Minwax	
CoSeal	Baxter	
Tisseel™	Baxter	Fibrinogen and thrombin
Evarrest™	Johnson & Johnson	
Vitagel™	Gelish	

7.1.3 Need for Adhesive Hydrogels

Adhesive hydrogels are hydrogels which have the inherent ability to adhere to a surface, without the presence of any external glue. They are extremely useful in engineering and medicine, and are used in various applications, including tissue adhesives [14], drug delivery [15], bioelectronics and biosensing [16], soft robotics [17], or additive manufacturing [18]. Specifically, there is a profound need for incorporating adhesive property into hydrogels for their usage in wound closure [19], organ sealing, and in fabrication of stretchable electronics. Moreover, adhesive hydrogels are very much required to develop adaptive and responsive devices.

In this chapter, we will comprehensively review adhesive hydrogels and their need in the medical world. To give the readers a better understanding, we will first briefly discuss the adhesion mechanisms of hydrogel adhesives, and how the adhesion property can be incorporated into the hydrogels. Then we will review the different types of adhesive hydrogels used in biomedicine and will conclude with the current challenges or future trends in hydrogel adhesives, hoping to inspire readers for research in this field.

7.2 Mechanisms of Adhesion

The synthesis of hydrogels is one of the main factors that affect the adhesiveness of hydrogels. They are usually produced by chemical or physical crosslinking. Chemical crosslinking refers to the process by which the polymer molecules within the hydrogels interact among each other covalently, mainly in the presence of an external chemical crosslinking agent. Chemically crosslinked hydrogels have high mechanical strength and are extremely stable [20]. Physical crosslinking is the linking between the polymer molecules by physical interactions, like hydrogen bonds or van der Waals forces [21]. Physical crosslinking is weaker than chemical crosslinking and is reversible in nature. However, physically crosslinked hydrogels are more biocompatible in nature. A new kind of crosslinking, dynamic crosslinking, which combines the stability and mechanical strength of chemical crosslinking and the reversibility of physical crosslinking, has been explored by various researchers [22–24]. Figure 7.1 shows the schematic diagrams for the three types of crosslinkings.

Various theories are proposed by researchers as the dominant adhesion mechanisms of hydrogels. The following section outlines the different kinds of mechanisms put forward by scientists, and their schematic diagrams are shown in Figure 7.2.

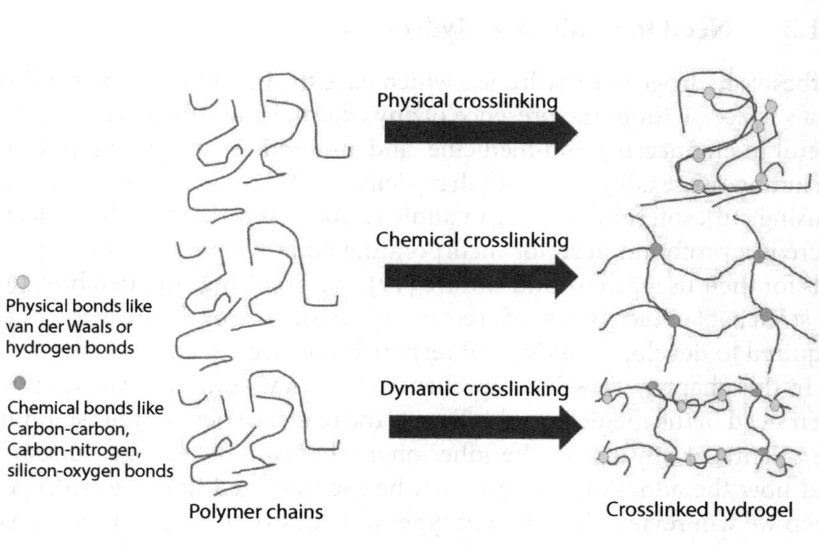

Figure 7.1 Picture depicting the three kinds of crosslinkings.

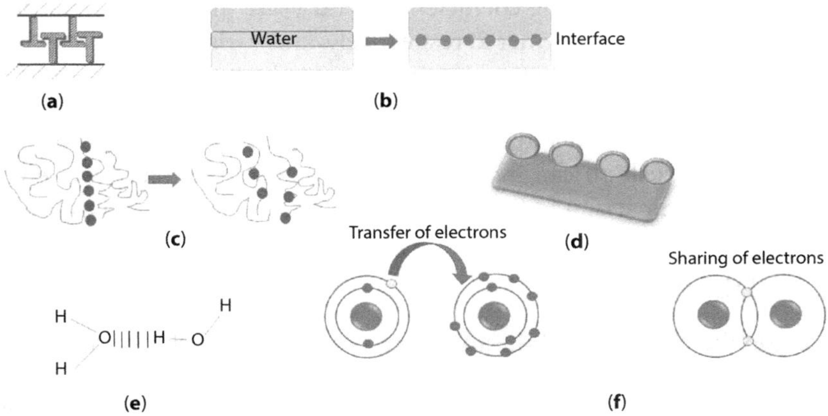

Figure 7.2 The schematic diagram showing (a) mechanical interlocking, (b) wet adhesion, (c) diffusion, (d) van der Waals force, (e) hydrogen bonding, and (f) ionic and covalent bondings.

7.2.1 Mechanical Interlocking

This is the oldest theory of adhesion, put forward by McBain and Hopkins [25] in 1925. Mechanical interlocking is the process by which the adhesive hydrogel and the rough surface of the substrate are interlocked together by filling the pores of the rough surface to achieve the bonding effect. Recently, Qiu *et al.* introduced graphene oxide and carboxymethyl cellulose onto carbon fibres to improve their interface adhesion by forming highly mechanically interlocked composite structures [26]. Wu *et al.* observed that presence of charged silicate nanoparticles in poly(ethylene glycol)-diacrylate hydrogels improved their contact with collagen by mechanical interlocking [27]. Lee *et al.* developed diatom silica reinforced polysaccharide hydrogels and observed that the surface pores provided sites for mechanical interlocking [28]. Also, bioinspired mechanical interlocks are now studied extensively [29].

7.2.2 Wet Adhesion

Wetting theory of adhesion is described as the extent of spreading of one polymer on another [30]. It can be represented as:

$$S = \gamma_1 - \gamma_2 - \gamma_{12} \tag{7.1}$$

Where S is the spreading coefficient, γ_1 and γ_2 are the surface free energies of polymer 1 and polymer 2, respectively, and γ_{12} is the interfacial free energy between the two phases. Wetting is said to be perfect when S>0. The adhesion between two polymers is maximum when polymer 1 completely wets polymer 2. Nguyen and Ho experimentally validated that the surface of a patterned cushion can increase the interfacial adhesion in wet environments in comparison to the surface of non-patterned substrates [31].

7.2.3 Diffusion

Diffusion theory states that the adhesion is established by the mutual diffusion of the molecules of the adhesive and the substrate surface, thus forming an interface phase. Sometimes, the interdiffusion of the component polymers in a hydrogel gives rise to its adhesion property [32]. Feng *et al.* exploited the intermolecular interaction between silk fibroin and konjac glucomannan to develop a highly biocompatible adhesive hydrogel for wound healing applications [33]. An interpenetrating adhesive hydrogel of fibrin and sodium alginate was developed by Vorwald *et al.* for cell targeting and delivery [34]. Zhang *et al.* embedded gold nanoparticles in gelatin-polyacrylamide to produce a tough adhesive hydrogel for its use as a strain sensor [35].

7.2.4 van der Waals Force

van der Waals force is a short-range attraction force between uncharged molecules, resulting from the dipole moment interactions. It is not considered to be the main driving force for adhesion in hydrogels, except in micro or sub-micro size materials, where it can generate sufficiently strong adhesion. Yi *et al.* concluded that the adhesion property shown by the poly(ethylene glycol)-dimethacrylate hydrogels under dry conditions was due to van der Waals forces [36]. These hydrogels have a promising application in the biomedical field owing to their adhesion under both dry and wet conditions.

7.2.5 Electrostatic Interaction

Electrostatic interactions also give rise to adhesive nature in hydrogels. Huang *et al.* observed a highly adhesive nature of a hybrid hydrogel to many substrates such as glass, polypropylene, wood, silicones, etc. and attributed it to electrostatic interactions [37]. A highly tissue adhesive hydrogel containing 2-acrylamido-2-methylpropane sulfonic acid, gelatin, chitosan, 2-methoxyethyl acrylate, and acrylic acid was developed by Tian *et al.* for recording electrophysiological signals [38].

7.2.6 π-π Interaction

Adhesive hydrogels are also developed by exploiting π-π interaction [39, 40]. This is the interaction between aromatic rings, which can be attached to different polymer chains. Famkar *et al.* developed a self-healing wound dressing hydrogel having strong adhesion characteristics by establishing π-π interaction between poly(vinyl alcohol) and graphene oxide [41]. Also, this approach has been adopted by others to crosslink the polymer backbones to form an adhesive hydrogel.

7.2.7 Hydrogen Bonding

A hydrogen bond represents a dipole-dipole attraction between a hydrogen atom and an electronegative atom having only one electron pair. This bond is one of the major mechanisms of adhesion in hydrogels. Hydrogen bond-based adhesion in poly(N-vinylpyrrolidone)/tannic acid hydrogel was achieved by Nam *et al.* [42]. Jiang *et al.* demonstrated that dual hydrogen bonding gave rise to robust adhesive hydrogels. Weak hydrogen bonds were observed between N,N-dimethyl acrylamide and acrylic acid, and strong multiple hydrogen bonds were achieved between 2-ureido-4[1H]-pyrimidinone (UPy) units [43].

7.2.8 Ionic and Covalent Bonding

Ionic bonds are formed by electron transfer between two oppositely charged ions, and they are highly strong bonds. Promising adhesive hydrogel strain sensors were fabricated by Huang *et al.* by the action of ionic bonding [44]. Covalent bonds are formed by sharing of electrons by adjacent atoms, giving rise to a strong interaction force. Yang *et al.* demonstrated that the main driving force for the adhesion of chitosan-polyacrylamide hydrogels to glass and other substrates was due to covalent bonds formed between the polymer and the substrate molecules [45]. Also, a self-healing ionic hydrogel used both ionic and metal coordination bonds to develop high transparency and strong adhesion.

Table 7.3 summarises some of the prominent adhesive hydrogels and their dominant mechanism of adhesion.

Table 7.3 The mechanisms of adhesion of some prominent adhesive hydrogels.

Components of the hydrogel	Dominant mechanism	Reference
Polyacrylamide, calcium alginate	Mechanical interlocking	Pan *et al.* [46]
Polyacrylamide, sodium alginate	Mechanical interlocking	Yuan *et al.* [47]
Propylene glycol, monomethyl ether acetate	Wet adhesion	Wang *et al.* [48]
N-(3,4-Dihydroxyphenylethyl) methacrylamide	Covalent bond	Gao *et al.* [49]
Poly(N-isopropylacrylamide), chitosan	Covalent bonds	Gao *et al.* [50]
Oxidized starch, n-Hydroxyethyl acrylamide	Hydrogen bonding	Zhang *et al.* [51]
Ethylacrylate,[2-(acryloyloxy)ethyl] trimethylammonium chloride, sodium tripolyphosphate	Electrostatic interaction	Huang *et al.* [52]
Polydopamine, silicate	Wet adhesion	Chen *et al.* [53]
Mussel protein, dihydroxyphenylalanine	Ionic bonds	Kim *et al.* [54]
Poly(ethylene glycol)	Diffusion	Sahlin and Peppas [55]

7.3 Design Principles for Adhesive Hydrogels

A hydrogel is made up of water and a polymer, both of which are crucial in increasing or decreasing the adhesive nature of the hydrogel.

If the hydrogels are meant for tissue repair, they must be biocompatible, have strong adhesion to living tissues, antibacterial in nature, tough, and they need to cope with the dynamic environment of human body. The average working temperature and pH of the hydrogels should be like those of wounded skin. The cohesion strength of the hydrogels determines the maximum load the hydrogels can bear, while good adhesion strength ensures the hydrogels do not get separated from the living tissues by blood or plugging.

The adhesive hydrogels must be non-toxic and safe, must be easily available commercially and easy to replace.

7.4 Commonly Used Adhesive Hydrogels

All types of materials that are currently being researched for their potential use as adhesive hydrogels can be broadly divided into four categories. These are:

a) Natural polysaccharides,
b) Poly(amino acid)s,
c) Adhesive proteins,
d) Synthetic polymers.

Also, polyphenols are now being widely investigated for their application in tissue adhesion.

7.4.1 Natural Polysaccharides

Natural polysaccharides which can be derived from plant/animal sources are ideal for developing biomaterials because of their high biocompatibility and biodegradability. Common natural polysaccharides which have been used for adhesive hydrogels are hyaluronic acid, alginate, cellulose, dextran, and chitosan. Table 7.4 summarises some of the work done in this field and their applications.

Natural polysaccharides possess chemical structure which can mimic extracellular matrix of human tissues and thus create a good environment for cell proliferation.

7.4.2 Poly(amino acid)s and Their Analogues

Poly(amino acid)s and their analogues such as poly-γ-glutamic acid, ε-poly-l-lysine, and thioctic acid are widely used for applications in biomedicine owing to their high biocompatibility and ease of functionalization due to the high content of amino or carboxyl groups. Some of the investigations carried out on these polymers as adhesive hydrogels are listed in Table 7.5.

Table 7.4 Some of the adhesive hydrogels fabricated from natural polysaccharides.

Polysaccharide used	Additives used	Uses	Reference
Hyaluronic acid	Catechol	Tissue adhesive	Shin *et al.* [56]
	Pyrogallol, hydroxyapatite	*In-vivo* bone formation	Choi *et al.* [57]
	Polypeptide, zinc oxide	Antimicrobial tissue adhesive	Shirzaei Sani *et al.* [58]
	Serotonin	Adhesive haemostatic agent	An *et al.* [59]
Alginate	Gallic acid	Strain responsive transdermal delivery	Abebe *et al.* [60]
	Gelatin	Cell encapsulation	Sarker *et al.* [61]
	Hydrazide-modified poly(L-glutamic acid)	Antibleeding and tissue repair	Yan *et al.* [62]
Cellulose	Polydopamine, poly(acrylic acid)	UV protection	Tang *et al.* [63]
	Silver, tannic acid	Wound healing	Chen *et al.* [64]
		Cigar wrapping	Dong *et al.* [65]
Dextran	Aminoethyl methacrylate	Tissue engineering	Levesque and Shoichet [66]
	Poly-l-lysine	Surgical tissue adhesive	Hyon *et al.* [67]
	Poly(ethylene glycol)	Tissue sealants	Artzi *et al.* [68]

(*Continued*)

Table 7.4 Some of the adhesive hydrogels fabricated from natural polysaccharides. (*Continued*)

Polysaccharide used	Additives used	Uses	Reference
Chitosan	2-Acrylamido-2-methylpropane sulfonic acid	Physiological signal monitoring	Jin *et al.* [69]
	Poly(ethylene glycol), tyramine	Wound dressing	Lih *et al.* [70]
	Nanocellulose	Wound healing agent	Patel *et al.* [71]

Table 7.5 Some of the work done on adhesive hydrogels made from poly(amino acid)s and their derivatives.

Poly(amino acid) used	Additives used	Use	Reference
Thioctic acid	Tannic acid, phytic acid	Wound dressing	Shao *et al.* [72]
	Gentamicin	Antibacterial wound dressing	Gao *et al.* [73]
Poly-γ-glutamic acid	Poly(3,4-ethylenedioxythiophene), poly (styrene sulfonate)	Bioelectronics	Zhang *et al.* [74]
	3,4-dihydroxyphenylalanine, glycidyl methacylate	Antibacterial bandage	Wang *et al.* [75]
	Ascidian, tannic acid	Wound healing	Xue *et al.* [76]
ε – Polylysine	Gallic acid, acrylic acid	Wound healing	Tang *et al.* [77]

7.4.3 Adhesive Proteins

Adhesive proteins such as elastin, gelatin, keratin, or silk fibroin are excellent candidates for development of adhesive hydrogels for biomedical applications. They are usually derived from animal tissues and possess excellent biocompatibility and high mechanical strength. They are also naturally adhesive in nature. Keratin hydrogels derived from human hair have been developed by researchers and have shown prominent applications in wound healing [78, 79] and nerve regeneration [80]. Kim *et al.* developed a methacrylated silk fibroin hydrogel which showed high adhesion and biocompatibility [81]. This hydrogel is promising for its use as a photocuring laparoscopic medical glue. Biocompatible gelatin methacryloyl hydrogels have been fabricated by Huang *et al.* [82] mimicking the principle of octopus suction cups for wound healing and other biomedical applications. A tough, self-adhesive, antifreezing organic hydrogel is developed by Cao *et al.*, composed of gelatin and MXene nanosheets, for use as wearable electronics [83].

7.4.4 Synthetic Polymers

Various synthetic polymers exhibit excellent biocompatibility and they are widely used in biomedical research. Most common polymers used in the field of adhesive hydrogels are poly(ethylene glycol), poly(vinyl alcohol), polyvinylpyrrolidone, polyurethane, polyacrylamide, poly(acrylic acid) and poly(lactic acid). These polymers exhibit high water solubility, high antifouling, and they are found to help in protein adhesion. Table 7.6 lists a few of the most important studies on this aspect.

Table 7.6 Some notable works done on adhesive hydrogel synthesized from synthetic polymers.

Polymer used	Additives used (if any)	Uses	Reference
Poly(ethylene glycol)	Gelatin	Wound healing	Hwang *et al.* [84]
		Sealants for dural closure	Zhu *et al.* [85]
	Lysozyme	Eye bandage	Zhang *et al.* [86]
	Poly(glycerol sebacate)	Joint wound treatment	Chen *et al.* [87]
	Chondroitin sulphate	Tissue adhesive	Strehin *et al.* [88]

(*Continued*)

Table 7.6 Some notable works done on adhesive hydrogel synthesized from synthetic polymers. (*Continued*)

Polymer used	**Additives used (if any)**	**Uses**	**Reference**
Poly(vinyl alcohol)	Cellulose nanocrystals	Tissue adhesive	Li *et al.* [89]
		Contact lenses	Gupta *et al.* [90]
	Magnetic nanohydroxyapatite	Osteoblast adhesion	Hou *et al.* [91]
	Ferric chloride	Smart devices, logic gates.	Tie *et al.* [92]
Polyvinylpyrrolidone	Poly(acrylic acid)	Drug delivery	Rac *et al.* [93]
	Polyurethane	Platelet adhesion	Kuźmińska *et al.* [94]
Poly(acrylic acid)	Reduced graphene oxide	Motion sensing	Wu *et al.* [95]
	Clay	Soft actuators	Zhao *et al.* [96]
	Laponite	Tissue adhesive	Shen *et al.* [97]
	Poly(ethylene glycol) diacrylate	Microfluidics	Valentin *et al.* [98]
Poly(lactic acid)		Wound dressing and sealants	Okamura *et al.* [99]
Polyurethane	Acrylamide monomer	Medical adhesive	Shen *et al.* [100]
		Bioelectronics	Li *et al.* [101]
Polyacrylamide	Tannic acid, kaolin	Hemostatic adhesive	Fan *et al.* [102]

7.5 Prospective Applications of Adhesive Hydrogels

7.5.1 Biomedicine

Hydrogels have a wide range of applications in biomedicine that include gene therapy, drug delivery, wound healing, tissue engineering, and as contact lenses. The adhesion property is incorporated in hydrogels for eliminating the need for external glues, to prevent the growth of anaerobic bacteria and for better biocompatibility.

7.5.2 Textiles

Textile research has adopted adhesive hydrogels to fabricate smart clothing, which will possess temperature and humidity regulation properties to maximise comfort. Addition of hydrogels on textiles can be done in two different ways: by grafting the monomer onto the textile, or by coating the textile with the polymer. These processes have led to the development of antibacterial, washable clothing with controlled release of fragrances. An interesting work has been done by Rai *et al.* [103] by loading a nanocellulose hydrogel with fine indigo particles and then applying this hydrogel on cotton fabrics.

7.5.3 Food Science

Hydrogels are utilized for their adhesion and other properties by food technologists in order to improve food nutrition, taste, detection, and packaging applications. Adhesive hydrogels have been used as adhesives between cigar paper and tobacco leaves, or as preservatives in refrigerated meat.

7.5.4 Bioelectronics

Conductive and adhesive hydrogels are widely used in bioelectronics due to their ability for tissue adhesion and they are excellent candidates for wearable implants owing to their high biocompatibility [104].

7.5.5 Cosmetics

Hydrogels contain high moisture, are non-irritable and safe to the skin, can release loaded drugs in a sustainable manner, and are easily degradable [105]. Taking advantage of these properties, active ingredients can be loaded into cosmetics for their better efficacy. Adhesive hydrogels have found their potential in the fields of anti-oxidation, anti-tanning, moisturizers, facial masks, and controlled release of skin serum (like hyaluronic acid). One main advantage of adhesive hydrogels over conventional hydrogels is their longer residence time on the skin, thus improving their performance.

7.6 Summary

Recently, adhesive hydrogels are one of the most researched soft materials with applications in biomedicine, e.g., in bioelectronics, signal monitoring, fabrication of soft robots, drug delivery, wound dressings, tissue engineering and many other fields. This is due to their unique properties such as viscoelasticity, biocompatibility, biodegradability, antibacterial nature, their ability to be engineered for favourable mechanical and sealing properties, and thus the hydrogels can overcome the drawbacks of many traditional treatments. In this chapter, we have discussed the dominant mechanisms by which hydrogels exhibit the adhesion property and the main types of polymers which can be used to synthesize adhesive hydrogels. Development of medical glues from polymer hydrogels is particularly beneficial because the polymers are already being extensively used in biomedicine. Also, with the advancement of 3D bioprinting technology, it is very convenient to fabricate adhesive hydrogels and thus their market is growing rapidly. However, development of hydrogels which have a perfect balance of adhesion, biocompatibility and mechanical strength is still a challenge due to the problems posed by residues of chemical crosslinkers, detachment and transfer of fragments, excessive swelling, etc. In future, adhesive hydrogels should be developed to attain controllable adhesion, repeated adhesion, and extensive adhesion.

References

1. O. Wichterle and D. Lím, Hydrophilic gels for biological use, *Nature* 185, 117–118 (1960).
2. A.D. Augst, H.J. Kong, and D.J. Mooney, Alginate hydrogels as biomaterials, *Macromolecular Bioscience* 6 , 623–633 (2006).
3. S.R. van Tomme and W.E. Hennink, Biodegradable dextran hydrogels for protein delivery applications, *Expert Rev Med Devices* 4, 147–164 (2014).
4. S.M.M. Quintero, R. v. Ponce F, M. Cremona, A.L.C. Triques, A.R. d'Almeida, and A.M.B. Braga, Swelling and morphological properties of poly(vinyl

alcohol) (PVA) and poly(acrylic acid) (PAA) hydrogels in solution with high salt concentration, *Polymer (Guildf)* 51, 953–958 (2010).
5. N.S.V. Capanema, A.A.P. Mansur, A.C. de Jesus, S.M. Carvalho, L.C. de Oliveira, and H.S. Mansur, Superabsorbent crosslinked carboxymethyl cellulose-PEG hydrogels for potential wound dressing applications, *Intl J. Biological Macromolecules* 106, 1218–1234 (2018).
6. B. Ramaraj and G. Radhakrishnan, Interpenetrating hydrogel networks based on gelatin and polyacrylamide: Synthesis, swelling, and drug release analysis, *J. Appl. Polym. Sci.* 52, 837–846 (1994).
7. R. Hafezi Moghaddam, S. Dadfarnia, A.M.H. Shabani, Z.H. Moghaddam, and M. Tavakol, Electron beam irradiation synthesis of porous and non-porous pectin based hydrogels for a tetracycline drug delivery system, *Mater. Sci. Eng.* C 102, 391–404 (2019).
8. A. Dey, P. Guha Ray, S. Dhara, and S. Neogi, Optically engineered ZnO nanoparticles: Excitable at visible wavelength and lowered cytotoxicity towards bioimaging applications, *Appl. Surface Sci.* 592, 153303 (2022).
9. P. Bhattacharya and S. Neogi, Antibacterial properties of doped nanoparticles, *Reviews Chem. Eng.* 35, 861–876 (2019).
10. I. Sengupta, P. Bhattacharya, M. Talukdar, S. Neogi, S.K. Pal, and S. Chakraborty, Bactericidal effect of graphene oxide and reduced graphene oxide: Influence of shape of bacteria, *Colloids Interface Sci. Communications* 28, 60–68 (2019).
11. R. Pinnaratip, M.S.A. Bhuiyan, K. Meyers, R.M. Rajachar, and B.P. Lee, Multifunctional biomedical adhesive, *Advanced Healthcare Mater* 8, e1801568 (2019).
12. L. Sanders and J. Nagatomi, Clinical applications of surgical adhesives and sealants, *Crit Rev Biomed Eng* 42, 271-292 (2014).
13. A.A. Pawar, M.A. Joshi, M. Gadhire, R. Shotriya, B. Phad, and J. Singh, Prospective randomized comparative study of skin adhesive glue (2-methyl-2-cyanopropionate or cyanoacrylate) versus conventional skin suturing by suture material/skin stapler in clean surgical cases, *Intl. Surgery J.* 5, 168-173 (2017).
14. J.H. Ryu, Y. Lee, W.H. Kong, T.G. Kim, T.G. Park, and H. Lee, Catechol-functionalized chitosan/pluronic hydrogels for tissue adhesives and hemostatic materials, *Biomacromolecules* 12, 2653–2659 (2011).
15. R. Narayanaswamy and V.P. Torchilin, Hydrogels and their applications in targeted drug delivery, *Molecules* 24, 603-621 (2019).
16. B. Ying, R. Zeyuan Chen, R. Zuo, J. Li, X. Liu, B. Ying, R.Z. Chen, R. Zuo, X. Liu, and J. Li, An anti-freezing, ambient-stable and highly stretchable ionic skin with strong surface adhesion for wearable sensing and soft robotics, *Adv. Funct. Mater.* 31, 2104665 (2021).
17. H. Jung, M.K. Kim, J.Y. Lee, S.W. Choi, and J. Kim, Adhesive hydrogel patch with enhanced strength and adhesiveness to skin for transdermal drug delivery, *Adv. Funct. Mater.* 30, 2004407 (2020).

18. X. Deng, R. Attalla, L.P. Sadowski, M. Chen, M.J. Majcher, I. Urosev, D.C. Yin, P.R. Selvaganapathy, C.D.M. Filipe, and T. Hoare, Autonomously self-adhesive hydrogels as building blocks for additive manufacturing, *Biomacromolecules* 19, 62–70 (2018).
19. C. Ghobril and M.W. Grinstaff, The chemistry and engineering of polymeric hydrogel adhesives for wound closure: a tutorial, *Chem. Soc. Reviews* 44, 1820–1835 (2015).
20. J. Xu, X. Liu, X. Ren, and G. Gao, The role of chemical and physical crosslinking in different deformation stages of hybrid hydrogels, *European Polymer J.* 100, 86–95 (2018).
21. W.E. Hennink and C.F. van Nostrum, Novel crosslinking methods to design hydrogels, *Advanced Drug Delivery Reviews* 64, 223–236 (2012).
22. A. Skardal, J. Zhang, L. Mccoard, S. Oottamasathien, and G.D. Prestwich, Dynamically crosslinked gold nanoparticle – hyaluronan hydrogels, *Advanced Materials* 22, 4736–4740 (2010).
23. M.C. Roberts, M.C. Hanson, A.P. Massey, E.A. Karren, and P.F. Kiser, Dynamically restructuring hydrogel networks formed with reversible covalent crosslinks, *Advanced Materials* 19, 2503–2507 (2007).
24. M.R. Arkenberg, H.D. Nguyen, and C.C. Lin, Recent advances in bio-orthogonal and dynamic crosslinking of biomimetic hydrogels, *J. Mater. Chem. B* 8, 7835–7855 (2020).
25. J.W. McBain and D.G. Hopkins, On adhesives and adhesive action, *J. Physi. Chem.* 29, 188–204 (1925).
26. B. Qiu, T. Sun, M. Li, Y. Chen, S. Zhou, M. Liang, and H. Zou, High micro-mechanical interlocking graphene oxide/carboxymethyl cellulose composite architectures for enhancing the interface adhesion between carbon fiber and epoxy, *Composites Part A:* 139, 106092 (2020).
27. C.J. Wu, J.J. Wilker, and G. Schmidt, Robust and adhesive hydrogels from cross-linked poly(ethylene glycol) and silicate for biomedical use, *Macromolecular Bioscience* 13, 59–66 (2013).
28. J. Lee, E. Park, A. Fujisawa, and H. Lee, Diatom silica/polysaccharide elastomeric hydrogels: adhesion and interlocking synergy, *ACS Appl. Mater. Interfaces* 13, 21703–21713 (2021).
29. M. Zhu, F. Zhang, and X. Chen, Bioinspired mechanically interlocking structures, *Small Structures* 1, 2000045-53 (2020).
30. A. Dey, P. Bhattacharya, and S. Neogi, Bioadhesives in biomedical applications: a critical review: in; *Progress in Adhesion and Adhesives* Vol: 6, K. L. Mittal (Ed.) pp.131-153, Wiley-Scrivener, Beverly, MA (2021).
31. V.P. Nguyen and V.A. Ho, Wet adhesion of soft curved interfaces with micro pattern, *IEEE Robotics Automation Letters* 6, 4273–4280 (2021)
32. E. Jabbari and N.A. Peppas, Polymer-polymer interdiffusion and adhesion, *J. Macromol. Sci. C* 34, 205–241 (2006).

33. Y. Feng, X. Li, Q. Zhang, S. Yan, Y. Guo, M. Li, and R. You, Mechanically robust and flexible silk protein/polysaccharide composite sponges for wound dressing, *Carbohydrate Polymers* 216, 17–24 (2019).
34. C.E. Vorwald, T. Gonzalez-Fernandez, S. Joshee, P. Sikorski, and J.K. Leach, Tunable fibrin-alginate interpenetrating network hydrogels to support cell spreading and network formation, *Acta Biomaterialia* 108, 142–152 (2020).
35. X. Zhang, Y. Zhang, W. Zhang, Y. Dai, and F. Xia, Gold nanoparticles-deranged double network for Janus adhesive-tough hydrogel as strain sensor, *Chem. Eng. J.* 420, 130447 (2021).
36. H. Yi, S.H. Lee, M. Seong, M.K. Kwak, and H.E. Jeong, Bioinspired reversible hydrogel adhesives for wet and underwater surfaces, *J. Mater. Chem. B* 6, 8064–8070 (2018).
37. G. Huang, Z. Tang, S. Peng, P. Zhang, T. Sun, W. Wei, L. Zeng, H. Guo, H. Guo, and G. Meng, Modification of hydrophobic hydrogels into a strongly adhesive and tough hydrogel by electrostatic interaction, *Macromolecules* 55, 156–165. (2022).
38. G. Tian, Y. Liu, M. Yu, C. Liang, D. Yang, J. Huang, Q. Zhao, W. Zhang, J. Chen, Y. Wang, P. Xu, Z. Liu, and D. Qi, Electrostatic interaction-based high tissue adhesive, stretchable microelectrode arrays for the electrophysiological interface, *ACS Appl. Mater. Interfaces* 14, 4852–4861 (2022).
39. Y.C. Lin, K.M. Liu, P.L. Chiu, C.M. Chao, C.S. Wen, C.Y. Wang, and H.H. Tseng, Enhancing the hydrophilicity and biofoulant removal ability of a PVDF ultrafiltration membrane via π-π interactions as measured by AFM, *J. Membrane Sci.* 641, 119874 (2022).
40. J. Yang, R. Bai, B. Chen, and Z. Suo, Hydrogel Adhesion: A supramolecular synergy of chemistry, topology, and mechanics, *Adv. Funct. Mater.* 30, 1901693 (2020).
41. E. Famkar, G. Pircheraghi, and H. Nazockdast, Effectively exerting the reinforcement of polyvinyl alcohol nanocomposite hydrogel via poly(dopamine) functionalized graphene oxide, *Composites Sci. Technol.* 217, 109119 (2022).
42. H.G. Nam, M.G. Nam, P.J. Yoo, and J.H. Kim, Hydrogen bonding-based strongly adhesive coacervate hydrogels synthesized using poly(N-vinylpyrrolidone) and tannic acid, *Soft Matter* 15, 785–791 (2019).
43. Z. Jiang, Y. Li, Y. Shen, J. Yang, Z. Zhang, Y. You, Z. Lv, and L. Yao, Robust hydrogel adhesive with dual hydrogen bond networks, *Molecules* 26, 2688-2698 (2021).
44. Y. Huang, L. Xiao, J. Zhou, T. Liu, Y. Yan, S. Long, and X. Li, Strong tough polyampholyte hydrogels via the synergistic effect of ionic and metal–ligand bonds, *Advanced Functional Materials* 31, 2103917 (2021).
45. Y. Yang, M. Zhou, J. Peng, X. Wang, Y. Liu, W. Wang, and D. Wu, Robust, anti-freezing and conductive bonding of chitosan-based double-network hydrogels for stable-performance flexible electronic, *Carbohydrate Polymers* 276, 118753 (2022).

46. S. Pan, F. Zhang, P. Cai, M. Wang, K. He, Y. Luo, Z. Li, G. Chen, S. Ji, Z. Liu, X.J. Loh, and X. Chen, Mechanically interlocked hydrogel–elastomer hybrids for on-skin electronics, *Advanced Functional Materials* 30, 1909540 (2020).
47. Y. Yuan, H. Wu, X. Ren, J. Wang, R. Liu, B. Hu, and N. Gu, Dual-network hydrogel based on ionic nano-reservoir for gastric perforation sealing, *Science China Materials* 65, 827–835 (2021).
48. Y. Wang, V. Kang, E. Arzt, W. Federle, and R. Hensel, Strong wet and dry adhesion by cupped microstructures, *ACS Appl. Mater. Interfaces* 11, 26483–26490 (2019).
49. L. Gao, S. Ma, L. Bao, X. Zhao, Y. Xiang, Z. Zhang, Y. Ma, Z. Ma, Y.M. Liang, and F. Zhou, Molecular engineering super-robust dry/wet adhesive with strong interface bonding and excellent mechanical tolerance, *ACS Appl. Mater. Interfaces* 14 12684–12692 (2022).
50. Y. Gao, X. Han, J. Chen, Y. Pan, M. Yang, L. Lu, J. Yang, Z. Suo, and T. Lu, Hydrogel-mesh composite for wound closure, *Proc. Natl. Acad. Sci. (USA)* 118, e2103457118 (2021).
51. Y. Zhang, J. Xue, D. Li, H. Li, Z. Huang, Y. Huang, C. Gong, S. Long, and X. Li, Tough hydrogels with tunable soft and wet interfacial adhesion, *Polymer Testing* 93, 106976 (2021).
52. G. Huang, Z. Tang, S. Peng, P. Zhang, T. Sun, W. Wei, L. Zeng, H. Guo, H. Guo, and G. Meng, Modification of hydrophobic hydrogels into a strongly adhesive and tough hydrogel by electrostatic interaction, *Macromolecules* 55, 156–165 (2022).
53. Y. Chen, H. Qin, A. Mensaha, Q. Wang, F. Huang, and Q. Wei, Biomimetic nanocomposite hydrogel networks for robust wet adhesion to tissues, *Composites Part B:* 222, 109071 (2021).
54. B.J. Kim, D.X. Oh, S. Kim, J.H. Seo, D.S. Hwang, A. Masic, D.K. Han, and H.J. Cha, Mussel-mimetic protein-based adhesive hydrogel, *Biomacromolecules* 15, 1579–1585 (2014).
55. J.J. Sahlin and N.A. Peppas, Enhanced hydrogel adhesion by polymer interdiffusion: Use of linear poly(ethylene glycol) as an adhesion promoter, *J. Biomater. Sci. Polym.* 8, 421–436 (2012).
56. J. Shin, J.S. Lee, C. Lee, H.J. Park, K. Yang, Y. Jin, J.H. Ryu, K.S. Hong, S.H. Moon, H.M. Chung, H.S. Yang, S.H. Um, J.W. Oh, D.I. Kim, H. Lee, and S.W. Cho, Tissue adhesive catechol-modified hyaluronic acid hydrogel for effective, minimally invasive cell therapy, *Advanced Functional Materials* 25, 3814–3824 (2015).
57. S. Choi, J.S. Lee, J. Shin, M.S. Lee, D. Kang, N.S. Hwang, H. Lee, H.S. Yang, and S.W. Cho, Osteoconductive hybrid hyaluronic acid hydrogel patch for effective bone formation, *J. Controlled Release* 327, 571–583 (2020).
58. E. Shirzaei Sani, R. Portillo-Lara, A. Spencer, W. Yu, B.M. Geilich, I. Noshadi, T.J. Webster, and N. Annabi, Engineering adhesive and antimicrobial hyaluronic acid/elastin-like polypeptide hybrid hydrogels for tissue engineering applications, *ACS Biomaterials Sci. Eng.* 4, 2528–2540 (2018).

59. S. An, E.J. Jeon, J. Jeon, and S.W. Cho, A serotonin-modified hyaluronic acid hydrogel for multifunctional hemostatic adhesives inspired by a platelet coagulation mediator, *Materials Horizons* 6, 1169–1178 (2019).
60. M.W. Abebe, R. Appiah-Ntiamoah, and H. Kim, Gallic acid modified alginate self-adhesive hydrogel for strain responsive transdermal delivery, *Intl. J. Biological Macromolecules* 163, 147–155 (2020).
61. B. Sarker, J. Rompf, R. Silva, N. Lang, R. Detsch, J. Kaschta, B. Fabry, and A.R. Boccaccini, Alginate-based hydrogels with improved adhesive properties for cell encapsulation, *Intl. J. Biological Macromolecules* 78, 72–78 (2015).
62. S. Yan, W. Wang, X. Li, J. Ren, W. Yun, K. Zhang, G. Li, and J. Yin, Preparation of mussel-inspired injectable hydrogels based on dual-functionalized alginate with improved adhesive, self-healing, and mechanical properties, *J. Mater. Chem. B* 6 , 6377–6390 (2018).
63. Z. Tang, Y. Miao, J. Zhao, H. Xiao, M. Zhang, K. Liu, X. Zhang, L. Huang, L. Chen, and H. Wu, Mussel-inspired biocompatible polydopamine/carboxymethyl cellulose/polyacrylic acid adhesive hydrogels with UV-shielding capacity, *Cellulose* 28, 1527–1540 (2021).
64. Y. Chen, Y. Zhang, A. Mensaha, D. Li, Q. Wang, and Q. Wei, A plant-inspired long-lasting adhesive bilayer nanocomposite hydrogel based on redox-active Ag/Tannic acid-cellulose nanofibers, *Carbohydrate Polymers* 255, 11750 (2021).
65. S. Dong, S. Feng, F. Liu, R. Li, W. Li, F. Liu, G. Shi, L. Chen, and Y. Zhang, Factors influencing the adhesive behaviour of carboxymethyl cellulose-based hydrogel for food applications, *Intl. J. Biological Macromolecules* 179, 398–406 (2021).
66. S.G. Lévesque and M.S. Shoichet, Synthesis of cell-adhesive dextran hydrogels and macroporous scaffolds, *Biomaterials* 27, 5277–5285 (2006).
67. S.H. Hyon, N. Nakajima, H. Sugai, and K. Matsumura, Low cytotoxic tissue adhesive based on oxidized dextran and epsilon-poly-l-lysine, *J. Biomedical Mater. Res. A* 102 ,2511–2520 (2014).
68. N. Artzi, T. Shazly, C. Crespo, A.B. Ramos, H.K. Chenault, and E.R. Edelman, Characterization of star adhesive sealants based on PEG/dextran hydrogels, *Macromolecular Bioscience* 9, 754–765 (2009).
69. R. Jin, J. Xu, L. Duan, and G. Gao, Chitosan-driven skin-attachable hydrogel sensors toward human motion and physiological signal monitoring, *Carbohydrate Polymers* 268, 118240 (2021).
70. E. Lih, J.S. Lee, K.M. Park, and K.D. Park, Rapidly curable chitosan–PEG hydrogels as tissue adhesives for hemostasis and wound healing, *Acta Biomaterialia* 8, 3261–3269 (2012).
71. D.K. Patel, K. Ganguly, J. Hexiu, S.D. Dutta, T. V. Patil, and K.T. Lim, Functionalized chitosan/spherical nanocellulose-based hydrogel with superior antibacterial efficiency for wound healing, *Carbohydrate Polymers* 284, 119202 (2022).

72. X. Shao, X. Yang, Y. Zhou, Q. Xia, Y. Lu, X. Yan, C. Chen, T. Zheng, L. Zhang, Y. Ma, and S. Gao, Antibacterial, wearable, transparent tannic acid–thioctic acid–phytic acid hydrogel for adhesive bandages, *Soft Matter* 18, 2814–2828 (2022).
73. Y. Gao, X. Zhan, S. Huo, L. Fu, Z. Tang, K. Qi, C. Lv, C. Liu, Y. Zhu, S. Ding, and Y. Lv, A gentamicin-thioctic acid multifunctional hydrogel for accelerating infected wound healing, *J. Mater. Chem. B* 10, 2171–2182 (2022).
74. C. Zhang, M. Wang, C. Jiang, P. Zhu, B. Sun, Q. Gao, C. Gao, and R. Liu, Highly adhesive and self-healing γ-PGA/PEDOT:PSS conductive hydrogels enabled by multiple hydrogen bonding for wearable electronics, *Nano Energy* 95, 106991 (2022).
75. P. Wang, Y. Pu, Y. Ren, S. Liu, R. Yang, X. Tan, W. Zhang, T. Shi, S. Li, and B. Chi, Bio-inspired hydrogel-based bandage with robust adhesive and antibacterial abilities for skin closure, *Science China Materials* 65, 246–254 (2021).
76. W. Xue, R. Yang, S. Liu, Y. Pu, P. Wang, W. Zhang, X. Tan, and B. Chi, Ascidian-inspired aciduric hydrogels with high stretchability and adhesiveness promote gastric hemostasis and wound healing, *Biomaterials Sci.* 10, 2417–2427 (2022).
77. L. Tang, Y. Dang, Y. Wang, Y. Zhang, T. Hu, C. Ding, H. Wu, Y. Ni, L. Chen, L. Huang, and M. Zhang, Rapid fabrication of bionic pyrogallol-based self-adhesive hydrogel with mechanically tunable, self-healing, antibacterial, wound healing, and hemostatic properties, *Biomaterials Advances* 136, 212765 (2022).
78. T. Khaliq, M. Sohail, S.A. Shah, A. Mahmood, M. Kousar, and N. Jabeen, Bioactive and multifunctional keratin-pullulan based hydrogel membranes facilitate re-epithelization in diabetic model, *Intl. J. Biological Macromolecules* 209, 1826–1836 (2022).
79. J. Wang, S. Hao, T. Luo, Z. Cheng, W. Li, F. Gao, T. Guo, Y. Gong, and B. Wang, Feather keratin hydrogel for wound repair: Preparation, healing effect and biocompatibility evaluation, *Colloids Surfaces B* 149, 341–350 (2017).
80. P. Sierpinski, J. Garrett, J. Ma, P. Apel, D. Klorig, T. Smith, L.A. Koman, A. Atala, and M. van Dyke, The use of keratin biomaterials derived from human hair for the promotion of rapid regeneration of peripheral nerves, *Biomaterials* 29, 118–128 (2008).
81. S.H. Kim, Y.J. Lee, J.R. Chao, D.Y. Kim, M.T. Sultan, H.J. Lee, J.M. Lee, J.S. Lee, O.J. Lee, H. Hong, H. Lee, O. Ajiteru, Y.J. Suh, H.S. Choi, Y.J. Cho, and C.H. Park, Rapidly photocurable silk fibroin sealant for clinical applications, *NPG Asia Materials* 12, 1–16 (2020).
82. R. Huang, X. Zhang, W. Li, L. Shang, H. Wang, and Y. Zhao, Suction cups-inspired adhesive patch with tailorable patterns for versatile wound healing, *Adv. Sci.* 8, 2100201 (2021).
83. L. Cao, Z. Zhao, J. Li, Y. Yi, and Y. Wei, Gelatin-reinforced zwitterionic organohydrogel with tough, self-adhesive, long-term moisturizing and

antifreezing properties for wearable electronics, *Biomacromolecules* 23, 1278–1290 (2022).
84. J. Hwang, P. le Thi, S. Lee, E.H. Park, E. Lee, E. Kim, K. Chang, and K.D. Park, Injectable gelatin-poly(ethylene glycol) adhesive hydrogels with highly hemostatic and wound healing capabilities, *J. Indus. Eng. Chem.* 109, 372–383 (2022).
85. T. Zhu, H. Wang, Z. Jing, D. Fan, Z. Liu, X. Wang, and Y. Tian, High efficacy of tetra-PEG hydrogel sealants for sutureless dural closure, *Bioactive Mater.* 8, 12–19 (2022).
86. S. Zhang, H. Zhou, C. Huang, J. Sun, X. Qu, and Y. Lu, A novel corneal adhesive based on functionally coupled PEG-lysozyme hydrogel for wound closure after surgical eye surgery, *Chinese Chemical Letters* 33, 4321-4325 (2022).
87. K. Chen, Z. Wu, Y. Liu, Y. Yuan, C. Liu, K. Chen, Z. Wu, Y. Liu, Y. Yuan, and C. Liu, Injectable double-crosslinked adhesive hydrogels with high mechanical resilience and effective energy dissipation for joint wound treatment, *Advanced Functional Materials* 32, 2109687 (2022).
88. I. Strehin, Z. Nahas, K. Arora, T. Nguyen, and J. Elisseeff, A versatile pH sensitive chondroitin sulfate–PEG tissue adhesive and hydrogel, *Biomaterials* 31, 2788–2797 (2010).
89. Z. Li, D. Wang, H. Bai, S. Zhang, P. Ma, W. Dong, Photo-crosslinking strategy constructs adhesive, superabsorbent, and tough PVA-based hydrogel through controlling the balance of cohesion and adhesion, *Macromol. Mater. Eng.* 305, 1900623 (2020).
90. S. Gupta, T.J. Webster, and A. Sinha, Evolution of PVA gels prepared without crosslinking agents as a cell adhesive surface, *J. Mater. Sci.: Materials in Medicine* 22, 1763–1772 (2011).
91. R. Hou, G. Zhang, G. Du, D. Zhan, Y. Cong, Y. Cheng, and J. Fu, Magnetic nanohydroxyapatite/PVA composite hydrogels for promoted osteoblast adhesion and proliferation, *Colloids Surfaces B* 103, 318–325 (2013).
92. J. Tie, H. Liu, J. Lv, B. Wang, Z. Mao, L. Zhang, Y. Zhong, X. Feng, X. Sui, and H. Xu, Multi-responsive, self-healing and adhesive PVA based hydrogels induced by the ultrafast complexation of Fe^{3+} ions, *Soft Matter* 15, 7404–7411 (2019).
93. V. Rac, S. Lević, B. Balanč, B. Olalde Graells, and G. Bijelić, PVA cryogel as model hydrogel for iontophoretic transdermal drug delivery investigations. Comparison with PAA/PVA and PAA/PVP interpenetrating networks, *Colloids Surfaces B* 180, 441–448 (2019).
94. A. Kuźmińska, B.A. Butruk-Raszeja, A. Stefanowska, and T. Ciach, Polyvinylpyrrolidone (PVP) hydrogel coating for cylindrical polyurethane scaffolds, *Colloids Surfaces B* 192, 111066 (2020).
95. L. Wu, Y. Hu, P. Tang, H. Wang, and Y. Bin, High stretchable, pH-sensitive and self-adhesive rGO/CMCNa/PAA composite conductive hydrogel with

good strain-sensing performance, *Composites Communications* 24, 100669 (2021).

96. L. Zhao, J. Huang, Y. Zhang, T. Wang, W. Sun, and Z. Tong, Programmable and bidirectional bending of soft actuators based on janus structure with sticky tough PAA-clay hydrogel, *ACS Appl. Mater. Interfaces* 9, 11866–11873 (2017).
97. M. Shen, Y. Sun, J. Xu, X. Guo, and R.K. Prud'Homme, Rheology and adhesion of poly(acrylic acid)/laponite nanocomposite hydrogels as biocompatible adhesives, *Langmuir* 30, 1636–1642 (2014).
98. T.M. Valentin, E.M. Dubois, C.E. Machnicki, D. Bhaskar, F.R. Cui, and I.Y. Wong, 3D printed self-adhesive PEGDA–PAA hydrogels as modular components for soft actuators and microfluidics, *Polymer Chemistry* 10, 2015–2028 (2019).
99. Y. Okamura, K. Kabata, M. Kinoshita, D. Saitoh, and S. Takeoka, Free-standing biodegradable poly(lactic acid) nanosheet for sealing operations in surgery, *Advanced Materials* 21, 4388–4392 (2009).
100. J. Shen, H. Zhang, J. Zhu, Y. Ma, H. He, F. Zhu, L. Jia, and Q. Zheng, Simple preparation of a waterborne polyurethane crosslinked hydrogel adhesive with satisfactory mechanical properties and adhesion properties, *Frontiers in Chemistry* 10, 855352 (2022).
101. X. Li, E. Zhang, J. Shi, X. Xiong, J. Lin, Q. Zhang, X. Cui, L. Tan, K. Wu, X. Li, E. Zhang, J. Shi, X. Xiong, J. Lin, Q. Zhang, X. Cui, L. Tan, and K. Wu, Waterborne polyurethane enhanced, adhesive, and ionic conductive hydrogel for multifunctional sensors, *Macromolecular Rapid Communications* 42, 2100457 (2021).
102. X. Fan, S. Wang, Y. Fang, P. Li, W. Zhou, Z. Wang, M. Chen, and H. Liu, Tough polyacrylamide-tannic acid-kaolin adhesive hydrogels for quick hemostatic application, *Mater. Sci. Eng. C* 109, 110649 (2020).
103. S. Rai, R. Saremi, S. Sharma, and S. Minko, Environment-friendly nanocellulose-indigo dyeing of textiles, *Green Chemistry* 23, 7937–7944 (2021).
104. N.R. Chodankar, S.H. Ji, Y.K. Han, and D.H. Kim, Dendritic nanostructured waste copper wires for high-energy alkaline battery, *Nano-Micro Letters* 12, 1–12 (2020).
105. P. Bhattacharya and S. Neogi, Techniques for deposition of coatings with enhanced adhesion to bio-implants, in: *Adhesion in Pharmaceutical, Biomedical, and Dental Fields*, K. L. Mittal and F. M. Etzler, (Eds.) pp. 235-256, Wiley-Scrivener, Beverly, MA (2017).

8

Adhesives in Dermal Patches

Niharika Lal[1]*, Praveen Kumar Gaur[1] and Navneet Verma[2]

[1]*Department of Pharmacy, Metro College of Health Sciences and Research, Greater Noida, Uttar Pradesh, India*
[2]*Department of Pharmacy, IFTM University, Moradabad, Uttar Pradesh, India*

Abstract
The skin offers an available and advantageous site for the administration of medicaments. For millennia, individuals have used substances on the skin for remedial therapy, and in the modern era, a variety of dermal formulations have been developed to treat local and systemic diseases. A dermal patch, also known as a skin patch, is a medicated adhesive patch that adheres to the skin to deliver medication. Transdermal drug delivery could be an alternative for oral drug delivery and is poised to provide an alternative to hypodermic injection too. Dermal patches often are known as Transdermal Patches and with the constant development of new and innovative approaches, they have been proposed as a safe and effective way to deliver drugs through the skin. But, for an appropriate drug delivery from a dermal patch, its adhesion to the skin is the primary factor. This is crucial to the product's safety, potency, and quality. The US Food and Drug Administration (FDA) has received multiple reports of "adhesion failure" for transdermal drug delivery devices through the Drug Quality Reporting System (DQRS). It is expected that in transdermal patches the adhesive will maintain a bond with the skin once the release liner has been removed and deliver therapeutic action to the skin without causing any kind of skin irritation.

In this chapter, the focus is primarily on the utilization of adhesives in the fabrication of dermal patches, the mechanism of adhesion, tests to determine the efficacy of adhesives, and possible adhesion failures of transdermal patches. Moreover, this chapter includes a summary of the factors that influence skin adhesion, adhesive types adhesive properties, *in vivo* adhesive performance, regulatory aspects, as well as quality control analyses. This chapter concludes with an overview of the potential prospects of novel adhesives in Transdermal Drug Delivery.

**Corresponding author*: niharikaLaL24@gmail.com

K.L. Mittal and S. Neogi (eds.) Adhesives in Biomedical Applications, (177–212) © 2023 Scrivener Publishing LLC

Keywords: Dermal drug delivery, transdermal patches, adhesion, adhesive patch

8.1 Introduction

In the current era, dermal drug delivery is an area of interest for most pharmaceutical and cosmeceutical industries whether to provide drug delivery or to incorporate cosmetics to enhance skin appearance as well as to treat various conditions. Dermal medication application is a common method for treating skin problems on a local level [1]. However, the stratum corneum, the uppermost layer of the skin, acts as a physical barrier to prevent chemicals from penetrating the skin. The active agents' and vehicles' physicochemical properties are important limiting factors for drug penetration through the skin. The drug can be incorporated in transdermal patches as nanoparticles, microparticles, dispersions, emulsions, or in solubilized form. While all topical and transdermal compounds are present in dermal drug formulation, only the transdermal formulations also known as "patches," are intended to penetrate the top layer of the skin and exert their action on deeper or distant tissues [2].

It has been claimed during the last few decades that the transdermal method of medication administration is superior and an active approach in comparison to oral and parenteral drug delivery. Delivery of drugs via the skin is important because it circumvents hepatic first-pass metabolism (systemic) which is typically observed with the oral route [3]. The first-pass effect can lead to low bioavailability, adverse secondary effects, and an erratic drug dosing schedule. The presence of bacterial enzymes, gut wall enzymes, hepatic enzymes, and gastrointestinal lumen enzymes has been implicated in the first-pass effect [4].

The utilization of delivering drugs via percutaneous transport is important from a pharmacotherapeutic standpoint as drug can be easily self-administered without causing pain, has minimal chances of enzymatic drug degradation and can supply effective drug levels in systemic circulation [5]. The drug diffuses across the stratum corneum (SC), enters the viable epidermis/dermis, and then is available for systemic absorption via dermal blood capillaries.

Additionally, transdermal drug administration requires low dosage medications because of the low diffusional length necessary to reach the vasculatures [6]. Another benefit is that side effects induced by non-uniform drug absorption and metabolism in the gastrointestinal tract are avoided. Additionally, dermal patches are simple to apply to the skin, are well tolerated, and are a viable option when oral drug delivery is difficult [7].

From an economic point of view, it has been estimated that the global transdermal drug delivery market will grow to approximately $96.57 billion by the year 2025. The global transdermal drug delivery market is currently worth $32 billion [8]. Several Transdermal Drug Delivery Systems (TDDS) containing drugs such as estradiol, fentanyl, clonidine, nicotine, oxybutynin, nitroglycerine, and scopolamine are available and the industry has been focusing on the implementation of dermal patches in treating psychiatric and neurological disorders [9].

The utmost critical factor for dermal drug delivery is adhesion, and for an appropriate adhesion, the choice of adhesive is an important aspect. The US Food and Drug Administration (FDA) has received multiple reports of "adhesion deficient" for transdermal drug delivery devices through the Drug Quality Reporting System (DQRS) [10]. To begin with, the therapeutic effect of the drug is dependent on the adhesion performance of the TDDS. Patch lift, or even the patch falling off, reduces the surface area of contact with the skin, which can minimize the drug delivery from the patch. In other words, poor adhesion leads to incorrect patient dosage [11]. Second, patches that do not last for the prescribed time must be replaced more frequently, increasing the cost to the patient. Finally, lack of adhesion is a potential safety hazard. Children who pick up fallen patches may inadvertently receive a dose [12]. Accidental exposure to certain patches has resulted in death and other major medical complications (e.g. transfer of a patch from an adult to a child while hugging, accidentally sitting, or lying on a patch) [13].

Overall, It is expected that in transdermal or dermal patches, adhesive will maintain a direct bond with the skin once the release liner has been removed, provide therapeutic action and desired skin adhesion without causing skin irritation [14].

So, for appropriate adhesion, the choice of adhesive is an important aspect. The term 'adhesive' is used to describe any substance that tends to attach or stick to other materials [15]. The interaction of the adhesive with the substrate surface is referred as adhesion, and it includes two dissimilar entities being held close such that mechanical force or work can be transferred across the interface. Several theories have been proposed to explain the adhesion process, with the forces involved including van der Waals forces, chemical bonding, or electrostatic attraction [16–18].

Over the past few decades, adhesives in one form or another have been explored and studied by formulation experts. This chapter provides an overview of the significance of the role of adhesive in a transdermal drug delivery system and the necessity for adhesion testing [19]. The properties of and issues associated with the adhesive use in currently marketed products, such as variability, safety, and regulatory aspects are then described.

8.2 Types of Dermal Patches

8.2.1 Microreservoir Type Dermal Patches

This drug delivery method is a hybrid between a reservoir and matrix-dispersion systems (Figure 8.1). The drug reservoir is created by suspending the drug in an aqueous solution of liquid polymer and then homogeneously dispersing the drug suspension in a lipophilic polymer, such as silicone elastomer, using the high energy dispersion technique and shear mechanical force to form thousands of unleachable microscopic spheres of drug reservoirs [20]. This technique was used in the creation of the Nitro disc (Key Pharmaceuticals) transdermal patch. Depending on the proportional degree of solubility of the drug in the liquid compartment and the polymer matrix, the release of a drug from a microreservoir-type system can follow either a partitioning or diffusion process.

8.2.2 Adhesive Dispersion Controlled Dermal Patches

The system, as shown in Figure 8.1, consists of a drug-impermeable backing membrane, a drug reservoir made by directly dispersing the drug in an

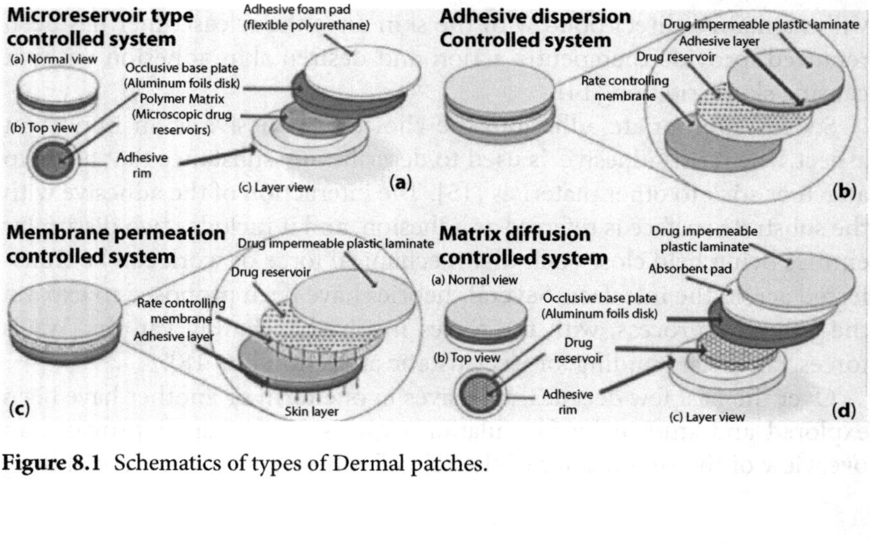

Figure 8.1 Schematics of types of Dermal patches.

adhesive polymer and then spreading the medicated adhesive onto a flat sheet of drug-impermeable backing by solvent casting or hot melting to form a thin drug adhesive layer. On top of this a constant-thickness layer of non-medicated, rate-controlling adhesive polymer is applied to create an adhesive diffusion-controlled drug delivery system with a detachable release liner which is removed and the patch is applied to the skin for the specified period. The adhesive substance must stick to the skin for an extended period without adhering to the release liner and must be drug compatible. The development and marketing of a Valsartan transdermal (Novartis) therapeutic system for once-daily application patch for angina pectoris, is an example of this type of system [21].

8.2.3 Membrane Permeation Controlled Dermal Patches

The drug reservoir is located in a compartment placed between a drug-impermeable backing laminate and a rate-controlling polymeric membrane in this system (Figure 8.1). The drug molecules are allowed to diffuse through the pores of the rate-controlling membrane, allowing them to cross it. The drug solids are dispersed homogeneously in a solid polymeric matrix (e.g. polyisobutylene) suspended in a viscous liquid medium (e.g. silicone fluid) to form a gel-like suspension, or dissolved in a suitable solvent (e.g. alkyl alcohol) to form a gel in solution within reservoir compartments. The rate-controlling membrane can be a microporous or non-porous polymer membrane with specified drug permeability, such as an ethylene-vinyl acetate copolymer. A thin layer of drug compatible adhesive polymer, such as silicone adhesives, can be added to the external surface of the polymeric membrane to provide close contact of the TDDS with the skin surface. The polymer composition, permeability coefficient, and thickness of the rate-controlling membrane and adhesive can all be altered to tailor the release rate of drug. TransdermScop (Scopolamine) for 3-day motion sickness protection and TransdermNitro (Nitroglycerine) for once-daily angina pectoris therapy are two examples of this technology [22].

8.2.4 Matrix Diffusion Controlled Dermal Patches

This approach is depicted in Figure 8.1. In this, drug reservoir is prepared by homogeneously dispersing drug particles in a hydrophilic or lipophilic polymer matrix. The medicated polymer is then molded into a medicated disc with a specified thickness and predetermined surface area. Drug particles can be dispersed in a polymer matrix by homogeneously mixing finely ground drug particles with a liquid polymer or a highly viscous base polymer and then crosslinking the polymer chains, or by homogeneously blending drug solids with a rubbery polymer at elevated temperatures and/or under vacuum. The drug reservoir is held on a polymer disc that is fixed to an occlusive base plate in a compartment made of a drug-impermeable backing. The adhesive polymer is then spread along the circumference to form a strip of adhesive rim around the medicated disc. The nitroglycerine releasing transdermal therapeutic system is the best example of this matrix transdermal medication delivery system. The absence of drug dumping is a benefit of matrix dispersion type transdermal systems [23].

8.3 Evolution of Adhesives in Medical Applications

Since man's earliest medical records, drugs have been applied to the skin to treat minor ailments, for transdermal administration of therapies to manage systemic ailments, and as cosmetics. Dermal delivery systems, on the other hand, did not become widely used until the last third of the twentieth century, when delivery technology was developed to allow precise and repeatable administration through the skin for systemic effects. The first transdermal patch that reached the market was Scopolamine [23, 24]. In 1944, 0.6 mg of Scopolamine was tested with other drugs for the prevention of seasickness in troops and was found to be effective in a larger dose of around 1.2 mg. But the therapeutic effect of Scopolamine lasts only for a short duration, as it has a short elimination half-life of 4.5 hr. It was found that Scopolamine had a considerable flux via human skin and in the late 1970s, Alza Corporation developed a transdermal therapeutic patch of Scopolamine for prevention and therapy of motion-induced nausea [25]. Today, dermal patches incorporating drugs like fentanyl, nicotine, buprenorphine, clonidine, oxybutynin, estradiol, nitroglycerine, and many others are quite popular in the global market. The formulation of dermal drug delivery compositions involves incorporating a medicament into a carrier such as a polymeric and/or pressure-sensitive adhesive to form a system that also serves as a reservoir for the medicament. If the polymer is a pressure-sensitive adhesive, the composition can adhere directly to the skin without the use of any additional adhesive. The pressure-sensitive adhesive must attach effectively to the skin and allow the medicament to migrate from the carrier to the application site and/or through the skin and into the living organism's bloodstream. The targeted distribution rate, the length of delivery, and the medications themselves all influence the design and performance of cutaneous drug delivery devices [26].

Due to the bulkiness of the reservoir system and leakage of the drug from the reservoir, matrix patches were introduced into the market. The early competitors Cygnus, LTS, and Theratech used the concept of matrix design for several drugs like estradiol, testosterone, and nitroglycerine to circumvent intellectual rights and summons linked with Alza technology. The matrix patches attracted the market as they are much thinner, flexible, comfortable to wear, have low cost, and have good adhesion. The matrix patches do not contain the liquid reservoir, the drug is incorporated into the polymer, and an adhesive is used to stick this polymer matrix to the skin [27]. Later, the alternative to matrix patches was well introduced by the drug-in-adhesive patches. In Drug–in-Adhesive (DIA) patches the drug is incorporated in the adhesive, which also functions to disperse the drug and controls its delivery rate. The adhesive used in DIA-based transdermal systems belongs to the category called 'Pressure-Sensitive Adhesives' (PSAs). DIA-based dermal patches are thin, light, and flexible. They are comfortable to wear and have better skin compatibility with marked patient acceptability [24]. The choice of an appropriate PSA for a drug that could improve its solubility, partition coefficient, and diffusion coefficient is a critical factor. DIA is composed of three layers in general: Backing membrane, pressure-sensitive adhesive (PSA) incorporating drug, and the release liner. Fabricating an adhesive-based system requires a profound knowledge of materials science, rheology of adhesive, and the biology of delivering drugs across the skin. The primary consideration is to assess the physicochemical compatibility of the drug with adhesive in transdermal technology [28]. The approach of controlled drug release plays a crucial role while fabricating the system depending on rational understanding of material selection and technique, making it feasible to control drug release. While talking about pressure-sensitive adhesives in fabricating DIAs, selection of adhesives, phenomenon of adhesion, and adhesive properties play important roles. It is anticipated that the transdermal patch must adhered to the surface of the skin up to the proposed time duration, for delivering therapeutic action. Over the past few decades, adhesives in one form or another have attracted formulation experts [29]. Generally speaking, adhesives are the nonmetallic substances utilized to create a bond with the surface of another material involving cohesion and adhesion. Adhesion and cohesion can be defined thermodynamically, although they are difficult to measure precisely. Cohesive bonding is a property of like molecules, i.e., of the same substance, to stick to each other due to mutual attraction [30]. Adhesive bonding is a technique in which adhesive acts as an intermediate layer for connecting the adhesive tape to the skin. In the late nineteenth century, the development of medical tapes and dressings resulted in the use of adhesives in the medical field. Medical adhesive tapes, or surgical tapes, are utilized to affix bandages and different dressings to skin around wounds. Most adhesive tapes are pressure-sensitive adhesives (PSAs) i.e., tapes that stick and stay firmly with substrate. There is no requirement for heat activation or use of solvent. PSAs are designed to form spontaneous adhesion when a small amount of pressure is applied [31].

PSAs are defined by their ability to form an immediate bond with the substrate without the need for activation. PSAs are viscoelastic materials and, therefore, their adhesive property is strictly related to the solid- and liquidlike behaviors that are dependent on frequency of the applied stress at a given temperature [32]. Moreover, it is important that a PSA should be removed easily (debonding) after desired therapeutic action from the applied skin surface by a light pull without leaving any residue on skin surface. PSAs should not change their physical properties since they should be easily removed (debonding) after achieving the desired therapeutic effect with a little pull without leaving any residue on the skin surface. They may form bonds due to their viscoelastic nature, and they could change their flow condition along the way or subsequently [33].

The three kinds of PSAs mostly utilized in fabricating transdermal devices are: polyisobutylene (PIB), polysiloxanes (silicones), and polyacrylate copolymers (acrylics). There is a slight possibility that these synthetic PSAs will cause allergic reactions. The choice of an appropriate PSA depends on numerous factors, together with patch composition, formulation design, expected wear time (up to 24 or 72 hours or for a week), and wear conditions (for example bathing, sweating, and humidity) [34].

8.4 Types of Adhesives Used in Dermal Patches

PSAs had their entry in the medical field in the 1850s when adhesive dressings based on the utilization of natural rubber were introduced. The discovery of synthetic tackifying resins enabled the adhesive's quality to be improved. Synthetic PSAs, such as oxidation-resistant acrylic copolymers, could only be used in the early stages of development if their purity was high enough to ensure skin biocompatibility [35].

Following are the characteristics that make a PSA acceptable for patch development: i) to adhere to the skin; ii) not to itch or irritate the skin in any way; iii) easy application; iv) to allow removal without undue trauma; v) to leave no residue on the skin after removal; vi) to be durable; and vii) be chemically and physically compatible with a variety of medications and excipients.

Because of its lower allergenicity, acrylic and silicone-based PSAs have mostly been replaced by polyisobutylene (PIB). Hydrocolloids or PSAs containing a substantial quantity of water, and matrices that acquire adhesive characteristics are two other materials commonly employed in patch formation [36].

8.4.1 PIB-Based PSAs

PIB is a vinyl polymer produced by cationic polymerization from the monomer isobutylene (IB). PIB is commonly categorized as a synthetic rubber or elastomer. PIB has extremely low air, moisture, and gas permeability, as well as superior thermal and oxidative stability, chemical resistance, and strong tack in adhesive composition [37]. It is a colorless to light-yellow elastomer that is viscous or semisolid, with no odor, taste, or toxicity. PIB has long been utilized as a base ingredient in pressure-sensitive adhesive formulations. PSAs can use high-molecular-weight PIB as an elastomeric base since it is strong and elastic. PIB is an extremely soft and liquid-like tackifier created from low-molecular-weight monomers. Since adhesion has two critical parameters: tack and holding power, different resins and tackifiers can be used to alter the tack of PIB-based PSAs [38]. The poor cohesive strength of PIB can be improved by adding high molecular-weight PIB or fillers which can improve tack, flexibility, and will provide good aging and chemical resistance. The utilization of PIB in PSA-based dermal patches needs desirable properties which can be achieved in three main ways.

1. A combination of low- and high-molecular-weight PIB is used first to obtain a balance of tack and cohesive strength. Pressure-sensitive adhesion requires a specific proportion of low-to high-molecular-weight PIB, which could be achieved simply by dry or wet mixing. This is generally thought to be around 80 wt% or less of low-molecular-weight PIB. PSAs with mild adhesive property are produced using this type of formulation [39].
2. A low-molecular-weight polyisobutene is blended with high- and medium-molecular-weight PIB. When added to the PIB mixture, the formulation range expands. It is possible to employ higher ratios of high-molecular-weight PIB to low-molecular-weight PIB. As a result, the formulation becomes a balance between an extremely sticky material (high PIB content) and material with low shear strength.

3. Tackifiers, plasticizers, fillers, waxes, oils, and other additives can be incorporated into the formulation to impart appropriate adhesive property and viscosity. For TDD applications, Adhesives Research, Inc. (Glen Rock, PA) offers fully formulated PIB adhesives, while other suppliers like Mactac (Moosic, PA) and National Starch and Chemical Company (Bridgewater, New Jersey) can provide custom formulations [40]. Examples of PIB- PSAs available in market are Vistanex PIB LM-MH, Hercolyn, Piccolyte S115, Parapol 950 , Hydrogenated carbon resin Ercorez 5320 [40].

8.4.2 Silicone-Based PSAs

The term "silicone" is not always used correctly, and it should only be applied to polymeric materials to avoid confusion with the metallic element silicon (Si). These are synthetic polymers with SiOSi linkage that are employed in a range of industries for their ability to wet-out surfaces, water repellency, high radiation absorption, and temperature and chemical resistance. Silicones have been used in a wide range of healthcare applications for over 60 years, along with in medical devices and as active pharmaceutical ingredients (APIs) and excipients in medicines, due to their various physical forms and physicochemical properties [41–43].

Silicone adhesives, notably those self-adhering that do not require any activation before use, are one type of silicone compounds that have sparked continuous interest and development. They are employed as excipients in transdermal patches as well as skin contact adhesives in the attachment of prosthetic and wound care devices [44]. Recent research supports the use of silicone-based pressure-sensitive adhesives in dermal patches, not only because they are skin-friendly, but also because they boost the efficacy of pharmacological APIs [45].

Silicone adhesives are widely known for their ease of removal from skin- an ability to remove cleanly from compromised skin without negatively impacting the wound healing process [46].

Silicone pressure-sensitive adhesives, also known as standard pressure-sensitive adhesives, form a network structure as a result of silanol condensation reaction between the polymer and resins. These polymers have sufficient cohesive strength for dermal applications, and the adhesive layer is removed intact from the skin [47]. These adhesives are usually packaged in a volatile solvent that is removed during the coating process. Silicone adhesives are commonly employed in medical devices in addition to cutaneous patches.

The Embrace® MINIMIZE Silicone Scar Aid, which consists of a silicone PSA coated onto silicone elastomer (rubber) sheet, is a recent novel example of silicone PSA's use in medical device attachment. The dressing can be placed with a special applicator to relieve tension on healing skin and reduce scar formation [48].

Dermal patches fabricated from silicone adhesives provide good skin adhesion, and controlled flux of the drug with better skin biocompatibility.

Silicone PSAs can be dispersed in several volatile polar and nonpolar hydrocarbon solvents, and the solvent in which the silicone PSA is dispersed as well as the concentration of the PSA in the solvent can be customized further [49]. It is crucial to note that exposure to amines and amino-functional medications and excipients will cause certain silicone PSAs to lose tack and their ability to rapidly cling to the skin [50]. The list of marketed silicone-based PSAs is presented in Table 8.1.

8.4.3 Acrylic PSAs

For nearly 50 years, acrylic or acrylate copolymers have been utilized commercially for PSA formulations. The following fundamental features must be present in an acrylic copolymer for it to be acceptable for usage as a PSA.

1. The copolymer should have a notable fraction of a monomer with a low glass transition temperature (Tg) as the first characteristic. Tg of 20° C is usual for this monomer, with a preferred range of 40° to 80° C. The PSA's soft, sticky quality is due to this low Tg monomer.
2. The copolymer's second property is that it will strengthen the PSA, preventing it from splitting and leaking during use.
   Adding high Tg monomers to the copolymer, adding monomers that generate intermolecular interactions between individual copolymers, covalent crosslinking of the copolymer, and physical crosslinking of the copolymer through graft or block copolymers all are options for providing reinforcement [41].

The tack property of an acrylic copolymer was first described in the 1930s; however, it was in the 1950s that with the discovery that incorporating a high Tg functional monomer into the acrylic copolymer provided the

Table 8.1 List of marketed silicone-based PSAs.

Product supplier name (Shin-Etsu Chemical Co., Ltd.)	**Cure system/ application**	**Solid wt%**	**Solvent**	**Crosslinker and amount**	**Viscosity at 25°C Pa-s**
KR-100	Peroxide (Adhesive Tape)	60	Toluene/ Xylene	--	100
KR 130		60	Toluene/ Xylene	--	100
KR 101-10		60	Toluene	--	100
KR 3700	Addition (Adhesive Tape)	60	Toluene	--	30
KR 3701		60	Toluene	--	30
X-40-3237		60	Toluene	X-92-122C(0.3)	150
X-40-3240		60	Toluene	X-92-122C (0.5)	20
X-40-3291-1		60	Toluene	X-92-122C (0.5)	120
KR-3704	Addition (Protective Film)	60	Toluene	--	100
X-40-3323		30	Toluene	--	3
X-40-3370-1		60	Toluene	X-92-226 (5.0)	100
X-40-3306		30	Toluene	--	15

Note: The amount of crosslinker mentioned is in % w/w.

reinforcement required for acceptable PSA performance, and thus true beginning of commercially useful acrylate PSAs began.

Acrylic acid (AA) and acrylamide as well as other reinforcing monomers are still frequently utilized in a variety of PSAs, for medical and pharmaceutical applications. Another reason for introducing functional monomers into a copolymer is to alter the copolymer's chemical characteristics.

Many PSA performance qualities, such as the PSA's ability to moisten a surface, and its ability to dissolve complex chemicals and excipients may influence the PSA characteristics [42].

Functional monomers, in particular, can have a substantial impact on how small-molecule additives like medications and pharmaceutical excipients dissolve or are complexed. The use of N-vinylpyrrolidone (NVP) as a reinforcing monomer in an acrylate-based PSA is a famous example of this. These monomers are employed in surgical drapes and many other medical applications that require antibacterial performance because iodine complexes can be coupled with NVP. However, a downside of integrating functional monomers to change chemical properties is that the functional monomers can have a significant impact on the copolymer's physical properties, particularly its ability to offer reinforcement [43].

Because excessive reinforcing would result in the loss of the soft, tacky qualities required in a PSA, this impact limits the quantity of functional monomer that can be added. Although an acrylate-based PSA with NVP can be created, increasing the amount of NVP results in a substance that behaves more like polyvinylpyrrolidone (PVP), a powder with no PSA qualities of its own [44]. List of marketed acrylic-based PSAs is provided in Table 8.2.

8.4.4 Other PSAs

Primary families of PSAs have several flaws due to their chemical structures. Because of their apolar character and the need for mineral oil as a plasticizer, PIB-PSAs are only available as dispersions in aromatic solvents, and their use is restricted to low polar drugs [51]. The manufacture of copolymers customized based on drug solubility and adhesive quality is enabled by the selection of suitable monomers in acrylic-based PSAs, but the novel excipient's safety must be verified in *in vivo* experiments. Silicone-based adhesives have several drawbacks, including low physical stability and high cost [52].

As a result, unique families of PSAs have been designed to address these challenges and overcome intellectual property constraints. Aside from the development of novel polymers, another method is to take advantage of the unique properties of pharmaceutical excipients used for various reasons. Because of their well-known biocompatibility and safety, polyvinylpyrrolidone (PVP), poly(methyl methacrylate), or mixtures thereof are commonly used, even if additives and/or chemical cross-linking agents are required to provide them with appropriate adhesive property [53].

Table 8.2 List of marketed acrylic-based PSAs.

Product supplier name (Henkel Adhesive Technologies)	Cure system/application	Solid % w/w	Solvent(s)	Crosslinker and amount	Viscosity (cP or mPa·s)
DURO-TAK 87-900A	Acrylate copolymer (Pressure-Sensitive Adhesive)	43	Ethyl acetate	--	1800
DURO-TAK 87-4098	Acrylate copolymer (Pressure-Sensitive Adhesive)	36.5	Ethyl acetate	--	6500
DURO-TAK 87-2074	Acrylate copolymer (Pressure-Sensitive Adhesive)	29.5	Ethyl acetate Isopropanol 2,4 Pentadione Toluene	Crosslinked acrylate	1500
GELVA GMS 3083	Acrylate copolymer (Pressure-Sensitive Adhesive)	38	Ethyl acetate	--	8500
GELVA GMS 788	Acrylate copolymer (Pressure-Sensitive Adhesive)	41	Ethyl acetate Toluene	--	5250
GELVA GMS 9073	Acrylate copolymer (Pressure-Sensitive Adhesive)	32	Ethyl Acetate Hexane Ethanol Isopropanol	Crosslinked acrylate	5000

(Continued)

Table 8.2 List of marketed acrylic-based PSAs. (*Continued*)

Product supplier name (Henkel Adhesive Technologies)	Cure system/application	Solid % w/w	Solvent(s)	Crosslinker and amount	Viscosity (cP or mPa·s)
DURO-TAK 387-2051/ 87-2051	Acrylate copolymer (Pressure-Sensitive Adhesive)	51.5	Ethyl acetate Heptane	--	4000
DURO-TAK 387-2054/ 87-2054	Acrylate copolymer (Pressure-Sensitive Adhesive)	47.6	Ethyl acetate Heptane 2,4 Pentadione Isopropanol Toluene	Crosslinked acrylate	2750
DURO-TAK 87-2196	Acrylate copolymer (Pressure-Sensitive Adhesive)	45	Ethyl acetate Heptane 2,4 Pentadione Isopropanol Toluene	Crosslinked acrylate	2100
DURO-TAK 87-4287	Acrylate copolymer (Pressure-Sensitive Adhesive)	39	Ethyl acetate	--	8000
DURO-TAK 387-2516/ 87-2516	Acrylate copolymer (Pressure-Sensitive Adhesive)	41.5	Ethyl acetate Heptane Methanol Ethanol	Crosslinked acrylate	4350

PSAs containing a considerable volume of water have recently gained popularity in transcutaneous immunization and protein administration via iontophoretic patches. These have also been used as a wound dressing and electrically conductive pad [54]. To boost the creep resistance of these hydrophilic PSAs made of swelling polymers, chemical or photo crosslinking is required. Only a few studies have looked into different approaches to avoid a complete cure, which could determine the physical or chemical instability of the loaded drugs [55, 56].

By combining PVP and lactide oligomers, Zhang and coauthors [57] presented a hydrophilic PSA containing about 25% w/w water suited for delivering ibuprofen and salicylic acid. A poly(methyl methacrylate) PSA with a higher water content (up to 40% w/w) has also been proposed for the administration of proteins or hydrophilic medicines.

In another study, Wilpiszewska and Czech [58] developed water-soluble pressure-sensitive adhesives (PSAs) based on acrylic acid and carboxymethyl starch (CMS). Tack and peel adhesion to various paper types (newsprint, sanitary, packaging, fax, and art paper) as well as dynamic shear adhesion at higher temperatures (70–240°C) were evaluated to use PSA as a double-sided splicing tape in the paper industry. Tack and peel adhesion values rose with CMS content (0.1–10 wt percent in PSA) for all paper types tested. Fax paper had the highest tack and peel adhesion values (for the system containing 10% CMS) of ca. 19.5 N/2.5 cm and 18.3 N/2.5 cm, respectively, and art paper had the lowest: 14.7 N/2.5 cm and 13.5 N/2.5 cm, respectively. The time it took for PSA to become entirely soluble in water varied depending on CMS content and pH value, although it did not surpass 5 minutes.

8.5 Testing Physical Properties of PSAs

Adhesives are nonmetallic materials that use adhesion and cohesion to join other materials. PSA's adhesive and end-use qualities necessitate a viscoelastic and non-Newtonian flow pattern due to the adhesive's macromolecular structure. The adhesives must have a distinct set of rheological characteristics. They must quickly wet the adherend surface and build up a desirable holding force. Wetting is determined not only by viscoelastic properties but also by the surface tension of adhesives and surface free energy of adherends. During bonding, a PSA must have sufficient flowability to conform to the adherend surface [59]. The bond formation can take place at a low deformation rate. However, adhesive joints may experience stress states with considerable component of compressive stress.

The completed bond must also be able to withstand the peel and shear stresses that occur during use.

In comparison to bond formation, bond disruption is a more dynamic process. The time it takes for adhesive to yield, for example, is on the order of a tenth of a second. The qualities of adhesives are tested by a shear test generally carried out to measure the ability of an adhesive tape to resist creep under a constant load, applied parallel to the surface of the tape and substrate.. Only PSAs combine these mutually exclusive qualities attributed to their complex viscoelastic behaviour. PSA's precise synthesis or formulation provides excellent physical qualities, allowing for a balanced combination of tack, peel resistance, and cohesiveness [60]. Table 8.3 lists several parameters that influence the adhesive's peel property

Table 8.3 Influence of viscoelastic parameters on adhesive properties.

Type of test	Description	Principle	Affecting factors
Peel Adhesion	Force required to detach a pressure-sensitive adhesive from its backing material or a test plate	It implies high strain rate. It is used to assess how much force is needed to separate two components that are bonded by an adhesive.	Nature of substrate Adhesive thickness Dwell time Peel rate Rigidity
Shear resistance	It refers to an adhesive's capacity to resist shear stress.	It is a type of stress in which the external force acting on a surface acts in parallel to the surface.	Dwell time Type of test panel
Tack	The maximal force required to break the bond	Test signifies adhesive that, with brief contact and light pressure, forms a bond with the surface of another material. The implication is that the adhesive separates from the surface cleanly, leaving no macroscopic residue.	Applied pressure Environmental factors Contact area Dwell time

8.5.1 Tack Test

Tack is a term used with PSA that refers to how quickly a bond between the adherend and the adhesive is formed. The adhesive and adherend are briefly brought together and pulled apart under a light pressure. The higher the tack, the more effort is required to separate it. Low tack, on the other hand, permits a sticky film or tape to be repositioned. The initial attachment of a PSA to an adherend takes only a few seconds and is influenced not only by rheological qualities at low frequencies but also by molecular interactions at the adhesive/adherend interface [59]. When the PSA's surface tension is much lower than the surface free energy of the adherend, the PSA's performance is generally ruled by the adhesive material's viscoelastic properties (rate of wetting), whereas for low-energy surfaces, the substrate's wetting is governed by the PSA critical surface tension, which is typically in the 28-32 mN/m range [60].

The lower the adherend's critical surface tension value is, less wetting and tack will be. The hydrolipidic nature of the skin is connected to its surface free energy, which ranges from about 25 mJ/m^2 for dry skin to 56 mJ/m^2 as the relative humidity and temperature of the skin increase. As a result, gender and age as well as physicopathological circumstances have a significant impact on PSA tack efficacy. A PSA must be removed from the surface gently and without leaving any residue [61]. The PSA is required to have high deformability and should act as a liquid while forming the bond with the adherend. When the bond is broken, it should have a high modulus and behave like a solid. A patch, on the other hand, requires low tack value because it is applied to the skin. Because determining the time required to achieve optimal adhesion is difficult, tack methods measure the force for debonding after a brief contact period and light pressure [62].

For pressure- sensitive adhesives, several standard tack test procedures have been developed. These usually fall under one of the following groups: 1) Ball (Rolling) 2) Loop Tack and 3) Probe Tack.

The rolling ball tack tester works on the principle of ball method where a steel ball is allowed to roll down from a specific starting point. This ball is inclined at a specific angle to the base of the test platform The pressure-sensitive tape is affixed just beneath the end of an inclined grooved path of the platform. The rolling ball attains maximum velocity at touchdown and rolls above the tacky surface which retards its motion by virtue of resistance offered by the adhesive till it stops the ball. The shorter the distance travelled the higher is the tackiness of the adhesive coating. The test is useful for quality control because the findings are affected by both adhesive stickiness and coating thickness [63, 64].

The probe tack test is designed for use with adhesives attached to both flexible and rigid backings. The probe tack test is designed to determine how much force is required to detach the probe from the adhesive layer after a light pressure is applied. As a result, tack is defined as the maximum force required to break the bond after a short period of contact [65].

The loop tack test defines tack as the force required to separate the loop produced by clamping the ends of a patch strip that has been placed in contact with a specific area of a specified surface at a specific speed [66]. A schematic diagram of tack tests is depicted in Figure 8.2.

8.5.2 Shear Adhesion or Creep Test

Shear adhesion or creep is another key feature of a PSA. It describes a PSA's resistance to tangential stresses and, as a result, the cohesion of PSA matrix. This feature is useful in the application area.

The force required to pull a standard area of a patch from a standard adherend plate in a parallel direction to the surface to which it has been affixed is measured as shear adhesion. By measuring the time it takes to remove a standard area of the patch from the adherend plate under a standard load, the shear test determines the force required to glide the patch on the adherend [76].

The adhesive should fail cohesively in the holding power tests, leaving an adhesive film on either the adherend plate or the backing layer. The holding power test is a test of how long the adhesive holds a constant weight at a constant temperature . If this pattern of failure occurs, it can be used to determine the adhesive's internal strength [77]. The greater the adhesive thickness, the greater the number of shearing layers, and the lower the

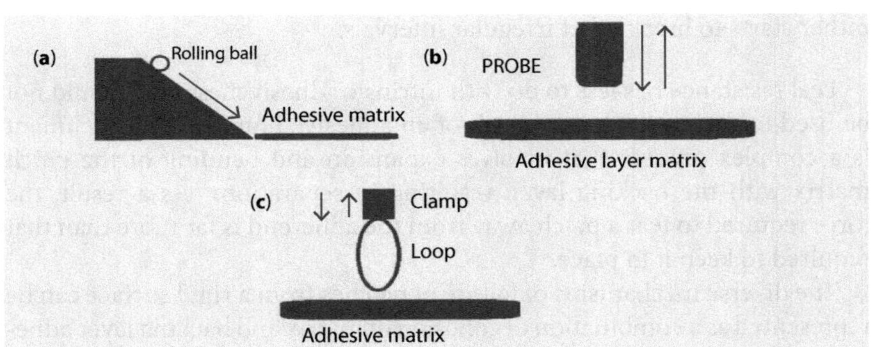

Figure 8.2 Schematic Diagrams of (a) Rolling ball tack, (b) Probe tack, (c) Loop tack tests.

cohesiveness of adhesive matrix. As a result of the exponential relationship, holding power is equivalent to patch matrix thickness [78].

8.5.3 Peel Resistance Test

The force required to break the bond between an adhesive tape and the surface to which it has been applied is measured in the peel test. This determines the bond strength of adhesive to the applied substrate. Peel test involves applying sticky tape to a surface, allowing it to sit for some time, and then pulling it away [67]. The parameters such as peel angle or direction, applied pressure, and the time for which the surfaces stay bonded are all standardized to ensure consistency of results. Peel-off patches are an important feature of transdermal medication administration. The higher the peel adhesion, the more difficult is to remove these. It is crucial in establishing the force required to peel back an adhesive tape or patch from a solid surface at a predetermined rate and peel angle [68]. Peel resistance is determined by adhesive adhesion to the surface, adhesive viscoelastic behavior, support material, peel rate, and temperature.

Peel force-peel rate curve can be divided into three portions.

1) When the peel rate is very low, the failure mode is cohesive within the adhesive.

2) If the peel rate is raised, the peel failure mechanism shifts from adhesion failure to cohesive failure within the adhesive. This transition begins with a decrease in peel force, but it then increases as the peel rate increases.

3) The third segment of the curve shows that as the peel rate increases, the peel force will decrease. Adhesives have a glassy response in this portion of the curve. In this location, the slip-stick effect can be seen, which occurs when the adhesive is not uniformly peeled from the surface but rather starts to break off at irregular intervals.

Peel resistance has less to do with intrinsic adhesiveness and should not be used to determine the strength of an adhesive bond [69]. Detachment is a complex process that involves expansion and bending of the patch matrix with the backing layer, resulting in separation. As a result, the force required to tear a patch away from the adherend is far more than that required to keep it in place.

The diverse mechanisms of failure of patches from a rigid surface can be represented as a combination of cohesion of matrix and backing layer adhesion. When a patch is peeled away from the substrate, it is often assumed to peel completely from the adherend, leaving no visible residues [70].

The method for determining adhesion is chosen based on the type of stress encountered in a specific test. In summary, the standardized test methods require the application of a patch strip to a rigid classic test plate (adherend plate), typically made of stainless steel, using pressure to ensure contact; after that, the strip is separated from the plate at a predetermined angle (i.e., 180° or 90°) and speed (i.e., 300 mm/min) for a predetermined time [71].

The stress is imparted to the backing layer through the adhesive matrix during the detachment process. As a result, the formulation composition, which can influence flexibility and elongation, or the thickness of the backing layer can cause differences in behaviour during the detachment process [72]. Obviously, as the thickness and/or rigidity of the backing layer increase, so does the energy required to bend the backing layer. Simultaneously, the peel test strip creates an extraordinary long moment arm (a moment arm is the perpendicular distance from an axis to the line of action of a force), resulting in a lower peel force and counteracting the energy released in the backing layer deformation.

The influence of the backing layer characteristics is less pronounced in a 90° peel adhesion test as compared to a 180° peel test, since a lower energy is required to deform the backing layer. Moreover, tests carried out at two peel angles provide different information, since the 180° peel is a combination of tensile and shear processes, while the 90° peel depends only on tensile event; therefore, the standard deviations in the data obtained by measuring the peel force at 90° is lower than at 180° [73].

The adhesion results at two peel angles are not completely equivalent. As an example, in the case of patches prepared using a neoprene rubber PSA, the 90° test allowed to discriminate the influence of the PSA thickness and the peel strength was about an order of magnitude higher than the experiments performed at 180°. The thicker the adhesive matrix, the higher the peel strength [23–25] levelling off at some thickness [74]. The higher the adhesive matrix thickness, the higher the peel strength, which levels off at a certain thickness. This feature can be explained considering the double effect of thickness on the peel force.

Thicker PSAs, by deforming the adhesive matrix, increase the volume of the matrix and, as a result, the peel force increases. However, as the angle and moment arm increase in patches with increased thickness, the peel force drops, resulting in less detachment effort. At certain critical value of the adhesive matrix thickness, the PSA deformation in the crack is large enough to form a further increase in thickness no longer noticeable [75].

The influences of various factors affecting shear adhesion, peel adhesion and tack tests are shown in Tables 8.3–8.6.

Table 8.4 Factors affecting shear adhesion.

Factor	Comments
Tg of adhesive	As the Tg rises to a temperature below room temperature, shear adhesion increases. The Tan δ value at the test temperature—room temperature—increases as Tg increases, as per rheological analysis. The better the flow or wetting of the tested adhesive, the higher the Tan δ value. When Tg is too high, however, G' (storage modulus) rises quickly in the glass transition zone. As a result, an adhesive may lose a certain amount of contact area with the bonding substrate under light pressure, resulting in decreased shear adhesion.
Tan δ minimum value of the adhesive	Empirically, it is found that shear adhesion improves as the Tan δ minimum value (in rubbery plateau zone) lowers to a certain value. This is because the disentanglement of molecular chains is easier to proceed for those adhesives exhibiting higher Tan δ values. However, the shear force is decreased again when the Tan δ minimum value is lower than a certain value. The adhesive becomes very elastic and loses the capability to dissipate shear stress within molecular chains by gradual disentanglement. As a result, the adhesive tends to delaminate from the bonded substrate more easily accompanied by an adhesion failure (no adhesive residues on the bonded surface) instead of presenting a cohesive failure mode. There may be a critical Tan δ minimum value that indicates the fracture mode change from cohesive failure to adhesion failure.

(Continued)

Table 8.4 Factors affecting shear adhesion. (*Continued*)

Factor	Comments
Rubbery plateau modulus	According to most adhesion research findings, M_e (Molecular weight between two entanglements) value is the most important parameter affecting shear adhesion. This is because the energy dissipation between entanglements will be higher for those having greater M_e and therefore will impart higher shear adhesion. This hypothesis may apply to those adhesives having very similar Tg and Tan δ minimum values. According to rheological measurements, when an adhesive possesses a higher M_e, it will exhibit higher storage modulus. Accordingly, for the same SBC (Styrenic Block Copolymer)-based formulations, if we simply select different SBCs having different styrene contents and the same percentage of di-block, we may easily obtain different levels of rubbery plateau modulus (G_n^0) without affecting Tg and Tan δ minimum. Through series of formulations, we may be able to discover the highest shear adhesion which may present a certain value of G_n^0.
Adhesive Polarity	Adhesive polarity is another key consideration that can influence all adhesion properties, including tack, peel, and shear. All adhesion properties should theoretically improve when more polar components are added to the formulation. Because of the polarity of the adhesive, physical adsorption between the adhesive and the bonded substrate can be improved. The presence of a polar component will undoubtedly alter the adhesive's rheological properties. To determine the extent of contribution from adhesive polarity to rheological qualities, a thorough investigation is required.
Curing of adhesive	Many cured solvent- and water-based PSAs exhibit very high shear adhesion, irrespective of Tg, Tan δ minimum value, and G_n^0. Obviously, the molecular weight of cured adhesives also affects shear adhesion significantly. However, this factor cannot be applied to conventional SBC-based HMPSAs which are not curable. A separate study for curable adhesives is needed [79–81].

Table 8.5 Factors affecting peel adhesion.

Factor	Comments
Peel angle	The peel angle used in the peel test has a significant impact on the measured peel strength value, according to a vast number of experimental results. When the peel angle is 0°, a conventional tensile shear test is performed, and the result is the tensile shear strength. In general, the tensile shear strength is substantially higher than the peel strength, and the peel strength value decreases as the peel angle increases. For standard testing, a 180° peel angle is used.
Peel rate	During testing, different forms of damage can occur at varying strip rates. The rubber layer progresses from cohesive failure to interface failure, and then a somewhat stable phase switches to the "stick-slip" stage as the peel rate increases. As a result, when using the 180° peel strength test, there are severe limitations on the peel rate, which is nominal 300mm/min.
Test temperature	All hot- melt adhesives, in theory, are viscoelastic and follow the temperature equivalence principle. When the 180° peel strength test is carried out at different temperatures, the measured peel strength changes significantly. The viscoelastic behavior of the adhesive is principally responsible for this shift in the measured peel strength. As a result, there are strict temperature limitations when developing 180° peel strength test criteria. The peel test is normally performed at room temperature, usually 23° ± 2° C.
Other test conditions	Other conditions in the peel strength test, in addition to the peel angle, peel rate, and test temperature, include the surface of the material to be bonded before preparation of the sample, the applied pressure to the sample, the manner and time of applied pressure. The peel strength test findings are affected by the time, the temperature, and environmental humidity during sample preparation, placement, and testing [82–84].

Table 8.6 Factors affecting tack behaviour.

Factor	Comments
Adhesive Application	It is vital to determine the weight and thickness of the adhesive coating. The techniques for applying adhesives in a repeatable manner while inspecting the adhesive layer visually are important; Tack is affected by the thickness and homogeneity of the adhesive layer; the drying time between application and bonding can have a considerable impact on tack.
Viscosity	Cohesion and elasticity are essential for tack property. The elements that affect viscosity and elasticity will have an impact on the tack characteristics of PSA in the end. The viscous and elastic behaviours of adhesives are influenced by the intrinsic properties of polymers as well as experimental and environmental conditions. Mingheeti *et al.* concluded that the tack characteristics of polymers could potentially be influenced by their molecular weight [85].
Temperature and Humidity	Tack behavior as functions of temperature and strain rate is more complex than the cohesion. Most adhesives are prepared to have tack at room temperature. The tack property of adhesives is evaluated at low temperatures using a temperature chamber; tack is sensitive to the adhesive's moisture content, and adsorbed moisture on the surfaces can interfere with sticking. As a result, a humidity environmental chamber is used to control humidity.
Experimental Parameters	Regardless of the material's inherent properties, the viscosity and elastic modulus of a PSA are influenced by the experimental conditions such as temperature, time, humidity, strain rate etc. The bonding surfaces are tack-sensitive [86–88].

8.6 Prediction of Patch *In Vivo* Adhesive Performances

It is required to assess a patch's adhesive performance by estimating the proportion of the dosage form that remains adhered to the skin throughout its use. Patch survival rate studies [89] are clinical investigations that examine the *in vivo* adhesive function of a patch.

In general, it is a scoring system based on observations of the patch's endurance, nature during detachment, and other subjective concerns of users/patients.

The FDA recommends an adhesion scoring system in which patients/volunteers receive one of the following scores. The following is an estimation of the transdermal system's adhesion:

Score 0 = 90% adhered (essentially no lift-off from the skin)

Score 1 = 75% to < 90% adhered (only some edges lift off from the skin)

Score 2 = 50% to < 75% adhered (less than 50% of the system lifting off from the skin)

Score 3 = < 50% adhered but not detached (more than 50% of the system lifting off from the skin without falling off)

Score 4 = patch detached (completely from the skin) [90].

The *in vivo* quantitative measurement of patch adhesion to the skin is commonly used to determine the peel force [86, 87] or the tackiness using modified probe tack tests [91]. The stress-strain curves obtained during patch removal are calibrated using a dynamometer that is attached to the patch sample, which is affixed to the forearm or dorsal side of the volunteers' hands. Raynaud *et al.* [92] made several observations about the testosterone transdermal patch's capability to adhere *in vivo*. The adhesion was qualitatively and quantitatively determined by a peel test performed at the standard peel rate of 300 mm/min and an angle of 135°. According to the following ranking order, the quantitative study reveals a strong relationship between peel strength and application site:

Thigh > Lower back > Arm [89].

The probable safety risks regarding drug absorption are of concern during *in vivo* evaluations. Therefore, test should be executed using optimized formulation and more thoroughly during clinical studies. To reduce the extent of *in vivo* studies, several efforts were made to propose *in vitro* assays drawing relationship between the *in vivo* performances and the *in vitro* quantitative determinations.

The necessity for materials with low energy surfaces rather than stainless steel, which can allow for data to be obtained closer to human skin, is a prevalent consideration. The effects of relative humidity on peel adhesion performance can also be investigated using collagen-coated plates [90].

8.7 Adhesive Properties and Formulation Studies

The goal of the formulation studies is to create stable patches that are able to assure maximum flux of drug through human skin while maintaining proper skin adhesion. Meanwhile, higher fluxes are required to deliver medications via patch with specific dimensions: the higher the flux, the lower the patch surface, the higher the patient compliance [93].

When making PSA-derived transdermal patches, it is important to consider the formulation variables that can affect the patch's adhesive characteristics. The following are some of them:

i) Composition and quantity of additives,
ii) The matrix thickness, which is correlated to the active ingredient dose [17] and, eventually, the drug penetration enhancement [57].
iii) A flexible backing layer that is compatible with the medicine or excipients.

The incorporation of drugs or excipients, such as drug crystallization inhibitors, penetration enhancers, and/or antioxidants might alter PSA's rheological properties and induce adhesive property variations [94, 95].

Generally speaking, when a substance is added to a PSA, an unpredictable alteration of the adhesive properties can occur due to its plasticizer/antiplasticizer effect [96]. Lin and colleagues [97] created a self-adhesive medication-loaded transdermal patch using piroxicam as a model drug. According to their findings, binding medication into a poly(amino methylmethacrylate) (PAMA)-based PSA decreases peel adhesion strength because the drug may induce an increase in the polymer's glass transition temperature. The amount of plasticizer supplied to the PAMA matrix was increased to counteract this impact. Another work by Gullick *et al.* found that after loading coumarin, the peel adhesion of patch matrices was not significantly altered [98]. Minghetti *et al.* found that adding miconazole nitrate to a methacrylic-based PSA reduced both peel and shear adhesion significantly [99]. When benzotropine was introduced in acrylate-based PSA, a progressive increase in the peel force was concomitant to the reduction in both tack and shear strengths [100]. These effects can be explained by variations in polymer chain entanglement, where the drug acts as a plasticizer [101].

8.8 Summary

The development of a transdermal medicine delivery system has made remarkable progress. As the partition coefficient between the patch matrix and the skin regulates medication absorption, changes in the contact area caused by displacement and segregation can result in variable bioavailability and therapeutic failure. The researchers are attempting to include novel pharmaceuticals into adhesives, which could lead to the development of promising technologies such as DIA matrix and reservoir TDDS. The critical role of adhesives in predicting safety and efficacy has been noted in numerous reports received by the US Food and Drug Administration's Drug Quality Reporting System. Aside from the wide range of adhesives used in the fabrication of transdermal systems, there are still a few issues to address, such as skin irritation, and drug crystallization, which could impair drug absorption through the rate-controlling membrane, and so on. However, lack of adhesion can occur owing to variations in ambient conditions such as humidity, heat, cold, or after showering or swimming. It might also be due to a quality issue, such as if the release liner was not completely removed from dermal patch.

To solve these concerns, methodologies for measuring adhesive performance must be developed. The use of adhesives in the medical and pharmaceutical fields has increased in recent years. The current scenario focuses on determining the impact of adding drugs or chemicals to a PSA to predict its *in vitro* and *in vivo* adhesive characteristics. Dynamic mechanical analysis studies provide a detailed understanding of the structure of a PSA and the viscoelastic properties of adhesives used in the patch. Regulatory bodies recommend that *in vitro* adhesive characteristics be assessed using the holding power and peel adhesion tests. Aside from adhesion property, the adhesives' effect on the skin should be investigated. An adhesive should be biocompatible with the skin structure, which could help TDDS become more commercially viable.

Acknowledgements

The authors are grateful to Prof Kanaklata, Director of MCHSR for providing moral support as well as necessary facilities for the completion of this chapter, and the Department of Pharmaceutical Technology, Metro College of Health Sciences & Research.

References

1. A Bolzinger, S Briancon, J Pelletier and Y Chevalier, Penetration of drugs through the skin, a complex rate-controlling membrane. *Curr. Opin. Colloid Interface Sci.*17, 156-165 (2012).
2. H Cui, P Quan, Z Zhou and L Fang, Development of a drug-in-adhesive patch combining ion pair and chemical enhancer strategy for transdermal delivery of zaltoprofen: pharmacokinetic, pharmacodynamic and *in vitro/in vivo* correlation evaluation. *Drug Deliv.* 23, 3461-3470 (2016).
3. N Lal and N Verma, The effect of combination of functional and nonfunctional acrylic polymers on transdermal patches of Carvedilol: *in vitro* permeation, *in vivo* evaluation using biochemical parameters, and stability studies. Future J Pharm Sci 7, 127 (2021).
4. T Igarashi, K Nishino and S Nayar, The appearance of human skin: a survey. Foundations and Trends in Computer Graphics and Vision, 3, 1-95 (2007).
5. S.H Jo, E.H Ka , H.S Lee, E Apostolidis, H.D Jang and Y Kwon, Comparison of antioxidant potential and rat intestinal α-glucosidases activities of quercetin, rutin and isoquercetin. Int. J. Appl. Res. Natural Products 2, 52-60 (2009).
6. M. Kakumoto, T Sakaeda,K. Takara,T Nakamura,T Kita and T Yagami. Effects of carvedilol on MDR1-mediated multidrug resistance: comparison with verapamil. *Cancer Sci.* 94, 81-86 (2003).
7. Y.N Kalia and R.H Guy, Modeling transdermal drug release. Adv Drug Deliv Rev. 48, 159–172 (2001).
8. N Lal, P Yadav, V Rastogi, L Pandey, N Verma and A Verma, Development and evaluation of transdermal therapeutic system of metoprolol succinate using acrylic polymer. *Asian J. Pharm.* 10, 178-187 (2016)
9. A Melero, M.T Garrigues, P Almudever and M.A Villodre, Nortriptyline hydrochloride skin absorption: development of a transdermal patch. *Eur. J. Pharm. Biopharm.* 69, 588–596 (2008).
10. T Kokubo, K Sugibayashi and Y Morimoto, Interaction between drugs and pressure-sensitive adhesives in transdermal therapeutic systems. *Pharm. Res.*11, 104-107 (1994).
11. E.J Lai, J Cheong and A McGrath, Structure and function of skin, hair and nails. *Medicine.* 37, 223-226 (2009).

12. A Mehdizadeh, M.H Ghahremani, M.R Rouini and T Toliyat. Effects of pressure sensitive adhesives and chemical permeation enhancers on permeability of fentanyl through excised rat skin. *Acta Pharmacia.* 56, 219-229 (2006).
13. E Limpongsa and K Umprayn, Preparation and evaluation of diltiazem hydrochloride diffusion-controlled transdermal delivery system. *AAPS PharmSciTech.* 9, 464-470 (2008).
14. C Liu, P Quan and L Fang, Effect of drug physicochemical properties on drug release and their relationship with drug skin permeation behaviors in hydroxyl pressure sensitive adhesive. *Eur J Pharm Sci.* 10, 437-446 (2016).
15. K Ahmed Saeed Al-Japairai, S Mahmood, S Hamed Almurisi, J Reddy Venugopal, A Rebhi Hilles, M Azmana and S Raman, Current trends in polymer microneedle for transdermal drug delivery. *Int J Pharm.* 587, 119673 (2020).
16. N Higo, Recent trend of transdermal drug delivery system development. Yakugaku Zasshi. 127, 655-662 (2007).
17. J. Schultz and M. Nardin, Theories and mechanisms of adhesion, *in: Handbook of Adhesive Technology,* Vol.2, A.Pizzi and K.L. Mittal(Ed.) pp.19-33, Marcel Dekker, New York (1994).
18. K..L Mittal(Ed.) *Acid – Base Interactions: Relevance to Adhesion Science and Technology,* Vol.2, CRC Press, Boca Raton, FL (2000).
19. J Park, H Lee, G.S Lim, N Kim, D Kim and Y.C Kim, Enhanced transdermal drug delivery by sonophoresis and simultaneous application of sonophoresis and iontophoresis. *AAPS PharmSciTech.* 20, 1224-1243 (2019).
20. N. Lal and N. Verma. Development and evaluation of transdermal patches containing Carvedilol and effect of penetration enhancer on drug release. *Res. J. Pharm. Technol.* 11, 1-8 (2008).
21. M Lopez-Cervantes, J Escobar-Chavez, N Casas-Alancaster, V Turlier and C Lauze, Development and characterization of a transdermal patch and an emulgel containing kanamycin intended to be used in the treatment of mycetoma caused by *Actinomadura madurai. Drug Dev. Ind. Pharm.* 35, 1511-1521 (2009).
22. M.B Magnusson, A.K Walters and M Roberts, Veterinary drug delivery: potential for skin penetration enhancement. *Adv. Drug Deliv. Rev.* 50, 205-227 (2001).
23. M Taghizadeh, A Soroushnia and F Mohamadnia, Preparation and *in vitro* evaluation of a new fentanyl patch based on functional and non-functional pressure sensitive adhesives. AAPS Pharm Sci Tech. 11, 278-284 (2010).
24. S Rhee, Y Kwon, W Park, Y Choi, J Byun, C Chi and S Park, Characterization of monolithic matrix patch system containing tulobuterol. *Archives Pharm Res.* 31, 1029- 1034 (2008).
25. A Moraes Teixeira, A Felix, C Fernande, A.S Moura, C.A Mandarim deLacerda and J.J De Carvalho, Exercise training enhances elastin, fibrillin and nitric oxide in the aorta wall of spontaneously hypertensive rats. *Expl. Mol. Pathol.* 89, 351-357 (2010).

26. M Panchaxari,S Pampana, T Pal, B Devabhaktuni and K Aravapalli, Design and characterization of diclofenac diethylamine transdermal patch using silicone and acrylic adhesives combination. *Daru.* 21, 2231-2236 (2013).
27. T Murakami, M Yoshioka and R Yumoto, Topical delivery of keloid therapeutic drug, tranilast, by combined use of oleic acid and propylene glycol as a penetration enhancer: evaluation by skin microdialysis in rats. *J Pharm Pharmacol.* 50, 49-54 (1998).
28. K Patel, H Patel and V Patel, Formulation and characterization of drug in adhesive transdermal patches of diclofenac acid. *Int. J. Pharmacy Pharm. Sci.* 4, 296-299 (2012).
29. S Neuhoff, P Langguth, C Dressler, T.B Andersson, C.G Regardh and H Spahn-Langguth, Affinities at the verapamil binding site of MDR1-encoded P-glycoprotein: drugs and analogs, stereoisomers and metabolites. *Int J Clinical Pharmacol Therapy,* 38, 168-79 (2000).
30. K Sato, N Mitsui, T Hasegawi, K Sugibayashi and Y Morimoto, Potential usefulness of solubility index for prediction of the skin permeation rate of 5-ISMN from pressure-sensitive adhesive tape. *J Controlled. Release.* 73, 269-277 (2001).
31. H Savoji, A Mehdizadeh and S Abadi, Transdermal nitroglycerin delivery using acrylic matrices: Design, formulation, and *in vitro* characterization. *ISRN Pharmacology.* Article ID 493245, 1-9 (2014).
32. B Ongpipattanakul, R.R Burnette, O.R Potts and L.M Francoeur, Evidence that oleic acid exists in a separate phase within stratum corneum lipids. *Pharm Res.* 8, 350-354 (1991).
33. I Nesseem, S Eid, and S El-Houseny, Development of novel transdermal self-adhesive films for tenoxicam, an anti-inflammatory drug. *Life Sci.* 89, 430-438 (2011).
34. K Paudel, M Milewski, C Swadley, N Brogden, P Ghosh and A Stinchcomb, Challenges and opportunities in dermal/transdermal delivery. *Therapeutic Delivery* 1, 109-131 (2010).
35. P.J Raynaud, M Auges, L Liorzou, V Turlier and C Lauze, Adhesiveness of a new testosterone-in-adhesive matrix patch after extreme conditions. *Int. J. Pharm.* 375, 288-334 (2009).
36. S Penckofer, D Schwertz and K Florezak, Oxidative stress and cardiovascular disease in type 2 diabetes: the role of antioxidants and prooxidants. *J Cardiovasc Nursing 16* (2), 68-85 (2002).
37. R Potts and R Guy, Predicting skin permeability. *Pharm. Res.* 9, 663-669 (1992).
38. C Ren, L Fang and L Ling, Design and *in vivo* evaluation of an indapamide transdermal patch. *Int. J. Pharm.* 370, 129-135 (2009).
39. R.M Prausnitz, S Mitragotri and R Langer, Current status and future potential of transdermal drug delivery. *Nature. Rev. Drug Discovery.* 3, 115–124 (2004).

40. S.D Roy, G.L Flynn, M Gutierrez and G.W Cleary, Characterizations of pressure-sensitive adhesives for matrix patch design. *J Pharm Sci.* 85, 491-495 (1996).
41. S Santoyo , A Arellano, P Ygartua and C Martın, Penetration enhancer effects on the *in vitro* percutaneous absorption of piroxicam through rat skin. *Int J Pharm.* 117, 219-224 (1995).
42. H Kalluri and K.A Banga, Formation and closure of microchannels in skin following microporation. *Pharm. Res.* 28, 82-94 (2011).
43. A Naik, N.Y Kalia and H.R Guy, Transdermal drug delivery: overcoming the skin's barrier function. *Pharm. Sci. Technol. Today.* 3, 318-326 (2000).
44. P Kotiyan and P Vavia, Eudragits: role as crystallization inhibitors in drug-in-adhesive transdermal systems of estradiol. *Eur. J. Pharm.* 52, 173-180 (2011).
45. M Naziroglu and P Butterworth, Protective effects of moderated exercise with dietary vitamin C and E on blood antioxidative defense mechanism in rats with streptozotocin-induced diabetes. *Canadian J. Appl. Physiol.* 30, 172-185 (2005).
46. R Parhia and S Padilamb, *In vitro* permeation and stability studies on developed drug-in-adhesive transdermal patch of simvastatin. *Bull. Fac. Pharm. Cairo Univ.*56, 26-33 (2018).
47. Minghetti, F Cilurzo and L Montanari, Evaluation of adhesive properties of patches based on acrylic matrices. *Drug Dev. Ind. Pharm.* 25, 1–6 (1999).
48 S Mitragotri, Y Anissimov, A Bunge, H Frasch, R Guy and J Hadgraft, Mathematical models of skin permeability: an overview. *Int. J. Pharm.* 418, 115-129 (2011).
49. M.A Wokovich, S Prodduturi, W.H Doub, A Hussain and F.H Buhse, Transdermal drug delivery system (TDDS) adhesion as a critical safety, efficacy and quality attribute. *Eur. J. Pharm. Biopharm.* 64, 1–8 (2006).
50. H Wolff, M Irsan and K Dodou, Investigations on the viscoelastic performance of pressure sensitive adhesives in drug-in-adhesive type transdermal films. *Pharm Res.*, 31, 2186-2202 (2014).
51. H Xi, Y Yang , D Zhao, L Fang, N Zhao, Y Zhao and Z Zheng, Transdermal patches for site-specific delivery of anastrozole: *in vitro* and local tissue disposition evaluation. *Int. J. Pharm.* 391, 73–78 (2010).
52. D Yang, X Wan, P Quan , C Liu and L Fang, The role of carboxyl group of pressure sensitive adhesive in controlled release of propranolol in transdermal patch: Quantitative determination of ionic interaction and molecular mechanism characterization. *Eur. J. Pharm. Sci.* 115, 330-338 (2018).
53. N Lal, P Yadav, V Rastogi V, A Verma and N Verma, Aspects of pressure sensitive adhesives in fabricating drug-in-adhesive transdermal therapeutic systems. Drug Delivery Letters, 7, 1-13 (2017).
54. G. Wypych, PIB polyisobutylene, in: *Handbook of Polymers,* Second Edition, pp 285-300, ChemTec Publishing, Ontario, Canada (2016).

55. A Yasar, F Erdemir and B Parlaktas. The effect of carvedilol on serum and tissue oxidative stress parameters in partial ureteral obstruction induced rat model. *Kaohsiung J Med Sci.* 29, 19-25 (2013).
56. X Zhan, J Mao, S Chen and L Wang, Formulation and evaluation of transdermal drug- delivery system of isosorbide dinitrate. *Brazilian J. Pharm. Sci.* 51,120-125 (2015).
57. J Zhang, L Deng and M Zhao, Pressure sensitive properties of poly (N-vinyl pyrrolidone)/D.L lactic acid oligomers/glycerol/water blend for TDDS. *J Biomater Sci Polym* Ed.2, 1-15 (2010).
58. K Wilpiszewska and Z Czech, Water-soluble pressure-sensitive adhesives containing carboxymethyl starch with improved adhesion to paper. *J Polym Environ* 26, 1453–1458 (2018).
59. S Kandavilli, V Nair and R Panchagnula. Polymers in transdermal drug-delivery systems.*Pharm. Technology. 26*, 62-81 (2002).
60. Y Morimoto, T Kokubo and K Sugibayashi, Diffusion of drugs in acrylic-type pressure- sensitive adhesive matrix. Influence of interaction. *J. Controlled. Release.* 18, 113-122 (1992).
61. K Moser, K Kriwet, A Naik, Y Kalia and R Guy, Passive skin penetration enhancement and its quantification *in vivo*. *Eur. J. Pharm. Biopharm.* 52, 103-112 (2001).
62. S Mudshinge, A Deore, S Patil and C Bhalgat, Nanoparticles: emerging carriers for drug delivery. *Saudi Pharm. J.* 19, 129-141 (2011).
63. I Som, K Bhatia and M Yasir, Status of surfactants as penetration enhancers in transdermal drug delivery. *J Pharm Bioallied Sci.* 4, 2-9 (2012).
64. C Valenta, B.G Auner and I Loibl, Skin permeation and stability studies of 5 aminolevulinic acid in a new gel and patch preparation. *J Controlled Release.* 107, 495-501 (2005).
65. F.J Stefano and A.F Scasso,Transdermal delivery system with two superimposed adhesive layers having different affinities to the active substance comprised. United States Patent 20050208116 A1 (2005).
66. N Vora and L Madan, Development and *in-vitro* evaluation of an optimized carvedilol transdermal therapeutic system using experimental design approach. *Asian J. Pharm.* 8, 28-38 (2013).
67. Y Krishnaiah, Y Yang, R Hunt and M Khan, Cold flow of estradiol transdermal systems: Influence of drug loss on the *in vitro* flux and drug transfer across human epidermis. *Int. J. Pharm.* 77, 73-80 (2014).
68. C Valenta and G.B Auner, The use of polymers for dermal and transdermal delivery. *Eur. J. Pharm. Biopharm.* 54, 279-289 (2004).
69. J Wang, Z Li, F Sun, S Tang, S Zhang, P Lv, J Li and X Caoa, Evaluation of dermal irritation and skin sensitization due to vitacoxib. *Toxicol Reports.*, 4, 287–290 (2017).
70. W Weng, P Quan, C Liu, H Zhao and L Fang, Design of a drug-in-adhesive transdermal patch for risperidone: effect of drug-additive interactions on the

crystallization inhibition and *in vitro/in vivo* correlation study. *J Pharm Sci.* 105, 3153-3161 (2016).

71. Niharika and N Verma, Adhesive polymers in fabrication of transdermal drug delivery. Res J Pharm Technol 9, 945-956 (2016).
72. S Wiedersberg and R Guy, Transdermal drug delivery: 30+ years of war and still fighting!. *J. Controlled. Release.* 190, 150-156 (2014).
73. Quality Working Party. Guideline on quality of transdermal patches European Medicines Agency, Committee for Medicinal Products for Human Use, EMA/CHMP/QWP/608924. Science Medicines Health. 12–19 (2014).
74. C A Dahlquist. Tack, in: *Adhesion Fundamentals and Practice: A Report of an International Conference*. pp. 143–151, Maclaren & Sons Ltd. London (1966).
75. F Cilurzo, C.G Gennari and P. Minghetti, Adhesive properties: a critical issue in transdermal patch development. *Expert Opin Drug Deliv.* 9, 33–45 (2012).
76. K Ulman and R Sweet, The correlation of tape properties and rheology. Dow Corning Corporation white paper. Form no. 51-979-01 (1998).
77. S Venkatraman and R Gale, Skin adhesives and skin adhesion: Transdermal drug delivery systems. *Biomaterials.* 19, 1119–1136 (1998).
78. C Fauth, S Wiedersberg, R.H.H Neubert and M Dittgen, Adhesive backing foil interactions affecting the elasticity, adhesion strength of laminates, and how to interpret these properties of branded transdermal patches. *Drug Dev Ind Pharm.* 28, 1251–1259 (2002).
79. H.S Tan and W.R Pfister, Pressure-sensitive adhesives for trandermal drug delivery systems. *Pharm Sci Technol Today*. 2, 260–269 (1999).
80. D.E Nedberge, R.M Gale, N.M Crisologo, J Wen, D.C Imbert and A Luciano, Transdermal rosperidone delivery system. US Patent:20070134310 (2007).
81. R.R Jansen and D.J Coronado, Multilayer drug delivery system with barrier against reservoir material flow. US Patent:20060257462 (2006).
82. K .Y Ho and K Dodou, Rheological studies on pressure sensitive silicone adhesives and drug-in-adhesive layers as a means to characterise adhesive performance. *Int. J Pharm.* 333, 24–33 (2007).
83. F Sun, Y Hu and H.G Du, Synthesis and characterization of MQ silicone resins. *J Appl Polym Sci.* 125, 3532-3536 (2012).
84. A Colas and J Curtis, Silicone biomaterials: History and chemistry, in: *Medical Applications of Silicones*, BD Ratner , AS Hoffman, FJ Schoen, and JE Lemons (Eds.), pp 80-86, Elsevier (2004).
85. P. Minghetti, F. Cilurzo and C. Gennari, Adhesive properties: A critical issue in transdermal patch development. *Expert Opin Drug Deliv*. 9, 33-45 (2011).
86. G Schalau and H Aliyar. Silicone excipients in pharmaceutical drug delivery applications, in: *Excipient Applications in Formulation Design and Drug Delivery*, A Narang and S Boddu (Eds.), pp 423-462, Springer, Cham (2015).
87. H.A Aliyar and G.K Schalau. Recent developments in silicones for topical and transdermal drug delivery. *Therapeutic Delivery.* 6, 827-839 (2015).
88. Noven Pharmaceuticals., Innovations in passive transdermal drug delivery: high doses in a small patch. In: Proc. 33rd Annual Meeting of the Controlled

Release Society, Vienna, Austria (2006) https://www.controlledreleasesociety.org/sites/default/files/docs/2018/v23i3.pdf

89. D Kanios, L Nartker, T Mitchell, C Evans, T Menjoulet and R Huber, The effect of silicone/acrylic hybrid pressure sensitive adhesive in controlling *in-vitro* permeation and delivery profile from transdermal drug delivery systems. In: Proc. American Association of Pharmaceutical Scientists Annual Meeting and Exposition; Washington D.C. USA. Poster T2238 (2011).
90. S.H Imam , S.H Gordon, L Mao and L Chen, Environmentally friendly wood adhesive from a renewable plant polymer: characteristics and optimization. *Polym Degrad Stabil* 73, 529-533 (2001).
91. 3M™ Scotch® Industrial Box Sealing Tape 371 Technical Data. (2020).
92. J.P Raynaud, M Auges, L Liorzou, V Turlie and C Lauze, Adhesiveness of a new testosterone-in-adhesive matrix patch after extreme conditions. *Int. J. Pharm.*, 375, 28-32 (2009).
93. P.N Kotiyan and P.R Vavia, Eudragits: role as crystallization inhibitors in drug-in-adhesive transdermal systems of estradiol. *Eur. J. Pharm.* 52, 173-180 (2011).
94. R.D Toddywala, K Ulman, P Walters and Y.W Chien, Effect of physicochemical properties of adhesive on the release, skin permeation and adhesiveness of adhesive-type transdermal drug delivery systems (a-TDD) containing silicone-based pressure-sensitive adhesives. *Int J Pharm.* 76:77-89 (1991).
95. S Diridollou, F Patat and F Gens, *In vivo* model of the mechanical properties of the human skin under suction. *Skin Res Technol.* 6, 214-222 (2000).
96. C Zhao, P Quan, C Liu, Q Li, and L Fang, Effect of isopropyl myristate on the viscoelasticity and drug release of a drug-in-adhesive transdermal patch containing blonanserin. Acta Pharm Sinica B. 6, 623-628 (2016).
97. S Lin, C Lee and Y Lin, Drug-polymer interaction affecting the mechanical properties, adhesion strength and release kinetics of piroxicam-loaded Eudragit E films plasticized with different plasticizers. *J. Controlled. Release.* 33, 375-381 (1995).
98. D.R Gullick, W.J Pugh, M.J Ingram, P.A Cox and G.P Moss, Formulation and characterization of a captopril ethyl ester drug-in-adhesive-type patch for percutaneous absorption. *Drug Dev. Ind. Pharm.* 36, 926-932 (2010).
99. P Minghetti, F Cilurzo, A Casiraghi and L Montanari, Application of viscometry and solubility parameters in miconazole patches development. *Int. J. Pharm.* 190, 91-101 (1999).
100. N.T Hai, J Kim, E.S Park and S.C Chi, Formulation and biopharmaceutical evaluation of transdermal patch containing benztropine. *Int. J. Pharm.* 357, 55-60 (2008).
101. J Renvoise, D Burlot, G Marin and C Derail, Peeling of PSAs on viscoelastic substrates: a failure criterion. *J Adhesion.* 83, 403-416 (2007).

9

Medical Adhesives from Extracted Mussel Adhesive Proteins

Yuvaraj Dinakarkumar*, Annushrie Arravind and Niranjana Murali Mohan

Department of Biotechnology, Vel Tech High Tech Dr. Rangarajan and Dr. Sakunthala Engineering College, Avadi, Chennai, Tamil Nadu, India

Abstract

Medical adhesives that are physically and mechanically strong, easy to apply and biologically safe are considered to be an effective alternative to sutures and staples in different surgical and clinical procedures. Even though progress in the development of adhesives has been observed, adhesion to wet tissues still remains challenging. A vast variety of marine organisms have the ability to adhere firmly to wet surfaces under water, hence a diverse set of highly effective adhesion systems have been developed which has inspired the design of biological medical adhesives. These adhesives have proven to be biocompatible and effective. In this chapter we provide an overview of the mussel adhesive proteins, their chemical and physical adhesion strategies and their applications in medical field.

Keywords: Medical adhesives, surgical, adhesion, biocompatible, mussel adhesive proteins

9.1 Introduction

The oceans of our world are home to many different organisms utilizing different adhesion mechanisms for their survival. Starfish employs adhesion bonding for locomotion; mussels, barnacles and oysters produce adhesive materials for surface attachment and stability; marine worms use an adhesive composite to dwell underwater by gluing onto sand grains and biomineral particles [1]. The adhesion or attachment mechanism in marine

**Corresponding author*: yuvarajdinakarkumar@gmail.com

K.L. Mittal and S. Neogi (eds.) Adhesives in Biomedical Applications, (213–252) © 2023 Scrivener Publishing LLC

organisms is mainly due to cross-linking of proteins, a complex process that takes place during post-translational modifications [2]. Mussels are marine organisms that have the ability to attach themselves firmly to dry and wet surfaces including glass, plastics and metal oxides [3]. They show strong adhesion capability and hence they can also stick to non-sticky surfaces like Teflon [4].

Mussels are usually found in the rocky, wind-swept seacoasts and they secrete glue-like sticky material known as byssus which is responsible for strong adhesion in the turbulent marine environment [5]. The byssus is a bundle of thread-like materials consisting of four parts, namely plaque, thread, stem and root [6]. High concentrations of polyphenolic proteins, collagen and catechol oxidase were among the first biochemical observations of byssal attachment [6]. Figure 9.1 depicts the anatomy of the mussel species *Mytilus edulis* and the structure of the byssus threads [7]. Mussel byssus is a proteinaceous material and does not contain any living cells. Plaque is the part of the byssal thread that attaches to the foreign surface, whereas the stem is connected to the tissues of the organism by means of a root, and the threads attach the stem and plaque together. Byssal threads are attached to the root at the base of mussel foot; during adhesion the mussel foot emerges from the shell, finds a suitable surface to attach and becomes almost motionless after adhesion [8]. The adhesion property of the mussels is mainly due to the catecholic functional group 3,4-dihydroxyphenylalanine (DOPA) present in the plaque [9–11]. Mussel-based adhesives are prepared by covalently linking catechol groups

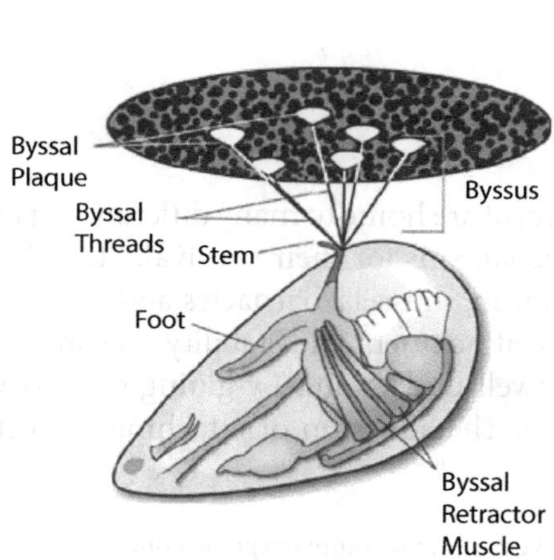

Figure 9.1 Structure of the Byssus thread and anatomy of *M. edulis* [7].

to end or side groups of existing polymers or by polymerization of monomers with existing or modified catechol functionality [12].

9.2 The Mussel Byssus

The byssus consists of a large number of threads secreted from the mussel foot to provide a stable attachment to different substrates in the intertidal zone [13, 14]. There are three mussel foot glands, namely the core gland, the cuticle gland and the plaque gland which are involved in the production of mussel byssus proteins [14–18].

The secreted thread proteins from the core gland are released into the ventral groove followed by the release of plaque proteins from the plaque gland into the distal depression at the foot tip [14, 19, 20]. A microemulsion fluid flows into the distal depression through the ciliated longitudinal ducts [21]. Then the accessory gland provides a protective cuticle covering the structure before disengagement from the groove [14, 19, 22]. Each byssal thread can be divided into proximal section and distal section, the proximal section is connected to the stem of the mussel foot whereas the distal section expands to form an adhesive plaque which is attached to the substrate surface [14, 15, 23, 24].

9.2.1 Collagenous Thread Protein

The collagenous proteins make almost 96% and 66% proteins in the proximal and distal regions respectively [24]. They are denoted as preCols, derived from the word pre-pepsinized collagens, originating from the core gland and they are considered as block co-polymers with dominant and central collagenous domains which are the basis for the formation of collagen triple helixes [24–28].

The preCols are divided into three variants, -D, -P, -NG, among which the flanking domains are considered to be a major contributor to mechanical properties [24, 29–32]. PreCol –D is a silk-like flanking domain found in the stiff distal sections, whereas the preCol –P is an elastic-like flanking domain found in the proximal part of the byssal thread [24, 33–35]. The preCol –D and preCol – P are distributed on each of the byssal thread providing stiffness and elasticity; they consequently show gradient properties from distally stiff to proximally elastic region [25, 26]. The preCol –NG functions as a mediator between preCol –D and –P.

9.2.2 Thread Matrix Protein

The byssal collagens are arranged in fibrils which are surrounded by the thread matrix proteins (TMPs). The TMPs also originate from the core gland and are distributed along the whole byssal thread among which one of the TMPs is found in the proximal section known as proximal thread matrix protein 1(PTMP1) [36, 37]. The TMPs show significant deamidation which contributes to the rapid maturation of solid byssal threads. The thread matrix proteins are considered to be the connector, lubricator, spacer and mediator between the collagenous proteins in the byssal threads [25, 36, 38, 39].

9.2.3 Mussel Foot Proteins (MFPs)

The mussel foot proteins are the major contributors towards adhesion process. The byssus thread proteins are comprised of 25-30 different mussel foot proteins (MFPs) among which only 5 types are found in the plaque.

As depicted in Figures 9.2 and 9.3 the five proteins can be roughly divided into two groups, namely DOPA-rich mfp-3 and mfp-5 proteins at the surface and mfp-6, mfp-2 and mfp-4 proteins that are located in the plaque. The mfp-3 and mfp-5 are mainly responsible for adhesion of the plaque to the substrate [40]. The mfp-1 and -6 originate in the cuticle

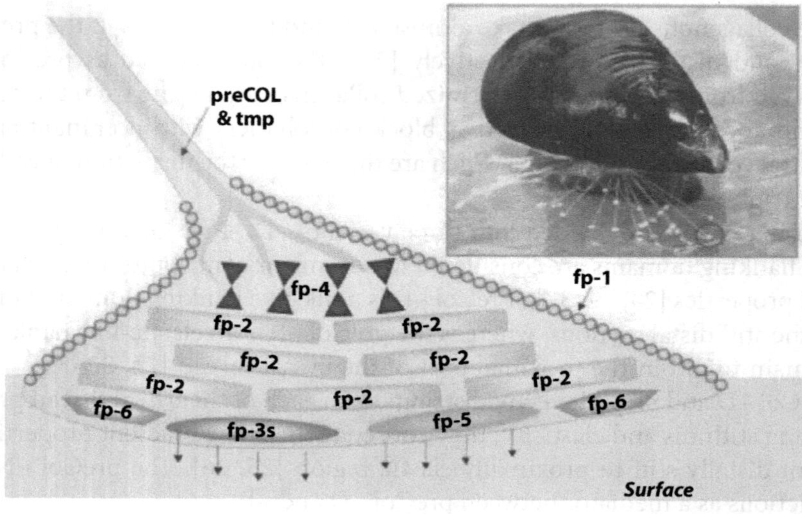

Figure 9.2 Representation of mussel foot proteins [40] (fp – foot protein in the mussel byssal thread).

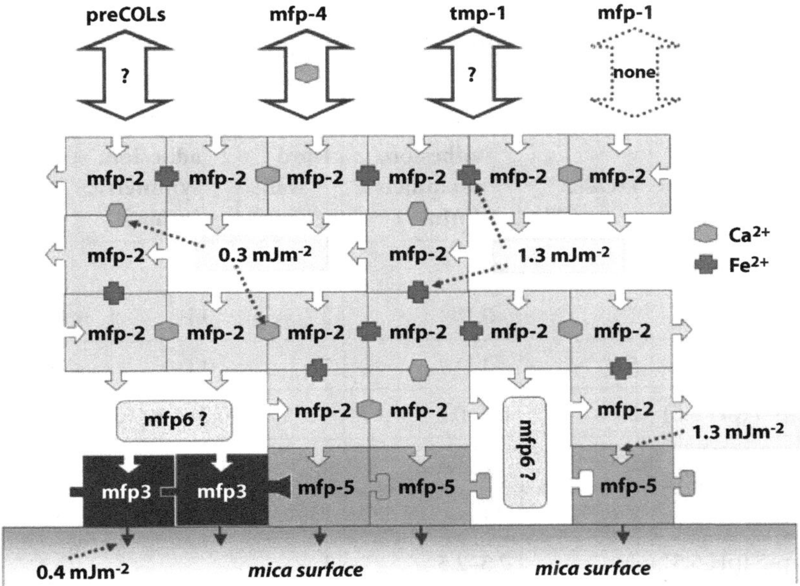

Figure 9.3 Protein interactions in adhesive plaques. The picture also depicts the strong surface interactions, self-interactions and specific adhesion [40].

gland and mfp-1 behaves as a protective coating in the cuticle layer of the byssus thread [38]. The mfp-2 is the most abundant protein found within the plaque with relatively low DOPA content; it is proposed to be a stabilizer in the byssus [41, 42]. The mfp-4 is located at the junction between distal section and the adhesive plaque and mediates the mechanical translation from thread to plaque and also contributes to the interconnection with collagens [43, 44]. The mfp-3 are polymorphous polar proteins and are subdivided into two groups, mfp-3f (fast) and mfp-3s (slow) based on their distinct elution in electrophoretic analysis, the fast mfp-3 contains more DOPA as compared to the slow mfp-3. The mfp-5 is similar to mfp-3 but is comparatively larger in size and contains the highest level of DOPA and the mfp-3 and -5 function as a primer for adhesion to different substrates [45]. The mfp-6 is rich in glycine and aromatic residues and exhibits low post-translational modifications of Tyr residues. The mfp-6 functions as antioxidant and reverses the oxidised DOPA residues in mfp-3 and -5 enabling sufficient adhesion of -3 and -5 to substrates [46–48]. The mfp-2 covers about 25% to 40% of the total protein content in the plaque. So in a byssus thread the mfp-5 and mfp-3 proteins are found near the surface

Table 9.1 Mussel foot protein (mfp) interaction strength (work of adhesion) determined from thin films deposited on mica using the surface forces apparatus (SFA) [6].

Mfp (mussel foot protein)	DOPA (mol%)	Work of adhesion, asymmetric (mJm^{-2})	Hard wall (nm)	Work of adhesion, symmetric (mJm^{-2})	Hard wall (nm)
Intrinsic					
Mefp-1	15	<0.1	8	<0.1	8
Mefp-2	5	<0.1	5	<0.1	10
Mcfp-3 (pH 5.5)	20	0.3–0.5		0.3–0.5	10–40
Mcfp-3 (pH 3)	20	1.2–1.4	4	1.2–1.4	5
Mcfp-4	2	<0.1	13	<0.1	50
Mcfp-5 (pH 5.5)	30	2.3–2.5	3		
Mcfp-5 (pH 3)	30	6–7	1–3		
Mcfp-6	3	<0.5	8	<0.5	12
Mcfp-3 ↔ Mcfp-5	20/30	2.3–2.6	3–8		
Mcfp-2 ↔ Mcfp-5	5/30	1.2–1.5	12		
Metal mediated					
Mefp-1 + Fe^{3+} [a]	15	0		4–5	8
Mcfp-2 + Fe^{3+} [a]	5	0		2–3	12
Mcfp-2 + Ca^{2+} [a]	5	0		0.2–0.4	15

*Mcfp –*Mytilus californianus* foot protein.
*Mefp –*Mytilus edulis* foot protein.
*Hard Wall – The hard wall distance is defined as the mica–mica separation distance or thickness of confined proteins, which does not appear to change with increase in the normal compressive load or pressure.
*Fe^{3+} and Ca^{2+} at 10 μM.
*Mussel foot protein interaction strengths are determined from the thin films deposited on mica both symmetrically and asymmetrically under conditions that include a contact time of 60 min and buffer of 0.3 M KNO3, 0.1-M Acetate with a pH 5.5 at room temperature.

where the attachment occurs and mfp-6 and mfp-2 are found away from the surface and thus DOPA levels are high in mfp-5 and mfp-3 [6, 7, 49].

The molecular interactions between specific mfps and between mfps and substrate surfaces in aqueous solutions are investigated using the surface forces apparatus (SFA). SFA is an instrument with angstrom distance and nanonewton force precision [6]. It can measure A) the repulsive or attractive forces during approach; B) the attractive adhesion forces; C) the attractive forces between a surface and liquid water [6, 50]. The measured forces reflect the nature of interation between the mfps or between mfps and a substrate surface such as mica, Table 9.1 is the summary of known mfp interactions as measured by the SFA on mica [50].

Study shows a strong adhesion between mfp-2 andmfp-5 in the presence of Fe^{3+} and Ca^{2+} metal ions [51]. Protein behavior on various materials of distinct hydrophobicities was studied using sum frequency generation (SFG) vibrational spectroscopy which helped to understand the chemical interactions that mainly involve mfp-3 which acts as a primer in the adhesive plaque formation [45].

9.3 Mussel-Inspired Adhesion

9.3.1 History

The polyphenolic proteins reported in 1981, were the first to be extracted from mussel foot that showed adhesive property. Since then the structure and properties were studied and variety of proteins showing adhesivity were found in mussels, barnacles and sandcastle worms [52, 53]. All the extracted proteins from various kinds of organisms showed the presence of 3.4-dihydroxy-L-phenylalanine (DOPA), tyrosine, phenylalanine to be responsible for the adhesive behavior [52]. The marine organisms showed large number of proteins controlling the adhesion mechanism; hence addressing the contribution of each protein in the nanoscale adhesion phenomenon was difficult.

The proteins were qualitatively analysed using the traditional techniques such as Quartz Crystal Microbalance with Dissipation (QCMD). Attenuated total internal reflectance Fourier transform infrared spectroscopy (ATR-FTIR) and resonance Raman spectroscopy characterize proteins adsorption by the change in the mass or via sensing certain target chemical species but these do not provide a quantitative explanation for the physics between the foot proteins and the substrate [54]. Thus researchers began to study the mussel adhesive proteins qualitatively and

Table 9.2 Properties of the mussel foot proteins involved in the adhesion process [54, 56].

Protein	Mass (kDa)	Isoelectric point	DOPA (%)	Location	Features of the MAPs (mussel adhesive proteins)	Post-translational modifications of MAPs (mussel adhesive proteins)	Repeat number	Species
Mfp-1	108	10.5	10-15	Cuticle (Exclusively)	Protective covering	Trans-2,3-cis-3,4-dihydroxyproline	75	*Mytilus edulis*
Mfp-2	42-47	9.10	2-3	Within the plaque	Plaque Protein	Cross-linked Cystine	11	*Mytilus edulis*
Mfp-3	6	> 10.5	25	Plaque-substrate interface	Mussel Adhesion	4-Hydroxyarginine	1	*Mytilus edulis*
Mfp-4	93	10.5	2	Within the plaque	Connects the thread to the plaque		36	*Mytilus californianus*
Mfp-5	9.5	9	28	Plaque-substrate interface	Enhances Mussel Adhesion	O-Phosphoserine	1	*Mytilus edulis*

(*Continued*)

Table 9.2 Properties of the mussel foot proteins involved in the adhesion process [54, 56]. (*Continued*)

Protein	Mass (kDa)	Isoelectric point	DOPA (%)	Location	Features of the MAPs (mussel adhesive proteins)	Post-translational modifications of MAPs (mussel adhesive proteins)	Repeat number	Species
Mfp-6	11.6	9.5	<5	Plaque-substrate interface	Reduces oxidized DOPA-quinone to DOPA	O-Phosphoserine	1	*Mytilus californianus*
Tmp-1	57	9.5	4	Byssal threads	Intimately associated with the preCOLs within the secretory granule		5	*Mytilis galloprovincialis*
PreCOL-D	240	10.5	< 1	Fibrous core of byssal threads	Provides tenacity to byssal threads	Trans-4-hydroxyproline	173	*Mytilus edulis*
PreCOL-NG	230	7.5	< 1	Fibrous core of byssal threads	Provides tenacity to byssal threads	Trans-4-hydroxyproline	143	*Mytilus edulis*

quantitatively but integrating modern analytical techniques with nanomechanical techniques such as the surface forces apparatus (SFA) and atomic force microscopy (AFM) methods [54]. The growth of nano-science enabled and encouraged the scientists to study the molecular interaction and influence of each protein in the adhesion process of the organism [55]. Table 9.2 shows the properties of the mussel foot proteins involved in the adhesion process [56].

The technological advancement saw a steep growth in the study of mussel adhesion mechanism and a way to quantify the intermolecular adhesion force of DOPA using atomic force microscopy.

In 1949, synthetic tissue adhesives were developed such as the cyanoacrylate glues which were used to close small skin cuts and lacerations, but these had certain side effects like chronic inflammatory reactions and poor mechanical properties; hence further attempts were made to develop better synthetic tissue adhesives like n-2-butylcyanoacrylate which also was not successful as it showed a low tensile strength and was highly brittle [57]. The biological tissue adhesives as drug delivery agents were first used in 1947 to deliver penicillin using a mucoadhesive formulated with tragacanth and a dental adhesive powder, this product was named as 'Orabase' which targeted the oral mucosa to treat mouth ulcers [58]. Orabase is commercially available in the market and is produced by Discus Dental, Inc., Culver City, CA, USA; Bristol-Myers Squibb Co., Princeton, NJ, USA; and Taro Pharmaceuticals, Hawthorne, NY, USA.

9.3.2 Bioadhesives

The conventional methods of reconnecting tissues and organs involve the use of sutures, staples and wires which mechanically hold the tissues to stop bleeding or fluid leakage; it also promotes faster healing and provides resistance to tensile loads applied on the tissue. These methods have many limitations such as damaging the surrounding tissues, requiring secondary surgeries for removal, and interfering with tissue healing processes; these also require high level experienced operating surgeons.

The adhesive biomaterials are an alternative that facilitate surgical reconnection of injured tissues and organs to restore structure, function and hemostasis [57–61]. These biomaterials are called tissue adhesives, bioadhesives or bioglues and are defined as substances that can polymerise *in-situ* and help bond two surfaces, one of which is a tissue [60]. Tissue adhesives offer a simple, quick and a sutureless way of reconnecting tissues and restoring their functionality, Depending on the application, a bioadhesive can function as a glue to keep tissues closely attached until healing,

Table 9.3 Properties that a material should have for use as a tissue adhesive [56, 60, 62].

Desired intrinsic property	Comments
Adhesion	Surgical adhesive must be able to hold two tissues together without external support until the wound heals, after which the adhesive should degrade into nontoxic biocompatible components.
Curing	The adhesive must have the ability to cure itself in the tissue environment, especially in a moist environment.
Flowability	The adhesive must have the ability to flow. Flowability of an adhesive relates to the movement of polymer chains and controls the extent of interpenetration between polymers and tissue surfaces.
Hydrophilicity	Hydrophilic polymers are preferred as they can form strong bonds with highly hydrophilic tissues.
Hydrogen bonding	The presence of strong hydrogen bonding groups like OH and COOH in the monomer can promote hydrogen bonding between polymer chains affecting the flowability and cohesive strength of the adhesive. A minimum cohesive strength is required for a good adhesive.
Molecular weight	High molecular weight polymers (>100 kDa) make better adhesives as they provide a high degree of entanglement.
Surface tension (wettability)	For any adhesive to adhere to a substrate, its surface tension must be equal to or less than that of the substrate. Lower surface tension promotes polymer spreading across the substrate for better adhesion.
Crosslinking	Crosslink formation in an adhesive promotes mechanical strength required by the adhesive to transfer shear stress across the adhered surface.

as a hemostat to prevent bleeding and blood loss from tissues, or as a sealant to avoid fluid leakage. Bioadhesives also have the potential to work as drug delivery carriers in cases where antibiotics and growth factors are released slowly covering a specific area to prevent infections and encourage cell growth and wound healing [58, 61]. Table 9.3 lists the desired properties that a specific material should fulfill for use as a surgical bioadhesive in different clinical settings.

9.3.2.1 Mechanisms of Biological Tissue Adhesion

Adhesion is related to the forces that attach two dissimilar materials; and when two similar materials attach, the process is referred to as cohesion. Adhesion is caused by intermolecular interactions at the interface between the materials and cohesion is due to the intramolecular attraction.

Thus bioadhesion can be defined as the state where biological materials are held together using a material which need not specifically be biological; this is accomplished by the forces acting at the interface of the two materials [58]. The term "bioadhesion" was first used in 1970 to describe the capability of different molecules of biological or synthetic origin to attach to biological tissues [57]. The mechanisms of bioadhesion are classified into four categories, namely intermolecular bonding, chain entanglement, mechanical interlocking, and electrostatic bonding. Surface wetting is also an important contributor to the adhesion mechanism, the wetting theory is related to the capability of the adhesive to wet and create molecular bonding with tissue surface [29, 56]. Figure 9.4 is a representation of the four main categories of biological adhesion mechanisms which are explained below.

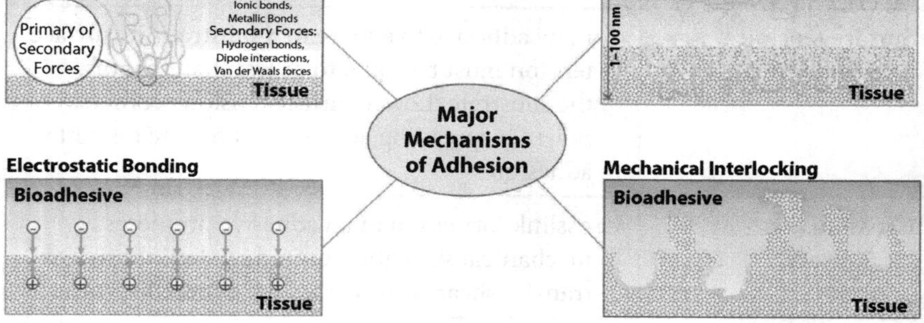

Figure 9.4 The four main categories of biological adhesion mechanism [60, 62, 67].

9.3.2.1.1 Electrostatic Bonding

When the surfaces of two materials with different electronic band structures are brought in contact, the possible transfer of some electrons which occurs to equalize the Fermi levels forms a double layer of electronic charge at the interface. This charge induces electrostatic forces, which may play a significant role in the intrinsic adhesion of the two contacting surfaces. However, in the case of insulator substrates the charge buildup would be very slow and the number of available electrons might be limited, hence it would require a long time to build up charge concentration. Although the presence of electrostatic forces from the charged double layers has been observed in some metals and semi-conductors, this mechanism does not play a significant role in adhesion of nonmetallic systems [63–65].

9.3.2.1.2 Intermolecular Bonding

This is a major mechanism of adhesion between adhesive materials and substrates, which arises from interatomic/intermolecular forces and the bonds between atoms/molecules of adhesive on one side and the surface atoms/molecules of substrates on the other [63]. These surface forces are classified as primary and secondary.

Examples of primary forces are cationic, ionic, and metallic chemical bonds, which are high energy bonds that form at the interface and result in strong adhesion. Secondary forces form low energy bonds, but they have a significant effect resulting from the large number of bonding sites that may exist at a given interface. Some of these secondary forces are hydrogen bonding, van der Waals forces, London dispersion, and dipole–dipole and dipole-induced interactions [56].

9.3.2.1.3 Chain Entanglement

Semi-permanent bonds between the adhesive and tissue surface are formed when polymer chains from the adhesive penetrate into the tissue surface to form a bond. The interpenetration of the polymer chains is driven by concentration gradients, the polymer chains penetrate the tissue and in turn the glycoprotein chains of the tissue penetrate the adhesive until an equilibrium depth is attained [66, 67]. This type of mechanism can be seen in the drug delivery systems [63].

9.3.2.1.4 Mechanical Interlocking

In this mechanism the adhesive interacts with the tissue at micro-irregularities or with the surface roughness to create bonding [68]. Here, the tissue irregularities or the tissue surface roughness plays a major role as the adhesive interlocks itself with the tissue on the sites where microscopic roughness can be noticed [69].

9.3.3 Role of DOPA in Adhesion Process

Mussels are marine organisms that secrete a mixture of proteins called mussel adhesive proteins (MAPs). These MAPs show the presence of at least six mussel foot proteins mfp1 to mfp6. The MAPs have a particular catecholic amino acid 3, 4-dihydroxyphenylalanine (DOPA), which is a key factor behind mussel's strong adhesion in underwater conditions. The foot proteins are known to contribute to the extreme wet adhesion [70, 71]. DOPA plays a vital role in the interfacial attachment and curing of the adhesive plaque proteins. It is also capable of binding to many different surfaces which can be organic or inorganic by establishing reversible non-covalent and irreversible covalent bonds [71].

DOPA can form strong hydrogen bonds via its dihydroxyfunctionality, increasing its adsorption on tissue surfaces. Catechol containing polymers have improved cohesive property due to the presence of benzene ring, which can interact with other aromatic rings via π–π interactions. In addition, the benzene ring can interact with positively charged ions through cation–π interactions, which promotes binding to positively charged surfaces and to substrates rich in cationic functional groups [72, 73]. Additionally, catechol can form robust reversible bonds with metal ions such as Cu and Zn, as well as interfacial bonds with metal oxide surfaces. Furthermore, the highly reactive oxidized form of catechol can be involved in intermolecular covalent cross-linking that causes curing of DOPA containing adhesives. Catechol is usually oxidized into semiquinone or quinone by one or two electron oxidation through enzymatic and chemical (periodate) reactions. Quinone is capable of reacting with several nucleophilic functional groups found in tissues ($-NH_2$, –SH, imidazole) to form interfacial covalent bonds, which can promote tissue adhesion [71].

9.3.3.1 DOPA Oxidation

Catechol exists in an acidic pH and as the pH changes to around pKa 9.3, which is the first dissociation constant of its –OH group, the catechol side chain autoxidizes to quinone, which exhibits a limited adhesion strength [74–76]. Under seawater conditions, DOPA oxidizes to DOPA-quinone, which can tautomerize to α, β-dehydro-DOPA. DOPA-quinone does not have the ability to form hydrogen bonds with surfaces [40]. Adhesion studies showed a higher work of adhesion on DOPA- modified gels compared to unmodified forms [77]. The mfp-3 and mfp-6 in the plaque behave as an antioxidant against the oxidation of the DOPA. The mfp-3 also creates a hydrophobic environment shielding the DOPA from oxidation and subsequently enhancing adhesion. The mussel foot proteins, with the exception of mfp-1, have the non-polar amino acid glycine adjacent to DOPA which can inhibit its oxidation through hydrophobic shielding and electrostatic shielding effects [40, 78, 79].

9.3.3.2 Coacervation

Coacervation is the phase separation of fluids driven by Columbic forces and charge neutralization between the chains of opposite charges. This process influences underwater adhesion by increasing the polymer concentration in the coacervates which, in turn, improves surface wettability via decreasing interfacial free energy and reducing viscosity of the phases to enhance adhesive delivery. In mussels, hydrophobic, π–π and cation–π interactions cause coacervation and subsequent increase in adhesion [71, 40, 19, 81].

9.3.3.3 Surface Drying

In underwater adhesion process, the interfacial molecular contact between the substrate and the adhesive is necessary; this contact is inhibited due to the presence of a layer of adsorbed water, which gives rise to strong repulsive hydration forces that obstruct an effective adhesion. This can be overcome with the mfp-3 due to its hydrophobic tryptophan side chains, which help in removal of the surface hydration layer. In addition to the mfp-3s, the cationic residues that are close to DOPA such as lysine also help in the removal of the hydration layer, promoting effective adhesion to the underlying surface [40, 82]. Hence the coacervates containing DOPA exhibit the capability to remove interfacial water molecules and form molecular hydrogen bonds with the substrates, which increases the adhesion strength [83, 84].

9.3.3.4 Catechol Modification

Catechol modification significantly changes its adhesive property, i.e., the reactivity of the catechol can be enhanced by the addition of a side electron-withdrawing nitro-group at the para position, which also increases bonding strength [85]. The nitro functionalization leads to higher resistance at higher temperatures as compared to an unmodified catechol [86–88]. The modified catechol exhibits higher rate of covalent cross linking and binds to biological substrates over a wide pH range [85]. The catechols that are nitro- and chloro-functionalized show characteristics that allow them to resist light-induced degradation and improve their antimicrobial property [89].

9.4 Mussel-Inspired Tissue Adhesives

9.4.1 Extraction of Mussel Protein

Marine adhesives secreted by micro- and macro-organisms can bind to any substrate and has the capability to displace water, this has led to an in-depth evaluation on their characterization and functions [90]. The Lewis bases simplify and restore the aromatic rings of the mussel foot proteins extracted by dilute chemical treatments [19].

Initially ring opening polymerization (ROP) of N-carboxyanhyride (NCA) monomers derived from α-amino acid was conducted to prepare DOPA and L-lysine copolymers. The cross-linking of these copolymers with an oxidizing agent was found to form water-resistant bonds with various substrates depending on the oxidizing agent used to provide large quantities of adhesive polymers consistent in quality [90]. A small model peptide extracted from *M. edulis* cultured using an expression system *E. coli* was studied to understand the adhesion mechanism of mussels after the peptide undergoes post-translation modification by hydroxylation of tyrosine residue on DOPA. The study revealed that the structural and functional properties of the adhesive depend on the post-translational hyroxylation of specific amino acids on DOPA. In this case the tyrosinase modification allows the Tyr- and Lys-residues to react and form O-quinone linkage on polymerization, this strong linkage allows solidification and precipitation. The genetically altered adhesive model peptide was purified using acetic acid and ion exchange chromatography; analysis of this modified peptide showed high adhesivity when compared to bovine serum albumin (BSA) [91]. The mfp-3 and mfp-5 proteins were found to contain high levels of DOPA correlating with the adhesion strength of MAPs which prompted the use of recombinant techniques. The recombinant mfp-151 a novel type of hybrid MAP, which is a fusion protein comprising of six *M. galloprovincialis* mfp-1 decapeptide added to the N- and C-termini of *M. galloprovincialis* mfp-5; it was constructed and prepared to overcome few of the limitations posed by recombinant mfp-3 and mfp-5. This provided a high yield showing efficient biocompatibility and cell adhesion to cell types like anchorage dependent and independent cells to form a practical bioadhesive [92]. *M. edulis and M. californianus* were initially studied to understand their large repeat sequences that are involved in strong adhesion and compared with other organisms that are similar in characteristics to gain insights in the adhesion mechanism. Guanidinium chloride gel filtration treatment elutes DOPA present in the mussel foot by 1.05 M (molar) after extraction from byssus thread [93]. The DOPA is mostly extracted using an acid based urea like allantoin, hydantoinetc [94]. The recombinant mussel adhesion proteins were tested with substrates like glass, plastics, Teflon, metals, etc. as an alternative to the mfps directly from organisms [95, 96]. Recombinant DNA technology is one of the most preferred cloning methods rather than chemical synthesis of the mfp and MAP based recombinant production mainly due to the high throughput of DOPA containing peptides [97]. Earlier studies on recombinant mfp-3 and mfp-5 produced in *E.coli* expression systems although expressed adhesivity their low production yield and complexity in purification posed as the limiting factors [98, 99].

9.4.2 Synthesis of Catechol-Functionalized Polymers

Different methods are applied to modify synthetic and natural polymers to incorporate catechol functional groups. In any case the additional protection and deprotection of the catechols along the synthesis route is vital to preserve their adhesive property in the end product [71]. There are various strategies to create functional DOPA polymersrelevant to healthcare: polymerization of catechol-modified monomers, synthesis of DOPA-containing peptides, and genetic engineering [56].

9.4.2.1 Direct Functionalization

Direct functionalization is a process where the polymers are directly linked to DOPA, for instance a linear or branched poly(ethylene glycol) (PEG) polymer grafted to DOPA. Here the polymer provides an inert atmosphere and a biocompatible molecular support to impart cohesive strength and the DOPA molecules impartadhesivity to the crosslinked hydrogels [56, 57]. DOPA and dopamine can also be directly conjugated onto polymeric backbone via the formation of amide, urethane, and ester linkages with the functional groups $-NH_2$, $-COOH$, and $-OH$ [100, 101]. In addition, mussel inspired PEG-based polymers show adhesive and mechanical properties (such as degradation, swelling and curing rate) that can be easily manipulated. This manipulation can be achieved by varying the PEG molecular weight, PEG architecture, number of DOPA residues used in the conjugation, and the type of linkage [56, 57].

Researchers have also investigated the drug delivery property of the PEG-DOPA polymers. A mixture of dipalmitoyl and dimyristoylphosphatidylcholine was used to create lipid vesicles with capability to release their contents at 37 °C. These liposomes were then combined with the PEG-DOPA polymers to form fast crosslinking hydrogels with an adhesion strength of 35.1 ± 12.5 kPa measured through lap shear test, this is much higher than the one achieved with fibrin glue [102]. Recent research shows that the PEG-catechol hydrogels can adhere to skin covered with blood with adhesion strength of above 40kPa. These were prepared by reinforcing the PEG-catechol material with collagen fibers and hydroxyapatite nanoparticles [103]. Nowadays researchers are focusing on developing injectable PEG and citrate based bioadhesives inspired by the mussel adhesive property. The injectable citrate based mussel bioadhesives (iCMBAs) are used for internal organs surgeries as these materials exhibit a strong wet adhesion and have adequate elastomericity, degradability and cytocompatibility. Studies show that iCMBAs can stop bleeding and help in closing the wounds without sutures [101]; Figure 9.5 shows the mechanism and application of iCMBA for sutureless wound closure. Through the application of click chemistry, iCMBAs were improved; a copper-catalyzed azide–alkyne Huisgen cycloaddition was used in combination with the chelating agent 4-(2-hydroxyethyl)-1-piperazineethanesulfonic acid (HEPES), to overcome the inhibitory effect between the catechol crosslinker and the copper reductant catalyst. The bioadhesive showed a higher wet adhesion strength than the fibrin glue. Additionally, it also showed antibacterial, antifungal, and haemostatic capabilities [104].

Further studies were done by combining catechol groups with different structures of PEG and also with other polymers to form block copolymers, which were then combined with catechol groups and conjugated onto polymer backbones. The results of thus derived products were compared to finalize a single material which proved to be a more efficient bioadhesive [100, 105, 106]. In the process, dopamine was combined with a 4-arm PEG and polycaprolactone (PCL) to prepare an adhesive polymer to coat hernia repair biologicalmesh and the mechanical strength tests were done on bovine pericardium that showed a shear strength of 107 ± 24.7 kPa [107]. DOPA has also been conjugated with poly(propylene oxide) PPO and poly(ethylene oxide) PEO. A study based on the above method was carried out in which the catechol groups were conjugated to a PEO-PPO-PEO block copolymer with a coupling efficiency about 80%. The product

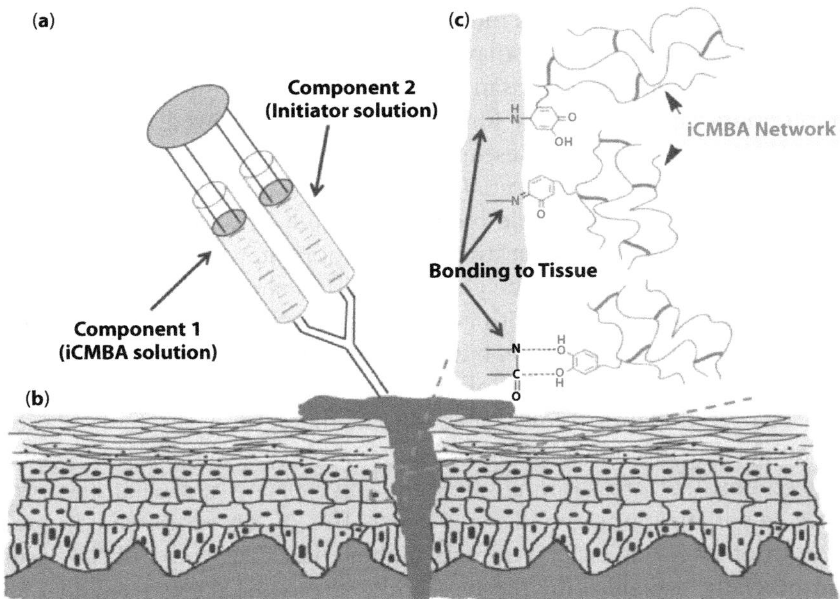

Figure 9.5 Representation of wound closure with an injectable citrate-based mussel-inspired bioadhesive (iCMBA). (a) preparation and application of a two-component adhesive consisting of iCMBA and oxidizing (sodium periodate) solution; (b) schematic representation of iCMBA for sutureless wound closure; (c) proposed mechanism of iCMBA adhesion to tissues [5].

exhibited temperature induced gelation; it also showed enhanced mucoadhesive characteristics which were demonstrated by viscosity measurements [100]. Other example of such conjugation is poly(ε-caprolactone) (PCL) with glycidylmethacrylate (GMA)and oligo (ethylene glycol) methacrylate (OEGMA) on the backbone which was utilized to combine with catechol grafts. Similarly, a DOPA-based mimetic adhesive with good adhesion to tissues and metals was formulated by immobilizing catechol groups on the epoxy rings of glycidylmethacrylate (GMA). The resultant product was less toxic but its adhesion strength was only 9kPa which is less than the strength shown by PEG-DOPA polymers [108].

Direct functionalization of DOPA with synthetic polymers and biopolymers also proved to be useful in producing hydrogels with strong adhesive property [57]. Biopolymers such as dextran, chitosan, hyaluronic acid, gelatin, and alginate have been covalently attached with acid- or amine-functionalized catechols and crosslinked to form hydrogels with strong tissue adhesive property [109–116].

Hyaluronic acid (HA) was modified by dopamine conjugation via carbodiimide chemistry. Using a layer-by-layer (LbL) technique, this material was combined with chitosan to form multilayer films with enhanced nanotopography [111]. These films exhibited an improved cell adhesion, proliferation, and viability. Researchers also developed a two-component bioadhesive system with potential to be used as a bone adhesive based on biocompatible and degradable biopolymers like chitosan, oxidized dextran, etc. On mixing in water, the two components covalently cross-link by Schiff's base formation. The characterization showed that the product derived was biocompatible with adhesion strength of 0.31 MPa, which is higher than that of fibrin glue, but less than that of cyanoacrylates [117]. Furthermore, a double-crosslink tissue adhesive (DCTA) was developed using a dopamine-conjugated gelatin macromer. The first gelation was attained using the formation of a catechol–Fe^{3+} complex and gelatin induced covalent crosslinking allowing a two-prong approach and produces an enhanced instant adhesive effect. The DCTA exhibited adhesion strength of 24.7 ± 3.3 kPa, which is higher than that of the fibrin-based commercially available bioadhesives, and it also showed elasticity, degradability, and biocompatibility [118].

9.4.2.2 Polymerization of Modified Monomers

This technique focuses on the polymerization of the monomers which are already modified with catechol groups. Dopamine methacrylamide (DMA) is a catechol containing monomer, which can undergo free radical polymerization when activated by heat or light to forman acrylate-based polymer [71]. DMA can also be combined with other monomers such as oligomeric ethylene glycol, monoacryloxyethyl phosphate, and methoxyethyl acrylate to form dopamine grafted polymers [119–121]. One study combined phosphate with DMA to mimic the natural glue proteins secreted by marine species. These materials formed coacervates at neutral pH and were used to bond wet cortical bone. The mechanical test on these products showed bond strength about 40% of that of cyanoacrylates [120]. Polyacrylamide (PAAm) was crosslinked with DMA to form nanocomposite hydrogels, which showed an increased adhesive property. They were also reported to be fracture resistant and also resisted repeated compression loads [122]. For example, a three-dimensional polymer network was formed by copolymerization of DMA with PAAm using a bi-functional crosslinker N,N′-methylenebisacrylamide. In this case, the DMA monomer was polymerized onto polyacrylamide (PAAm) hydrogels and crosslinked with different amounts of nano-silicate in this case LAPONITE®. The products obtained showed viscous dissipation properties as well as increased stiffness, rendering these able to release fracture energy[1] while reducing deterioration of the three-dimensional structure [122]. Similarly, catechol containing vinyl groups like 3,4-dihyroxystyrene, 4-vinylcatechol acetonide, and 3-vinylcatechol acetonide can also be polymerized with monomeric styrene to produce polystyrene based catecholic polymers [123, 124]. Scientists have also developed biomimetic polymeric adhesives based onstyrene–catechol copolymers. Catechol groups were crosslinked with polystyrene backbone to mimic L-DOPA. The resulting poly[(3,4-dimethoxystyrene)-costyrene] polymers showed curing properties similar to natural adhesive proteins and also exhibited an increased adhesive property [123]. A study showed that the poly(3,4-dihydroxystyrene)-co-styrene could also be formed by distributing pendant catechol groups (i.e., 3,4-dihydroxystyrene) at different concentrations onto polystyrene backbone and the derived bioglue had a balanced cohesive and adhesive characteristics. The adhesive property of the material was similar to cyanoacrylate-based commercial glues when tested on both low and high energy surfaces [124].

9.4.2.3 Methods to Synthesize Catechol Peptides

In this method the DOPA-containing peptides were synthesized to mimic the MAPs. The sequential polytripeptides and polydipeptides(X-Tyr-Lys) n, (X=Gly, Ala, Pro, Ser, Leu, Ile, Phe), (Y-Lys)n, (Y=Gly, Tyr), and (Gly-Tyr)n, which imitate a mussel adhesive protein were synthesized. These polypeptide chains were then crosslinked with tyrosinase at a pH of 7.4 and subjected to shear adhesion strength tests. It was demonstrated that the shear adhesion strengths of polytripeptides (X-Tyr-Lys)n were equal to those of the synthetic polydecapeptides, (Ala-Lys-Pro-Ser-Tyr-Pro-Pro-Thr-Tyr-Lys)n and (Gly-Pro-Lys-Thr-Tyr-Pro-Pro-Thr-Tyr-Lys)n, model polydecapeptides for blue mussel and Californian mussel [125]. In another case, monomers of DOPA and L-lysine were prepared using ring-opening polymerization. Then poly((Lys·HBr) x-(DOPA)y) was prepared as a synthetic mimic of natural MAPs and cured with ferric citrate. The adhesion strength of such polypeptide was 0.21 ± 0.10 MPa [126].

9.4.2.4 Recombinant Technology

Similarly to the above method of synthesing catechol-containing peptides, here these peptides are incorporated onto the replicating gene of the

[1]Fracture energy is defined as the amount of energy absorbed to create a unit area of a crack.

microorganism to increase the production of polypeptides. *E. coli* is an organism that shows good replication property. Hence, this organism is used to replicate DOPA-containing peptide sequences similarly to those found in MAPs. The microorganisms secrete these sequences employing their naturally occurring tyrosine residues, which are then transformed into DOPA. For example, the gelation of a hydrogel system based on DOPA-containing recombinant MAPs was obtained by Fe^{3+} mediated non-covalent crosslinking. The hydrogel produced showed deformable and self-healing properties, with an adequate adhesion strength [127]. Recombinant technology is also used to produce adhesive products with hybrid characteristics by combining different polypeptide sequences. Scientists have used this concept to develop a bioadhesiveon a recombinant hybrid MAP mimicking mfp-1 and mfp-5. The resulting adhesive exhibited wet tissue adhesion strength of 72.2 ± 3.7 kPa and was also biocompatible [128].

9.4.3 Mussel-Inspired Medical Bioadhesives

The topic of bioadhesives is a fast growing area of research and has the potential to cover a vast range of applications and most importantly in the field of medicine. According to PubMed and Google Scholar search, it was just a few years ago that the research on mussel-inspired adhesive products began. Until now there are not many medical bioadhesives that are approved for usage in daily life as there are a vast range of properties of the mussel adhesive proteins (MAPs) that are yet to be studied thoroughly. Some of the applications of the mussel-inspired medical adhesives have been listed and explained below.

9.4.3.1 Bioglues

Bioglues are mostly used as surgical glues to stop bleeding and to ensure a fast sutureless wound closing. As seen in the previous sections of this paper, injectable citrate-based adhesives have been developed to promote wound healing and biodegradation without significant inflammation [101]. In another study the poly (γ-glutamic acid) was combined with dopamine to form hydrogels that exhibited 10-fold wet tissue adhesion strength as compared to commercially available fibrin glue [129]. Recombinant MAPs were also used to produce bioglues; they were photo-crosslinked using a dityrosine-based strategy inspired from structural proteins found in insects. The products showed light-controllable gelation and proved to have good adhesive property, tissue regeneration promotion, and less inflammation [128]. These bioadhesives have been employed to coat biological scaffolds used for Achilles tendon or hernia repair. There are two ways to secure a surgical graft, the first is where a MAP-based tissue adhesive is used as a coating to secure surgical graft material and the other way is where the MAPs are combined with PEG and PCL to form films that coat a hernia repair biological mesh. These were further tested for adhesiveness, mechanical strength, and biocompatibility [107]. Mussel-inspired bioadhesive can also be applied for orthopedic surgeries such as treating a fracture; researchers have developed a phosphate-based adhesive intended for fractured bone repair, with promising results [120]. A two-component bone glue was prepared by conjugating L-DOPA onto oxidized dextran and then mixing with chitosan in an aqueous medium. The system cross-linked via Schiff's base formation, which also helped the glue to bind to exposed amino groups in fractured bone. The product displayed a bond strength three times higher than that of fibrin glue and exhibited high biocompatibility [57, 113].

9.4.3.2 *Drug Delivery*

Bioadhesive-based drug delivery systems employ adhesive hydrogels in combination with therapeutic agents to reach specific body regions and deliver drugs for an extended period [59]. A muco-resistant platform based on mussel-inspired adhesive was developed as an oral mucosa drug delivery system. Hydrocaffeic acid (HCA), a carboxyl-terminated catechol, was conjugated onto the backbone of chitosan. The product showed a high potential to serve as a drug delivery carrier and could be applied to medical devices [130]. The regeneration of oral mucosa and healing of oral ulcer was tested in animal models using triamcinolone acetonide oral paste carried in a catecholmodified chitosan (Chi-C) oral patch [130]. The adhesive polysaccharide oral patch called 'Chitoral' could be considered as a versatile drug delivery system for wet adhesion. The results were positive, exhibiting a reconstructed mucosa within 7 days and suggesting the potential use of this material as a mucoadhesive drug delivery system [130]. One of the therapeutic approaches to ulcerative colitis includes the rectal administration of sulfasalazine (SSZ), but its therapeutic efficiency is reduced due to its limited residence time. Researchers have hypothesized that the use of a mussel-inspired bioadhesive hydrogel composed of chitosan functionalized with catechol and loaded with SSZ (SSZ-Cat-Chitosan) might be helpful to overcome this limitation. The material was evaluated in animal models with ulcerative colitis and compared against oral treatment with SSZ, showing an increased delivery and therapeutic efficiency. These results also point to the potential use of enhanced mucoadhesive rectal drug delivery to improve the outcomes of ulcerative colitis, colon cancer, or related diseases [131]. Similarly, catechol-containing tannic acid (TAC[2]) and gelatin were combined by a Michael addition reaction under oxidizing conditions. The product called Gel-TAc was tested for adhesion strength, cytocompatibility, and antimicrobial and antifungal properties. The lap shear strength for Gel-TAc hydrogels crosslinked by silver nitrate was much higher than that of fibrin glue. The antimicrobial and antifungal studies suggest a strong biocidal activity, pointing to a potential use of Gel-TAc for wound closure and drug delivery [132].

Drug delivery system design is taking advantage of mussel-inspired bioadhesives with increased delivery efficiency, extended retention time, and possible reduction in therapeutic doses [57].

9.4.3.3 *Tissue Engineering*

In this method the bioadhesive material should help in tissue regeneration as well as a carrier of biomolecules to induce cell growth. A mussel-inspired hydrogel was prepared by dopamine polymerization followed by complexation with sodium alginate (SA) in polyacrylamide (PAM) networks with a good adhesion strength of 24.5 kPa and also affording cell proliferation, attachment, and spreading [133].

9.4.3.3.1 Bioadhesives in Wound Closure

One of the most widely used applications of bioadhesives is in wound closure. It is utilized as an alternative to other practices of wound closure like plasters, suturing, wires, staples, etc., Bioadhesives are in high demand mainly due to their non invasiveness and ability to lessen pain as well as their healing effects with no formation of scar, inflammation and infection. A bioadhesive should have firm adhesion with good biocompatibility and biodegradability without hindering the natural healing process. In Figure 9.6, the applications of a bioadhesive on the wound A) within the wound edges, B) outside the wound, C) outside and within the wound are depicted in rats after incision. The bioadhesive on the closed wounds after application exhibited complete epidermis and dermis structures and high deposition of collagen than the untreated wounds [134].

The other uses of bioadhesives include A) fixation of hard brittle tissues such as bones which provide quick results especially in the case of small bone fragments, B) as tissue sealants to prevent leakage, C) as an

[2]TAC – Represents TA-C.

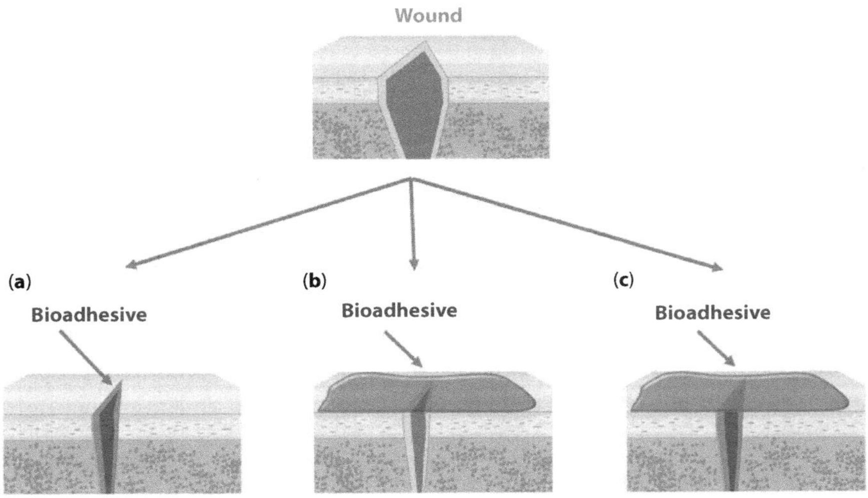

Figure 9.6 Applications of bioadhesives on the wound in rats [134].

immobilizing material that functions as an agent for wound dressing or drug delivery [134].

9.4.3.3.2 Prevention of Scar Formation

A bioadhesive based on recombinant collagen-binding MAPs and glycosaminoglycans was developed and tested. It was targeted to bind type I collagen and to regulate fibrillogenesis. This was tested on animal models and it was observed that the material showed re-epithelialization and neovascularization, as well as accelerated collagen synthesis. This new adhesive showed improved wound healing rate, and effective scar inhibition [135].

9.4.3.3.3 Tissue Repair and Reconstruction

Researchers have attempted to mimic the property of byssus using synthetic chemistry by focusing on the metal-coordination in threads resulting in hydrogels that could self–heal, polymer coatings and even surgical adhesives [136]. A mussel-inspired hydrogel which is enzymatically biodegradable and composed of a DOPA-mimetic modified PEG linked to analanine (Ala) dipeptide substrate was prepared. The material was designed with elastase-sensitive Ala-Ala links to degrade as a response to the enzyme secretion from neutrophils during the inflammatory phase. Rheological analysis showed that the hydrogel was adequate for soft tissue contact with an adhesion strength of 30.4 ± 3.4 kPa. This material was also tried in animal models to conclude that the biodegradation of the product took several months [137].

9.4.3.3.4 Cartilage Regeneration

In this type of tissue engineering, the researchers have focused on cartilage regeneration without the use of growth factors. This study started with the development of a mussel-inspired pDA-chondroitin sulphate-polyacrylamide hydrogel with improved adhesive and mechanical properties [138]. The catechol-enriched hydrogel demonstrated good cell affinity, increased cell adhesion and tissue integration as compared to chondroitin sulphate hydrogel. The derived product showed the necessary mechanical properties needed for an efficient cartilage repair.

9.4.3.4 Cellular Engineering

In cellular engineering, bioadhesives are used for cell division process and for improving cell-surface interactions. The advances in cellular engineering can be further improved by applying mussel-inspired bioadhesives as they have the capacity to attach, protect, and function as well as to control the cellular surroundings.

There are various studies where a mussel-inspired bioadhesive was used in cellular studies, one of which focused on improving the current hyaluronic acid (HA) based adhesive hydrogel for cell therapy. This was accomplished by conjugating dopamine onto HA using ethyl-dimethylaminopropyl-carbodiimide-N-hydroxysuccinimide (EDC-NHS) chemistry. This hydrogel was formed by oxidative crosslinking and was further studied on animal models. From the study it was inferred that the produced hydrogel showed an improved adhesiveness of the cell-containing hydrogel on wet surface. It also exhibited an enhanced cell compatibility when compared with the HA adhesive hydrogel, hence showed feasibility of this tissue adhesive for minimally invasive cell therapy. The study also concluded that the hydrogel produced could be used in paint materials [80, 114].

In another case, researchers have investigated the feasibility of protecting living cells and controlling their division process, this was carried out by encapsulating cells with polydiacetylene (PDA) [5]. In Figure 9.7, yeast cells are used as a model of encapsulation, which was found to be an efficient way not only for protecting and stabilizing the cells but also for fabricating artificial spores. It was demonstrated that this technique provided cells with resistance against harsh environments while allowing control of

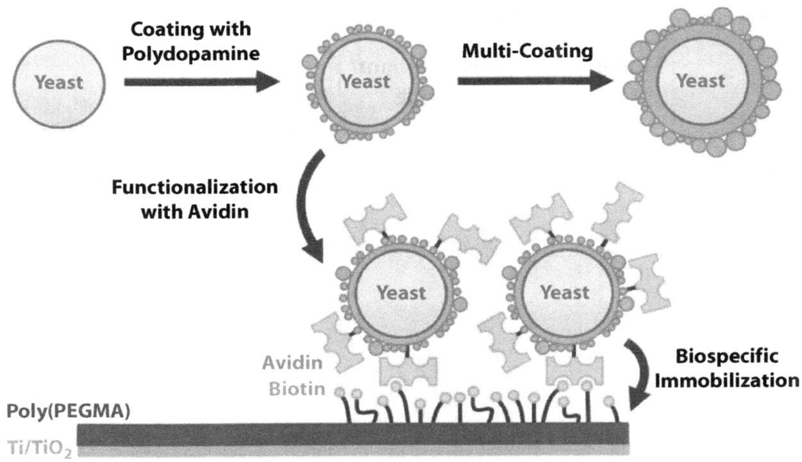

Figure 9.7 Representation of yeast cell encapsulation for protecting living cells and controlling their division process [5].

the cell cycle; and due to the catecholic nature of the encapsulating agent, the cell surfaces can be easily functionalized.

9.5 Summary

This chapter provides an understanding of bioadhesives, their history and also the current trends. Adhesives are materials that have always been in the list of essential commodities and the discovery of the mussel adhesive property paved the way for the development of bioadhesives that are environmentally friendly. Mussel adhesion is mainly due to the mfps (mussel foot proteins) which are composed of DOPA (3, 4-dihydroxyphenylalanine), but the various properties of mfps are yet to be studied. Some of the materials obtained through mussel-inspired proteins are listed in Table 9.4 [135]. Currently the focus in on the uses of the bioadhesives that are developed on the basis of mussel adhesion process, one of the main applications of these is in the medical field. Keeping the catechol groups as the base, these bioadhesives are modified appropriately to meet the needs in wound closures, drug delivery, tissue growth and reconstruction, protecting living cells and in inhibiting cell growth. The medical bioadhesives have shown a higher adhesion strength than the synthetic ones. Future studies are in progress to evaluate the efficacy of these adhesives in bone related therapies.

Table 9.4 Materials obtained through mussel-inspired proteins [135].

	Mussel thread protein	Recombinant/ peptide version	Resulting materials	Properties
Purified Thread Proteins	PreCols	NA	Hand-drawn Fiber	Similar structure and properties to native threads. E = 71 MPa; σ_{ult} = 40 MPa; ε_{ult} = 107%
	Byssal thread hydrolysate	NA	Water-insoluble films	pH-tunability; dye removal. E = 10 MPa; σ_{ult} = 2.5 MPa
	mfp-3	NS	Self-Coacervate	Dependent on pH and ionic strength
Recombinant thread proteins	PreCol-D	PreCol-D	Nanofibrils	Defined secondary structure, presence of triple helices
	Ptmp-1	Ptmp-1	Adhesive	Enhancing mouse osteoblast cell adhesion, collagen interaction
	mfp-1	rfp-1	Surface coating	Fe^{3+}mediated bridging. W_{ad} = 5.7 mJ/m^2

(Continued)

Table 9.4 Materials obtained through mussel-inspired proteins [135]. (*Continued*)

	Mussel thread protein	Recombinant/ peptide version	Resulting materials	Properties
	mfp-1	rfp-1	Adhesive Hydrogel	Flexible viscoelastic and self-healing behavior
	mfp-1	rfp-1	Polymeric nanoparticles	pH-responsive anti-breast-cancer drug release
	mfp-1	rfp-1	Coacervate	Cation-π stabilized. $W_{ad} = 4.8$ mJ/m^2
	mfp-1	rfp-1	Self-Coacervate	Salt-triggered. $W_{ad} = 5.0$ mJ/m^2
	mfp-5	rfp-5	Adhesive	Suitable for cell biological applications
	mfp-5	rfp-5	Adhesive	Immobilization of antibodies
	mfp-1, mfp-5	fp-151	Adhesive	Greater production yields
	mfp-1, mfp-5	fp-151	Adhesive	improved purification
	mfp-1, mfp-5	fp-151	Composite Nanofibers	Scaffold for bioactive molecules. $E = 44$ MPa; $\sigma_{ult} = 18$ MPa; $\varepsilon_{ult} = 107$ %

(*Continued*)

Table 9.4 Materials obtained through mussel-inspired proteins [135]. (*Continued*)

	Mussel thread protein	Recombinant/ peptide version	Resulting materials	Properties
	mfp-1, mfp-5	fp-151-RGD	Adhesive	Superior cell-adhesion and spreading abilities
	mfp-1, Mfp-5	fp-151-peptide	Adhesive	Enhanced adhesion, proliferation and spreading of mammalian cells
	mfp-1, mfp-3, mfp-5	Fp-131, fp-151	Complex Coacervate	Enhanced adhesion strength
Mussel-Inspired Peptides	PreCol-D	HRD-DN	Free-standing films	Metal-tunable architecture and mechanics. E = 3.5 MPa (w/o metals; 23 MPa (w/ Zn2+)
	mfp-3	mfp-35-pep	Coacervate adhesive	Improved adsorption kinetics and quantity
	mfp-5	PEP-pl-4, PEP-pl-6.5, PEP-pl-10, PEP-pl-4-dimer	Adhesive	Enhanced adhesion via amino acid synergy

E = stiffness; σ_{ult} = ultimate tensile strength; ε_{ult} = ultimate tensile extensibility; W_{ad} = work of adhesion.

HRD-DN = HRD peptides based on the N-terminal histidine-rich domain of preCol-D

fp = foot protein; mfp- mussel foot protein.

rfp = recombinant foot protein.

References

1. J. R. Burkett, J. L. Wojtas, J. L. Cloud, and J. J. Wilker. A method for measuring the adhesion strength of marine mussels. *J Adhesion*, 85, 601-615 (2009).
2. J. K. Román and J. J. Wilker. Cooking chemistry transforms proteins into high-strength adhesives. *J. Am. Chem. Soc.* 141, 1359-1365 (2018).
3. Q. Lu, E. Danner, J. H. Waite, J. N. Israelachvili, H. Zeng, and D. S. Hwang, Adhesion of mussel foot proteins to different substrate surfaces, *J Royal Soc Interface*, 10, 1–11 (2013).
4. H. Lee, S. M. Dellatore, W. M. Miller, and P. B. Messersmith, Mussel-inspired surface chemistry for multifunctional coatings, *Science*, 318, 426–430 (2007).
5. N. K. Kaushik, N. Kaushik, S. Pardeshi, J. G. Sharma, S. H. Lee, and E. H. Choi, Biomedical and clinical importance of mussel-inspired polymers and materials, *Marine Drugs*, 13, 6792–6817 (2015).
6. B. P. Lee, P. B. Messersmith, J. N. Israelachvili, and J. H. Waite, Mussel-inspired adhesives and coatings, *Annual Rev Mater Res*, 41, 99–132 (2011).
7. H. G. Silverman and F. F. Roberto, Understanding marine mussel adhesion. *Marine Biotechnol*, 9, 661-681 (2007).
8. S. A. Mian and Y. Khan, The adhesion mechanism of marine mussel foot protein: Adsorption of L-Dopa on α-and β-Cristobalite silica using density functional theory. *Journal of Chemistry*, Article ID 8756519, 1-6 (2017).
9. Y. Shao, L.F. Molnar, Y. Jung, J. Kussmann, C. Ochsenfeld, S.T. Brown, A.T. Gilbert, L.V. Slipchenko, S.V. Levchenko, D.P. O'Neill and R.A. DiStasio Jr., Advances in methods and algorithms in a modern quantum chemistry program package. *PhysChemChemPhys*, 8, 3172-3191 (2006).
10. B. P. Lee, J. L. Dalsin and P. B. Messersmith, Biomimetic adhesive polymers based on mussel adhesive proteins. *In Biological Adhesives*. A. M. Smith and J. A. Callow (Eds.) pp. 257-278, Springer, Berlin (2006).
11. H. Lee, N. F. Scherer and P. B. Messersmith, Single-molecule mechanics of mussel adhesion. *Proc Natl Acad Sci USA*, 103, 12999-13003 (2006).
12. A. Cholewinski, F. K. Yang and B. Zhao, Algae–mussel-inspired hydrogel composite glue for underwater bonding. *Materials Horizons*, 6, 285-293 (2019).
13. T. J. Deming, Mussel byssus and biomolecular materials. *Current Opinion Chem Biology*, 3, 100-105 (1999).
14. T. Priemel, E. Degtyar, M. N. Dean and M. J. Harrington, Rapid self-assembly of complex biomolecular architectures during mussel byssus biofabrication. *Nature Commun*, 8, 1-12 (2017).
15. C. H. Brown, Some structural proteins of *Mytilus edulis*. J Cell Sci 3, 487-502 (1952).
16. R. D. Lillie, Further experiments with the Masson trichrome modification of Mallory's connective tissue stain. *Stain Technol* 15, 17-22 (1940).

17. D. Miki, Y. Takeuchi, K. Inoue, and S. Odo, Expression sites of two byssal protein genes of *Mytilus galloprovincialis*. *Biological Bulletin* 190, 213-217 (1996).
18. A. Tamarin and P. J. Keller. An ultrastructutal study of the byssal thread forming system in Mytilus. *J Ultrastructure Res* 40, 401-416 (1972).
19. J. H. Waite. Mussel adhesion–essential footwork. *J Expl. Biology,* 220, 517-530 (2017).
20. A. Tamarin, P. Lewis and J. Askey. The structure and formation of the byssus attachment plaque in Mytilus. *Morphology* 149, 199-221 (1976).
21. J. H. Waite. The formation of mussel byssus: anatomy of a natural manufacturing process. *Structure, Cellular Synthesis and Assembly of bioPolymers* 19, 27-54 (1992).
22. N. Holten-Andersen and J. H. Waite. Mussel-designed protective coatings for compliant substrates. *J Dental Res,* 87, 701-709 (2008).
23. C.V. Benedict and J.H. Waite, Composition and ultrastructure of the byssus of Mytilusedulis. *J Morphology*, 189, 261-270 (1986).
24. J. H. Waite, X. X. Qin, and K. J. Coyne. The peculiar collagens of mussel byssus. *Matrix Biology* 17, 93-106 (1998).
25. J. H. Waite, E. Vaccaro, C. Sun, and J. M. Lucas. Elastomeric gradients: a hedge against stress concentration in marine holdfasts? *Philos Trans Royal Soc London. Series B:*357, 143-153 (2002).
26. E. Bell and J. Gosline. Mechanical design of mussel byssus: material yield enhances attachment strength. *J Expl Biology* 199, 1005-1017 (1996).
27. L. V. Zuccarello, The collagen gland of *Mytilus galloprovincialis*: an ultrastructural and cytochemical study on secretory granules. *J Ultrastructure Res.* 73, 135-147 (1980).
28. X. Qin and J. H. Waite. Exotic collagen gradients in the byssus of the mussel *Mytilusedulis. J Expl. Biology* 198, 633-644 (1995).
29. K. Pathak and R. Malviya. Introduction, theories and mechanisms of bioadhesion, in:*Bioadhesives in Drug Delivery*, K.L. Mittal, I. S. Bakshi and J. K. Narang (Eds.), pp. 1-27, Wiley-Scrivener, Beverly, MA (2020).
30. A. Hagenau, P. Papadopoulos, F. Kremer and T. Scheibel. Mussel collagen molecules with silk-like domains as load-bearing elements in distal byssal threads. *J Structural Biology* 175, 339-347 (2011).
31. A. A. Arnold, F. Byette, M. O. Séguin-Heine, A. LeBlanc, L. Sleno, R. Tremblay, C. Pellerin, and I. Marcotte. Solid-state NMR structure determination of whole anchoring threads from the blue mussel *Mytilusedulis. Biomacromolecules* 14, 132-141 (2013).
32. M. Heim, M. B. Elsner and T. Scheibel. Lipid-specific β-sheet formation in a mussel byssus protein domain. *Biomacromolecules* 14, 3238-3245 (2013).
33. X. X. Qin, K. J. Coyne and J. H. Waite. Tough tendons: mussel byssus has collagen with silk-like domains. *J Biological Chem* 272, 32623-32627 (1997).
34. K. J. Coyne, X. X. Qin and J. H. Waite. Extensible collagen in mussel byssus: a natural block copolymer. *Science* 277, 1830-1832 (1997).

35. X. X. Qin and J. H. Waite. A potential mediator of collagenous block copolymer gradients in mussel byssal threads. *Proc Natl Acad Sci USA* 95, 10517-10522 (1998).
36. J. Sagert and J. H. Waite. Hyper unstable matrix proteins in the byssus of *Mytilusgalloprovincialis. J Expl Biology* 212, 2224-2236 (2009).
37. C. Sun, J. M. Lucas and J. H. Waite. Collagen-binding matrix proteins from elastomeric extraorganismicbyssal fibers. *Biomacromolecules* 3, 1240-1248 (2002).
38. J. H. Waite Evidence for a repeating 3, 4-dihydroxyphenylalanine-and hydroxyproline-containing decapeptide in the adhesive protein of the mussel, *Mytilusedulis L. J Biological Chem* 258, 2911-2915 (1983).
39. R. L. Strausberg and P. R. Link. Protein-based medical adhesives. *Trends Biotechnology* 8, 53-57 (1990).
40. A. H. Hofman, I. A. Van Hees, J. Yang, and M. Kamperman. Bioinspired underwater adhesives by using the supramolecular toolbox. *Advanced Mater* 30, 1704640 (2018).
41. L. M. Rzepecki, K. M. Hansen and J. H. Waite. Characterization of a cystine-rich polyphenolic protein family from the blue mussel *Mytilusedulis* L. *Biological Bulletin* 183, 123-137 (1992).
42. K. Inoue, Y. Takeuchi, D. Miki, and S. Odo. Mussel adhesive plaque protein gene is a novel member of epidermal growth factor-like gene family. *J Biological Chem* 270, 6698-6701 (1995).
43. S. C. Warner and J. H. Waite. Expression of multiple forms of an adhesive plaque protein in an individual mussel, *Mytilusedulis. Marine Biology* 134, 729-734 (1999).
44. H. Zhao and J. H. Waite. Proteins in load-bearing junctions: the histidine-rich metal-binding protein of mussel byssus. *Biochemistry* 45, 14223-14231 (2006).
45. N. Bandara, H. Zeng and J. Wu. Marine mussel adhesion: biochemistry, mechanisms, and biomimetics. *J Adhesion Sci Technol.* 27, 2139-2162 (2013).
46. H. Zhao and J. H. Waite. Linking adhesive and structural proteins in the attachment plaque of *Mytiluscalifornianus. J Biological Chem* 281, 26150-26158 (2006).
47. J. Yu, W. Wei, E. Danner, R. K. Ashley, J.N. Israelachvili, and J. H. Waite. Mussel protein adhesion depends on interproteinthiol-mediated redox modulation. *Nature Chem Biology* 7, 588-590 (2011).
48. S. C. T. Nicklisch, S. Das, N.R. Martinez Rodriguez, J. H. Waite, and J. N. Israelachvili. Antioxidant efficacy and adhesion rescue by a recombinant mussel foot protein-6. *Biotechnology Progress* 29, 1587-1593 (2013).
49. J. Wang and T. Scheibel. Recombinant production of mussel byssus inspired proteins. *Biotechnology J.* 13, 1800146 (2018).
50. E.W. Danner, Y. Kan, M.U. Hammer, J.N. Israelachvili, and J.H. Waite, Adhesion of mussel foot protein Mefp-5 to mica: an underwater superglue. *Biochemistry* 51, 6511-6518 (2012).

51. M. A. Even, J. Wang and Z. Chen. Structural information of mussel adhesive protein Mefp-3 acquired at various polymer/Mefp-3 solution interfaces. *Langmuir* 24, 5795-5801 (2008).
52. B. K. Ahn, Perspectives on mussel-inspired wet adhesion. *J. Amer. Chem. Soc* 139, 10166-10171 (2017).
53. J. H. Waite and M. L. Tanzer. Polyphenolic substance of *Mytilusedulis*: novel adhesive containing L-DOPA and hydroxyproline. *Science* 212, 1038-1040 (1981).
54. L. Li and H. Zeng. Marine mussel adhesion and bio-inspired wet adhesives. *Biotribology* 5, 44-51 (2016).
55. S. Basak, Co-evolving with nature: the recent trends on the mussel-inspired polymers in medical adhesion. *Biotechnology Bioprocess Eng.* 26, 10-24 (2021).
56. K. L. Mittal, The role of the interface in adhesion phenomena. *Polym Eng. Sci* 17, 467-473 (1977).
57. N. Pandey, L. F. Soto-Garcia, J. Liao, P. Zimmern, K. T. Nguyen, and Y. Hong. Mussel-inspired bioadhesives in healthcare: design parameters, current trends, and future perspectives. *Biomaterials Sci.* 8, 1240-1255 (2020).
58. S. Khanlari and M. A. Dubé. Bioadhesives: a review. *Macromolecular Reaction Eng.* 7, 573-587 (2013).
59. K. L. Mittal, I. S. Bakshi and J. K. Narang (Eds.). *Bioadhesives in Drug Delivery*. Wiley-Scrivener, Beverly, MA (2020).
60. A. P. Duarte, J. F. Coelho, J. C. Bordado, M. T. Cidade, and M. H. Gil. Surgical adhesives: Systematic review of the main types and development forecast. *ProgPolymSci.* 37, 1031-1050 (2012).
61. T. B. Reece, T. S. Maxey and I. L. Kron. A prospectus on tissue adhesives. *AmerJ Surgery* 182, S40-S44 (2001).
62. P. Alves, P. Ferreira, and M. H. Gil. Biomedical polyurethanes-based materials. Polyurethane: properties, structure and applications, in; *PolymerSci and Technology*, L. I. Cavaco and J.A.Melo (Eds.), pp. 25-50, Nova Science Publishers, New York, (2012).
63. M. Mehdizadeh and J. Yang. Design strategies and applications of tissue bioadhesives. *Macromolecular Biosci.* 13, 271-288 (2013).
64. G. F. Hardy., A review of adhesion and adhesives: Science and technology *J. Adhesion* 28, 199-200 (1989).
65. S. Ebnesajjad and A. H. Landrock. *Adhesives Technology Handbook*, third edition, pp. 1-18. William Andrew, Norwich, NY (2014).
66. A. Ahuja, R. K. Khar and J. Ali. Mucoadhesive drug delivery systems. *Drug Development Industrial Pharmacy* 23, 489-515 (1997).
67. S. J. Marshal, S. C. Bayne, R. Baier, A. P. Tomsia, and G. W. Marshall. A review of adhesion science. *Dental Mater.* 26, e11-e16 (2010).
68. K. Modaresifar, S. Azizian, and A. Hadjizadeh. Nano/biomimetic tissue adhesives development: from research to clinical application. *Polym Reviews* 56, 329-361 (2016).

69. J. Yan, L. Yang, M. F. Lin, J. Ma, X. Lu, and P.S. Lee. Polydopamine spheres as active templates for convenient synthesis of various nanostructures. *Small* 9, 596-603 (2013).
70. B. Yang, N. Ayyadurai, H. Yun, Y.S. Choi, B.H. Hwang, J. Huang, Q. Lu, H. Zeng, and H.J. Cha. *In vivo* residue-specific DOPA-incorporated engineered mussel bioglue with enhanced adhesion and water resistance. *Angew. Chemie* 126, 13578-13582 (2014).
71. F. P. Kord and B. P. Lee. Recent approaches in designing bioadhesive materials inspired by mussel adhesive protein. *J Polym Sci Part A: Polym Chem.* 55, 9-33 (2017).
72. J. P. Gallivan and D. A. Dougherty. A computational study of cation– π interactions vs salt bridges in aqueous media: implications for protein engineering. *J Amer Chem Soc.* 122, 870-874 (2000).
73. Q. Lu, D. X. Oh, Y. Lee, Y. Jho, D. S. Hwang and H. Zeng. Nanomechanics of cation–π interactions in aqueous solution. *Angew Chemie 125*, 4036-4040 (2013).
74. B. P. Lee, C. Y. Chao, F. N. Nunalee, E. Motan, K. R. Shull, and P. B. Messersmith. Rapid gel formation and adhesion in photocurable and biodegradable block copolymers with high DOPA content. *Macromolecules* 39, 1740-1748 (2006).
75. H. Lee, N. F. Scherer and P. B. Messersmith. Single-molecule mechanics of mussel adhesion. *Proc Natl Acad Sci.* USA 103, 12999-13003 (2006).
76. J. Yu, W. Wei, M. S. Menyo, A. Masic, J. H. Waite and J. N. Israelachvili. Adhesion of mussel foot protein-3 to TiO2 surfaces: the effect of pH. *Biomacromolecules* 14, 1072-1077 (2013).
77. V. Bhagat and M. L. Becker. Degradable adhesives for surgery and tissue engineering. *Biomacromolecules* 18, 3009-3039 (2017).
78. S. Seo, S. Das, P. J. Zalicki, R. Mirshafian, C. D. Eisenbach, J. N. Israelachvili, J. H. Waite, and B. K. Ahn. Microphase behavior and enhanced wet-cohesion of synthetic copolyampholytes inspired by a mussel foot protein. *J Amer Chem Soc.* 137, 9214-9217 (2015).
79. W. Wei, J. Yu, C. Broomell, J. N. Israelachvili, and J. H. Waite. Hydrophobic enhancement of DOPA-mediated adhesion in a mussel foot protein. *J Amer Chem Soc* 135, 377-383 (2013).
80. S. K. Seidlits, Z. Z. Khaing, R. R. Petersen, J. D. Nickels, J. E. Vanscoy, J. B. Shear and C. E. Schmidt. The effects of hyaluronic acid hydrogels with tunable mechanical properties on neural progenitor cell differentiation. *Biomaterials* 31, 3930-3940 (2010).
81. W. Wei, Y. Tan, N. R. M. Rodriguez, J. Yu, J. N. Israelachvili, and J. H. Waite. A mussel-derived one component adhesive coacervate. *Acta Biomaterialia* 10, 1663-1670 (2014).
82. G. P. Maier, M. V. Rapp, J. H. Waite, J. N. Israelachvili, and A. Butler. Adaptive synergy between catechol and lysine promotes wet adhesion by surface salt displacement. *Science* 349, 628-632 (2015).

83. Y. Akdogan, W. Wei, K. Y. Huang, Y. Kageyama, E. W. Danner, D. R. Miller, N. R. M. Rodriguez, J. H. Waite and S. Han. Intrinsic surface-drying properties of bioadhesive proteins. *Angew Chemie* 126, 11435-11438 (2014).
84. L. M. Cox, J. P. Killgore, Z. Li, R. Long, A. W. Sanders, J. Xiao, and Y. Ding. Influences of substrate adhesion and particle size on the shape memory effect of polystyrene particles. *Langmuir* 32, 3691-3698 (2016).
85. M. Cencer, M. Murley, Y. Liu, and B. P. Lee. Effect of nitro-functionalization on the cross-linking and bioadhesion of biomimetic adhesive moiety. *Biomacromolecules* 16, 404-410 (2015).
86. E. Amstad, A. U. Gehring, H. Fischer, V. V. Nagaiyanallur, G. Hähner, M. Textor and E. Reimhult. Influence of electronegative substituents on the binding affinity of catechol-derived anchors to Fe3O4 nanoparticles. *J. Phys Chem C* 115, 683-691 (2011).
87. X. Ding, G. K. Vegesna, H. Meng, A. Winter and B. P. Lee. Nitro-group functionalization of dopamine and its contribution to the viscoelastic properties of catechol-containing nanocomposite hydrogels. *Macromolecular Chem Phys* 216, 1109-1119 (2015).
88. E. Amstad, T. Gillich, I. Bilecka, M. Textor, and E. Reimhult. Ultrastable iron oxide nanoparticle colloidal suspensions using dispersants with catechol-derived anchor groups. *Nano Letters* 9, 4042-4048 (2009).
89. Z. Shafiq, J. Cui, L. P. Pérez, V. S. Miguel, R. A. Gropeanu, C. Serrano and A. Campo. Bioinspired underwater bonding and debonding on demand. *Angew Chemie* 124, 4408-4411 (2012).
90. M. Yu and T. J. Deming. Synthetic polypeptide mimics of marine adhesives. *Macromolecules* 31, 4739-4745 (1998).
91. M. Kitamura, K. Kawakami, N. Nakamura, K. Tsumoto, H. Uchiyama, Y. Ueda, I. Kumagai and T. Nakaya. Expression of a model peptide of a marine mussel adhesive protein in *Escherichia coli* and characterization of its structural and functional properties. *J PolymSciPart A: Polym Chem.* 37, 729-736 (1999).
92. D. S. Hwang, Y. Gim, H. J. Yoo and H. J. Cha. Practical recombinant hybrid mussel bioadhesive fp-151. *Biomaterials* 28, 3560-3568 (2007).
93. J. H. Waite, D. C. Hansen and K. T. Little. The glue protein of ribbed mussels (*Geukensiademissa*): a natural adhesive with some features of collagen. *J. Comparative Physiology B* 159, 517-525 (1989).
94. R. Santonocito, F. Venturella, D. F. Piaz, M.A. Morando, A. Provenzano, E. Rao, M.A. Costa, D. Bulone, P.L. San Biagio, D. Giacomazza and A. Sicorello, Recombinant mussel protein Pvfp-5β: A potential tissue bioadhesive. *J. Biological Chem.* 294, 12826-12835 (2019).
95. H. J. Cha, D. S. Hwang and S. Lim. Development of bioadhesives from marine mussels. *Biotechnology J: Healthcare Nutrition Technology* 3, 631-638 (2008).
96. D. R. Filpula, M.L. Shwu, R. P. Link, S. L. Strausberg and R. L. Strausberg. Structural and functional repetition in a marine mussel adhesive protein. *Biotechnol Prog* 6, 171-177 (1990).

97. Y. K. Jo, H. J. Kim, Y. Jeong, K. I. Joo and H. J. Cha. Biomimetic surface engineering of biomaterials by using recombinant mussel adhesive proteins. *Advanced Mater. Interfaces* 5, 1800068 (2018).
98. D. S. Hwang, H. J. Yoo, J. H. Jun, W. K. Moon and H. J. Cha. Expression of functional recombinant mussel adhesive protein Mgfp-5 in *Escherichia coli. Appl Environmental Microbiology* 70, 3352-3359 (2004).
99. D. S. Hwang, Y. Gim and H. J. Cha. Expression of functional recombinant mussel adhesive protein type 3A in *Escherichia coli. BiotechnolProg* 21, 965-970 (2005).
100. K. Huang, B. P. Lee, D. R. Ingram and P. B. Messersmith. Synthesis and characterization of self-assembling block copolymers containing bioadhesive end groups. *Biomacromolecules* 3, 397-406 (2002).
101. M. Mehdizadeh, H. Weng, D. Gyawali, L. Tang and J. Yang. Injectable citrate-based mussel-inspired tissue bioadhesives with high wet strength for sutureless wound closure. *Biomaterials* 33, 7972-7983 (2012).
102. S. A. Burke, J. M. Ritter, B. P. Lee and P. B. Messersmith, Thermal gelation and tissue adhesion of biomimetic hydrogels. *Biomedical Materials*, 2, 203-210 (2007).
103. J. Feng, X. A. Ton, S. Zhao, J. I. Paez and A. Del Campo, Mechanically reinforced catechol-containing hydrogels with improved tissue gluing performance. *Biomimetics*, 2, 23-38 (2017).
104. J. Guo, G. B. Kim, D. Shan, J. P. Kim, J. Hu, W. Wang, F. G. Hamad, G. Qian, E. B. Rizk and J. Yang. Click chemistry improved wet adhesion strength of mussel-inspired citrate-based antimicrobial bioadhesives. *Biomaterials* 112, 275-286 (2017).
105. M. Brodie, L. Vollenweider, J. L. Murphy, F. Xu, A. Lyman, W. D. Lew, and B. P. Lee. Biomechanical properties of Achilles tendon repair augmented with a bioadhesive-coated scaffold. *Biomedical Materials* 6, 015014 (2011).
106. D. G. Barrett, G. G. Bushnell and P. B. Messersmith. Mechanically robust, negative-swelling, mussel-inspired tissue adhesives. *Advanced Healthcare Materials* 2, 745-755 (2013).
107. J. L. Murphy, L. Vollenweider, F. Xu and B. P. Lee. Adhesive performance of biomimetic adhesive-coated biologic scaffolds. *Biomacromolecules* 11, 2976-2984 (2010).
108. Y. Shi, P. Zhou, V. Jérôme, R. Freitag and S. Agarwal. Enzymatically degradable polyester-based adhesives. *ACS Biomater. Sci. Eng.* 1, 971-977 (2015).
109. J. H. Ryu, H. Seonki and H. Lee. Bio-inspired adhesive catechol-conjugated chitosan for biomedical applications: A mini review. *Acta Biomaterialia* 27, 101-115 (2015).
110. L. Foster, R. John and E. Karsten. A chitosan based, laser activated thin film surgical adhesive,' SurgiLux': preparation and demonstration. *J Visualized Experiments* 68, e3527–e3534(2012).
111. A. I. Neto, A. C. Cibrão, C. R. Correia, R. R. Carvalho, G. M. Luz, G. G. Ferrer, G. Botelho, C. Picart, N. M. Alves and J. F. Mano. Nanostructured

polymeric coatings based on chitosan and dopamine-modified hyaluronic acid for biomedical applications. *Small* 10, 2459-2469 (2014).
112. K. Yamada, T. Chen, G. Kumar, O. Vesnovsky, LD. T. Topoleski, and G. F. Payne. Chitosan based water-resistant adhesive. *Analogy to mussel glue. Biomacromolecules* 1, 252-258 (2000).
113. S. Hong, K. Yang, B. Kang, C. Lee, I.T. Song, E. Byun, K.I. Park, S.W. Cho and H. Lee. Hyaluronic acid catechol: a biopolymer exhibiting a pH-dependent adhesive or cohesive property for human neural stem cell engineering. *Advanced Functional Materials* 23, 1774-1780 (2013).
114. J. Shin, J. S. Lee, C. Lee, H. J. Park, K. Yang, Y. Jin, J. H. Ryu, K. S. Hong, S. H. Moon, H. M. Chung and H. S. Yang. Tissue adhesive catechol-modified hyaluronic acid hydrogel for effective, minimally invasive cell therapy. *Advanced Functional Materials* 25, 3814-3824 (2015).
115. C. Lee, J. Shin, J. S. Lee, E. Byun, J. H. Ryu, S. H. Um, D. I. Kim, H. Lee and S. W. Cho. Bioinspired, calcium-free alginate hydrogels with tunable physical and mechanical properties and improved biocompatibility. *Biomacromolecules* 14, 2004-2013 (2013).
116. J. Y. Park, J. Yeom, J. S. Kim, M. Lee, H. Lee and Y. S. Nam. Cell-repellant dextran coatings of porous titania using mussel adhesion chemistry. *Macromolecular Biosci.*13, 1511-1519 (2013).
117. B. Hoffmann, E. Volkmer, A. Kokott, P. Augat, M. Ohnmacht, N. Sedlmayr, M. Schieker, L. Claes, W. Mutschler and G. Ziegler. Characterisation of a new bioadhesive system based on polysaccharides with the potential to be used as bone glue. *J MaterSci: Mater Medicine* 20, 2001-2009 (2009).
118. C. Fan, J. Fu, W. Zhu and D. A. Wang. A mussel-inspired double-crosslinked tissue adhesive intended for internal medical use. *Acta Biomaterialia* 33, 51-63 (2016).
119. A. Pechey, C. N. Elwood, G. R. Wignall, J. L. Dalsin, B. P. Lee, M. Vanjecek, I. Welch, R. Ko, H. Razvi and P. A. Cadieux. Anti-adhesive coating and clearance of device associated uropathogenic *Escherichia coli* cystitis. J. Urology 182, 1628-1636 (2009).
120. H. Shao, K. N. Bachus and R. J. Stewart. A water-borne adhesive modeled after the sandcastle glue of *P. californica. Macromolecular Biosci.* 9, 464-471 (2009).
121. H. Lee, B. P. Lee and P. B. Messersmith. A reversible wet/dry adhesive inspired by mussels and geckos. *Nature* 448, 338-341 (2007).
122. S. Skelton, M. Bostwick, K. O'Connor, S. Konst, S. Casey and B. P. Lee. Biomimetic adhesive containing nanocomposite hydrogel with enhanced materials properties. *Soft Matter* 9, 3825-3833 (2013).
123. G. Westwood, T. N. Horton and J. J. Wilker. Simplified polymer mimics of cross-linking adhesive proteins. Macromolecules 40, 3960-3964 (2007).
124. P. Matos, R. Cristina, J. D. White and J. J. Wilker. Polymer composition and substrate influences on the adhesive bonding of a biomimetic, cross-linking polymer. *J Amer Chem Soc.* 134, 9498-9505 (2012).

125. H. Tatehata, A. Mochizuki, T. Kawashima, S. Yamashita and H. Yamamoto. Model polypeptide of mussel adhesive protein. I. Synthesis and adhesive studies of sequential polypeptides (X-Tyr-Lys) n and (Y-Lys) n. *J Appl Polym Sci.* 76, 929-937 (2000).
126. J. Wang, C. Liu, X. Lu and M. Yin. Co-polypeptides of 3, 4-dihydroxyphenylalanine and L-lysine to mimic marine adhesive protein. *Biomaterials* 28, 3456-3468 (2007).
127. B. J. Kim, D. X. Oh, S. Kim, J. H. Seo, D. S. Hwang, A. Masic, D. K. Han and H. J. Cha. Mussel-mimetic protein-based adhesive hydrogel. *Biomacromolecules* 15, 1579-1585 (2014).
128. E. Y. Jeon, B. H. Hwang, Y. J. Yang, B. J. Kim, B. H. Choi, G. Y. Jung and H. J. Cha. Rapidly light-activated surgical protein glue inspired by mussel adhesion and insect structural crosslinking. *Biomaterials* 67, 11-19 (2015).
129. W. Chen, R. Wang, T. Xu, X. Ma, Z. Yao, B. O. Chi and H.A. Xu. Mussel-inspired poly (γ-glutamic acid) tissue adhesive with high wet strength for wound closure. *J. Mater. Chem. B* 5, 5668-5678 (2017).
130. J. H. Ryu, J. S. Choi, E. Park, M. R. Eom, S. Jo, M. S. Lee, S. K. Kwon and H. Lee. Chitosan oral patches inspired by mussel adhesion. *J. Controlled Release* 317, 57-66 (2020).
131. J. Xu, M.Tam, S.Samaei, S. Lerouge, J. Barralet, M. M. Stevenson and M. Cerruti. Mucoadhesive chitosan hydrogels as rectal drug delivery vessels to treat ulcerative colitis. *Acta Biomaterialia* 48, 247-257 (2017).
132. J. Guo, W. Sun, J.P. Kim, X. Lu, Q. Li, M. Lin, O. Mrowczynski, E.B. Rizk, J. Cheng, G. Qian and J. Yang, Development of tannin-inspired antimicrobial bioadhesives. *Acta Biomaterialia* 72, 35-44 (2018).
133. M. Suneetha, K. M. Rao and S. S. Han. Mussel-inspired cell/tissue-adhesive, hemostatic hydrogels for tissue engineering applications. *ACS Omega* 4, 12647-12656 (2019).
134. W. Duan, X. Bian and Y. Bu. Applications of bioadhesives: a mini review. *Frontiers Bioeng Biotechnol* 9, 1-12 (2021).
135. E. Y. Jeon, C. H. Bong, D. Jung, B. H. Hwang and H. J. Cha. Natural healing-inspired collagen-targeting surgical protein glue for accelerated scarless skin regeneration. *Biomaterials* 134, 154-165 (2017).
136. M. J. Harrington, F. Jehle and T. Priemel, Mussel byssus structure-function and fabrication as inspiration for biotechnological production of advanced materials. *Biotechnology J.*, 13(12), (2018).
137. C. E. Brubaker and P. B. Messersmith. Enzymatically degradable mussel-inspired adhesive hydrogel. *Biomacromolecules* 12, 4326-4334 (2011).
138. L. Han, M. Wang, P. Li, D. Gan, L. Yan, J. Xu, K. Wang, L. Fang, C.W. Chan, H. Zhang and H. Yuan, Mussel-inspired tissue-adhesive hydrogel based on the polydopamine–chondroitin sulfate complex for growth-factor-free cartilage regeneration. *ACS Appl Mater Interfaces* 10, 28015-28026 (2018).

10

Dental Adhesives: State-of-the-Art, Current Perspectives, and Promising Applications

Lamia Sami Mokeem[1], Isadora Martini Garcia[2], Abdulrahman A. Balhaddad[3], Kevin Cline[2], Gabriel Rakovsky[2], Fabrício Mezzomo Collares[4] and Mary Anne S. Melo[1,2*]

[1]*Ph.D. Program in Biomedical Sciences, University of Maryland School of Dentistry, Baltimore, MD, USA*
[2]*Operative Dentistry Division, General Dentistry Department, University of Maryland School of Dentistry, Baltimore, MD, USA*
[3]*Department of Restorative Dental Sciences, College of Dentistry, Imam Abdulrahman Bin Faisal University, Dammam, Saudi Arabia*
[4]*Dental Materials Department, School of Dentistry, Federal University of Rio Grande do Sul, Porto Alegre, RS, Brazil*

Abstract

Secondary caries and material fractures have been recognized as the most common causes of the failure of tooth-colored restorations and the tooth-restoration bonding interface in dental practice. The imperfectly bonded tooth-restoration interface is susceptible to hydrolytic degradation and bacterial acid attack, severely jeopardizing bonded restorations longevity. This chapter covers the development and current status of dental adhesives and highlights the significant attempts to improve this class of dental materials properties and performance. In addition, we focus on the ongoing investigations of antibacterial compounds incorporated into dental adhesives, stressing the advantages of bioactive properties and how they impact clinical applications in restorative dentistry. Finally, close attention is paid to how developing contemporary antibacterial dental materials require extensive interdisciplinary collaboration to accomplish particular and complex biological tasks in the oral cavity with its challenging environment.

**Corresponding author*: mmelo@umaryland.edu

K.L. Mittal and S. Neogi (eds.) Adhesives in Biomedical Applications, (253–278) © 2023 Scrivener Publishing LLC

Keywords: Dental adhesives, nanotechnology, dental materials, composites, dental caries

10.1 Introduction

Worldwide, dental caries is a commonly spread disease, which exemplifies one of the most significant and predominant problems in oral health [1]. According to the World Health Organization (WHO) report, around 60-90% of adolescents and around 100% of adults have caries worldwide. Therefore, it is considered a threat to oral and systematic health. In addition, it causes a heavy financial burden globally [2, 3]. For example, in 2016, around $124 billion were spent on dental treatments in the United States alone [4, 5].

Dental caries is a biofilm-induced multifactorial disease [2]. Demineralization results from the imbalance in the oral flora, where an acidic attack initiates the cariogenic bacteria in dental plaque biofilm, establishing the mechanism of dental caries [6–8]. Cariogenic bacteria result from dysbiosis, which can multiply and survive in an acidic microenvironment (low pH levels) [9]. Oral biofilms consist of more than 700 bacterial species, which are embedded in a bacterial matrix, salivary proteins, peptides, and food debris [10, 11].

Once biofilm acids have caused dental caries, the treatment includes removing the carious tissue and replacing the missing tooth structure via a restorative material. In the United States, around 200 million restorations are placed yearly [12, 13]. Resin composites are broadly used in replacing missing tooth structures [14, 15] due to their superb aesthetic features, good mechanical properties, and biocompatibility [16]. Dental adhesives are used for bonding resin composites to the tooth structure [17, 18]. Unfortunately, composite resin restoration accumulates more biofilms than other restorative materials [19–21], negatively affecting its longevity with a failure rate ranging between 15% to 50% [22]. Studies showed that more than half of the restorations placed failed within ten years [23, 24], making replacement dentistry a significant problem to consider as its annual cost is $46 billion in the USA alone [25]. Restoration failure is caused mainly by secondary caries at the restoration-tooth margins [26–28].

In tooth bonding, resin-dentin bonds are less durable than resin-enamel bonds due to dentin's heterogeneous configuration and composition [28]. The resin-dentin interdiffusion zone is a hybrid layer created when an adhesive system penetrates the collagen matrix of the dentin, offering micromechanical retention for restoration [29]. Despite the significant advancements in dental adhesives, the resin-dentin interface is still the weakest part of composite restorations due to bond deterioration [30, 31]. The degradation of this weak phase leads to microleakage and gap formation creating a practical pathway for oral biofilm invasion and secondary caries formation [18, 32]. Several studies have been pursued to improve bonding property, durability, and longevity [17].

Since dental adhesives were introduced in the early 1950s, many developments have taken place, seeking efficiency and simplicity. Basically, dental adhesive systems consist of a volatile solvents mixture such as water, alcohol, and acetone, hydrophilic and hydrophobic monomers that infiltrate the demineralized dental substrates and polymerize, creating micromechanical bonds [33].

Adding antibacterial and remineralization features to resin composites and dental adhesives was favorable in refining the general properties (physical and mechanical) and overcoming the withdrawals by adding bioactive features (antibacterial and ion releasing), especially with the current treatment modalities, such as selective caries removal, where residual bacteria remain in the prepared tooth and the affected dental tissue is preserved [34, 35].

Involving nanotechnology was one of the innovative approaches to preventing and treating dental caries, particularly in controlling and managing dental plaque biofilms and remineralizing initial dental caries [36]. Nanoparticles' size ranges from 100 nm to smaller; as the particle size is reduced, the resultant properties can change dramatically [37]. Nanoparticles have emerged as a new class of inorganic agents due to their pronounced effect on reducing bacterial proliferation [38]. Their mode of action can be summarized as controlling mineral loss and reducing biofilm formation [39]. For example, metallic nanoparticles were known for their antibacterial effects, while calcium phosphate-based nanoparticles for their ion release that balances the mineral loss [40]. Nanoparticles properties dictate their functionality. For instance, the greater surface area enhances their contact with the bacteria to provide a broad-spectrum antibacterial efficacy [41]. It is essential to mention that incorporating an antibacterial property into dental adhesives is challenging due to the complexity of dentin's composition and the need to mimic the clinical scenario through accurate and robust *in vitro* and *in vivo* models. Two main methods are followed to achieve antibacterial effects through dental adhesives: tooth pre-treatment via etchants or disinfectants and incorporating antibacterial agents as leachables, polymerizable or fillers. Table 10.1 represents each method's most common antibacterial agents incorporated into dental adhesives.

Table 10.1 Most common antibacterial agents used in each anti-caries approach using dental adhesives.

Method	**Dental adhesive antibacterial agent**
Pre-Treatment [41, 42]	Benzalkonium chloride, chlorhexidine, sodium hypochlorite, urushiol, titanium tetrafluoride
Incorporating Leachable Agents [43, 44]	Benzalkonium chloride, glutaraldehyde, chlorhexidine, epigallocatechin-3-gallate
Incorporating Polymerizable Agents [45, 46]	12-Methacryloxydodecylpyridinium bromide (MDPB), benzotriazole-hydroxyphenyl-ethylmethacrylate, dimethylaminohexadecyl methacrylate
Incorporating Fillers [47, 48]	Nano-silver, copper iodide, zinc oxide

10.2 Brief History of Dental Adhesive Systems

The dental interface in adherend-adhesive combinations entails the quality of mechanical and physical mechanisms to create an attachment between the tooth and different materials [49]. Adhesive systems and dental tissue hybridization techniques have constantly been evolving since the introduction of the adhesive agent Sevriton Cavity Seal in the 1940s by Oskar Hagger [50]. The pioneer adhesive consisted of glycerophosphoric acid dimethacrylate and interacted with the tooth surface at a molecular level. However, the performance of this first adhesive agent was very limited and could not withstand the high polymerization shrinkage stress generated by the curing of the composite formulated at that time.

Michael Buonocore, in 1955, was the first to demonstrate that the application of phosphoric acid to enamel increased the bond strength between resin and enamel [51]. In addition, he believed that acid etching increased the microscopic surface area for resin retention. John Gwinnett, one of Buonocore's students, examined the interface under a microscope and reported that resins could penetrate enamel prisms and envelop the prisms after acid etching the enamel [52]. This approach was first described as hybrid layer, although this term was not yet identified in dentistry.

In 1979, Fusayama *et al.* [53] introduced the concept of "total etching," i.e., simultaneous acid etching of enamel and dentin. In this context, Nakabayashi *et al.* [54] were the first to demonstrate proper hybrid layer formation in acid-etched dentin. Nakabayashi's group demonstrated that resins could infiltrate acid-etched dentin to form a new structure composed of collagen fibers supported by a resin matrix. This new biocomposite came to be called the hybrid layer.

The first commercially available etch-and-rinse adhesive system was the Clearfil Bond System-F (Kuraray, Japan) in 1978, which used 40% phosphoric acid [18]. Initially, there were many reports of pulp reactions in the United States and Europe, which did not occur in Japan. It was found that the difference was due to how the dentinal tissue was removed and the way total etching was applied to the dental tissues.

In Japan, cavity preparation was more conservative, based on the removal of carious tissue, while in Europe and the United States, the removal was less conservative and extended to healthy tissues. In addition, in the United States and Europe, after complete etching with acid, dentists would dry the cavity walls intensively to ensure that the enamel margins were opaque and etched. However, it was noticed that the intense drying of dentin led to the collapse of its collagen fibers. This demineralized, fiber-collapsed dentin has less area for bonding agent retention, as it has lost interfibrillar spaces between exposed collagen fibrils. Consequently, while the bond strength to enamel was around 20 MPa, it was only about 5 MPa to dentin. With the dramatic decrease in bond strength, polymerization shrinkage was observed at the time. Although this effect was only described in 1992 by Kanca, bonding was already performed on wet dentin in Japan. These events, associated with resin detachment, bacterial infiltration, and recurrent caries, led to greater pulp sensitivity in patients treated in the United States and Europe compared to those treated in Japan [18].

Kanca, in 1992, observed that water was a great agent for rewetting dentin and introduced the concept of "wet adhesion" [55]. This technique increased the bond strength between bonding agents and dentin, allowing for good marginal sealing and lower postoperative sensitivity. From then on, the bond strength of the dentin/bonding agent interface equaled or exceeded the values found for the enamel/bonding agent interface.

With research advancements, dental adhesive systems have evolved over the decades. Therefore, studying adhesive systems for generations has led to understanding that the processes of formulating etch-and-rinse and self-etch systems occur differently. However, it can be said that the original material for the formulation of self-etch adhesive systems was the same for all adhesive system i.e. the self-curing adhesive created by Oskar Hagger based on glycerophosphoric acid dimethacrylate in the 1940s. Therefore, this chapter will use the classification proposed by Van Meerbeek *et al.* in 2003 [56], which categorizes the materials into etch-and-rinse and self-etch adhesive systems. Also, universal or multimode adhesive systems will be addressed.

10.3 Classification and Composition of Adhesive Systems

Dental adhesives are "amphiphilic" as they contain hydrophilic and hydrophobic monomers that facilitate bonding to the tooth structure and restorative material [49]. The dental adhesive systems are composed of three components: etchant, primer, and adhesive. Adhesive systems infiltrate the tooth structure via an acidic solution to physically match the adhesive solution's wetting ability [49]. Etchant contains the required acidic solution that minimizes the contact angle with the bonding solution. Primer is another component needed to create adequate substrate wetting [49].

The classification can be according to the way of conditioning and the number of steps into three-step etch-and-rinse, two-step etch-and-rinse, two-step self-etch, and one-step self-etch. Etch-and-rinse adhesive systems always involve using phosphoric acid separately, in a separate bottle, from the primer or adhesive. Self-etch adhesive systems depend on the use of acid primer instead of phosphoric acid in a separate bottle. In addition to these four types, the current classification uses " universal adhesive" or "multimode" to address an adhesive system that can be used as etch-and-rinse or self-etch.

10.3.1 Three-Step Etch-and-Rinse Conventional Adhesive System

The conventional three-step etch-and-rinse adhesive system consists of 37% phosphoric acid, primer, and adhesive. For dentin, these three materials are used. However, for enamel, there is no need to use a primer. Therefore, in a cavity with dentin and enamel, the step-by-step application of this adhesive system consists of:

- Acid etching: Application of phosphoric acid to the enamel for 15 s. Then, the acid is applied over the dentin for 15 s, totaling a conditioning time of 30 s for the enamel and 15 s for the dentin.
- Washing: Wash the acid etching solution for at least the same time as it was applied.
- Drying: While enamel must be completely dry, dentin should not be overly dry. This step is critical in the hybridization

process using etch-and-rinse adhesive systems. Excessive drying of the dentin leads to the collapse of the collagen network that has been exposed, causing the following materials (primer and adhesive) to be unable to penetrate through the surface of dentin into the tubules. Therefore, drying of the dentin must be moderate and can be performed with a sterilized filter paper. The paper is gently positioned over the dentin to protect it against the jet of air applied to the enamel. The enamel must be dried until its opacity is observed, indicating that the surface has been well-conditioned and is porous. After applying air to the enamel, the paper is removed from the dentin. Then it must be gently dried with an absorbent paper.

Excessive drying of dentin is a problem, as mentioned above. However, leaving too much water also harms the adhesion process, especially in the long term. In addition to dissolving the primer applied in next step, excess water makes the bonding interface more hydrophilic and prone to degradation, especially hydrolytic and polymeric degradation.

Therefore, when etch-and-rinse adhesive systems are used, dentin's "wet adhesion" concept is applied, which has a shiny and moist appearance but without excess water.

Etching with 37% phosphoric acid removes the topmost layer of enamel (about 10 μm), increasing its surface free energy. Acid also creates pores and increases surface free energy by selectively dissolving enamel prisms to a depth of about 20 μm [57]. In dentin, this process dissolves about 8 μm, and the collagen fibers are practically stripped of hydroxyapatite [56].

- Primer: Applied with a micro-applicator on the dentin for 20 s. The application must be made gently to prevent the breakage of collagen fibers.

The elemental composition of the primer is solvents and hydrophilic monomers. Solvents carry the bare collagen mesh through the hydrophilic monomers, usually represented by 2-hydroxyethyl methacrylate (HEMA). HEMA is a low molecular weight monomer (130.14 g/mol), with a hydroxyl group and a methacrylate group at the other end. The hydroxyl groups help in the process of diffusion through the collagen mesh and can bind to the organic tissue and water of the dentin. On the other hand, the methacrylate group covalently bonds with the other monomers of the adhesive, which will be applied in the next step. Therefore, the primer is essential for adequate adhesion between the dentin (hydrophilic substrate) and the adhesive (hydrophobic material).

- Solvent evaporation: Excess solvent from the primer is removed using gentle air. If excess solvent remains in the dentin, it is responsible for creating a more hydrophilic interface and more susceptible to degradation processes. In addition, subsequent adhesive polymerization is impaired when solvents are in the dentin, reducing the material's physical and chemical properties.
- Adhesive: a thin film must be applied to both enamel and dentin, covering the entire surface. After that, it must be light-cured for the time recommended by the manufacturer (generally 10 to 20 s).

Adhesives are composed of monomers, initiating agents, polymerization inhibitors, and, sometimes, inorganic fillers. The most used monomer is Bisphenol A-Glycidyl Methacrylate (Bis-GMA), which is a dimethacrylate aromatic ester with high molecular weight (512 g/mol) and high viscosity [57]. This monomer is responsible for good mechanical properties of the adhesives, providing a higher modulus of elasticity and greater hydrolytic stability than HEMA, for example. The percentage of Bis-GMA varies from about 50 to 70% by mass, depending on the commercial brand [58]. In addition to Bis-GMA, diluent monomers are added, such as triethylene glycol dimethacrylate (TEGDMA), with lower molecular weight (286 g/mol), and HEMA. These monomers help in material handling, reducing viscosity and increasing adhesive flow, and helping in the dentin hybridization process, as they are more hydrophilic, especially HEMA.

Currently, adhesives are light-cured; therefore, initiating agents are used that sensitize with light and initiate the polymerization process. Photoinitiators can be either type I or type II. As an example of type I photoinitiators, phosphine oxide stands out. These initiators undergo fragmentation after absorbing electromagnetic energy from the photoactivator, generating free radicals that initiate polymerization. As an example of a type II photoinitiator, camphorquinone can be highlighted. Type II photoinitiators need a co-initiator for the formation of free radicals. Co-initiators commonly used with camphorquinone are tertiary amines [59]. Currently, the initiation system is based on camphorquinone and amine at a molar mass of 0.5 to 1% [58].

Low concentrations (1 to 5%) are generally used so as not to increase the material's viscosity and impair the dental tissue's wetting when inorganic fillers are added to adhesives. Instead, fillers are added primarily to fine-tune viscosity, improve mechanical properties (such as higher modulus of elasticity and higher strength) and reduce hydrolytic degradation. Examples of fillers are silica, quartz, and glass based on barium, boron, aluminum, and silicon [58].

As an advantage, this adhesive system has a long clinical follow-up period and demonstrates satisfactory results *in vivo*, being the gold standard among adhesive systems. However, there are disadvantages: critical technique with multiple steps; extensive clinical protocol; aggressive demineralization of the dentin, as the acid dissolves the hydroxyapatite crystals and leaves the collagen fibers bare deep into the tissue; and part of the demineralized dentin is filled by the adhesive resin, leaving an area below the hybrid dentin layer unprotected by polymer and more susceptible to hydrolytic and enzymatic degradation.

10.3.2 Two-Step Etch-and-Rinse Conventional Adhesive System

The industry has released a simplified version of the three-step etch-and-rinse adhesive system to simplify the clinical procedure using the conventional three-step adhesive system. This strategy consists of using acid separately and combining the primer and adhesive. The step-by-step clinical procedure with this strategy is the same as the previous one, except for the steps of applying the primer and the adhesive separately. For the conventional two-step etch-and-rinse adhesive system, wet adhesion will be generated after carefully drying the dentin. First, a mixture of primer and adhesive material is applied with a micro-applicator. After applying this mixture, the solvent is volatilized with the etch-and-rinse adhesive system. Next, a layer of the combined primer and adhesive is applied again. The solvent is volatilized again. Finally, the material is photoactivated according to the manufacturer instructions.

The composition of this system is very similar to the un-simplified version with three steps, but the number of steps is reduced.

Although this strategy seems advantageous for reducing the operatory steps, it is debated whether the time difference is significant since this material is applied twice, so there is a significant layer of adhesive on the tissues. As a disadvantage, this system still presents difficulties related to operative sensitivity and aggressive dentin conditioning, as with the conventional 3-step adhesive system. Furthermore, this system has fewer hydrophobic monomers and forms a more hydrophilic hybrid layer, which is not sealed by a separate hydrophobic layer of adhesive.

Creating a more hydrophilic hybrid layer and the non-hydrophobic sealing by an adhesive layer over this hybrid layer causes simplified adhesives to act as semipermeable membranes [60]. As a result, water is entrapped at the adhesive interface [61], creating water channels called "water-tree" or "water trees" [62], which make the dental interface prone to hydrolytic degradation.

10.3.3 Two-Step Self-Etch Adhesive System

In self-etch adhesive systems, phosphoric acid is replaced by acidic functional monomers. As a result, washing and drying steps after acid conditioning are dispensed with. Therefore, self-etch systems have fewer operator errors by reducing the number of steps required for their application.

The basic composition of the 2-step self-etch adhesive system is a primer composed of solvents, hydrophilic monomers (HEMA), acidic monomers, and an adhesive composed of photoinitiator, polymerization inhibitor, and a mixture of hydrophobic monomers with other monomers. As the ionizing medium, water is present in self-etch adhesives, with rare commercial exceptions [17].

For most two-step self-etch adhesive systems, restoration of enamel and dentin must be accomplished by selective acid etching of enamel. First, the dentin must be protected (it can be done with sterilized paper), and 37% phosphoric acid in the gel must be applied to the enamel margins (the same acid as used in the etch-and-rinse adhesive system) for 30 s, washed, and dried. Next, the acid primer is applied via rubbing motion to the dentin with a micro-applicator for 20 s. Next, the solvent must be evaporated using an air jet. Next, a thin adhesive layer should be applied to both enamel and dentin with a micro-applicator. Finally, the adhesive is photoactivated for the time established by the manufacturer. As advantages, self-etch adhesive systems have lower operating errors and cause a less aggressive demineralization of dentin, bringing advantages to the dentin-adhesive interface compared to etch-and-rinse adhesive systems.

Dentin minimum invasion occurs in three ways: demineralization at a lower depth; monomeric hybridization simultaneously with the demineralization process, leaving virtually no exposed dentin layer at the base of the hybrid layer; higher stability of the bond between hydroxyapatite calcium and acidic monomers compared to phosphoric acid.

Before adhesion, the decalcification of the hydroxyapatite occurs. First, acids bind to calcium in the hydroxyapatite; then, phosphate and hydroxyl ions are released into the medium. Calcium and acid binding to hydroxyapatite can occur at any stage, and they are quickly released into the medium, altering the bond stability and leaving the fibers exposed. On the other hand, phosphoric acid does not affect bond stability. With acidic monomers, this bond can have varying degrees of stability depending on the pH of the monomer and its molecular structure. Among the main acidic monomers: 2-Hydroxyethyl methacrylate (HEMA), 4-Methacryloxyethyl trimelliatic acid (4-MET), 2-Methacryloyloxyethyl phenyl phosphoric acid (phenyl-P), and Methacryloyloxydecyl dihydrogen phosphate (10-MDP), 10-MDP has shown the best clinical results [17].

Some studies suggest that the best performance of 10-MDP adhesive occurs due to the formation of self-organized nanolayers, which are formed between 10-MDP and calcium in the tooth hard tissues. The first work to show the formation of nanolayers at the adhesive interface when using commercial adhesive systems with 10-MDP was in 2012 [63]. The authors suggested that the nanolayers increased the hydrophobicity of the hybrid layer, preventing water molecules from diffusing through the hybrid layer and causing collagen and polymer hydrolysis. Furthermore, these systems could create residual hydroxyapatite more resistant to acid dissolution [63]. However, although more recent studies still suggest that nanolayers constitute a critical phase in bond strength [63], there are studies demonstrating that nanolayers do not have high stability by themselves and that perhaps they are not the main ones responsible for the good results of adhesive systems with 10-MDP [64].

Self-etch adhesives can also be classified according to their acidity into moderately strong with pH ≥ 1.5, and strong with pH <1.5, according to Peumans *et al.* [65]. The thickness of the hybrid layer dictates the bond strength. For instance, it was found that the hybrid layer thickness for strong bonding is 4 μm, while when it drops to 300 nm, less bond strength is achieved, depending on the pH of the self-etch adhesive [17]. Strong pH adhesives create a thick and more densely demineralized hybrid layer, similar to phosphoric acid demineralization. Nevertheless, with the disadvantage of keeping the demineralization products at the interface, there is no washing step after using the monomer acid. When performing a restoration involving enamel and dentin, the ideal procedure is to perform the selective acid etching technique on enamel with phosphoric acid and applying a self-etch adhesive system with pH ≥ 1.5 and 10-MDP on dentin.

10.3.4 One-Step Self-Etch Adhesive System

In order to reduce clinical time, the industry created the one-step self-etch adhesive system. Two bottles are used to separate the reagents to maximize the adhesive's shelf-life. The composition is similar to the two-step self-etch system but has more hydrophilic monomers.

The step-by-step restoration involving enamel and dentin with this adhesive system should be performed with selective acid etching on enamel as described for the two-step self-etch system. Here, a drop from each vial must be mixed with a micro-applicator and applied to the dental tissues to volatilize the solvent and light-activate the material. Therefore, like the two-step self-etch systems, this adhesive system has advantages such as shorter clinical time and lower technical application errors. However, like a two-step etch-and-rinse adhesive system, it generates a more hydrophilic hybrid layer. There is no hydrophobic sealing by an adhesive layer over the hybrid layer, forming a more hydrophilic interface and a hybrid layer that acts as a semipermeable membrane, impairing the longevity of restorative treatments.

10.3.5 Universal or Multimode Adhesive System

This most recent class of dental adhesives was created to simplify the technique and facilitate its use by dentists. This material is a self-etch adhesive system that allows its application to dental tissues with or without phosphoric acid. In addition, some brands indicate their use in adhesive protocols involving metals and ceramics [66]. The universal adhesives present an elementary composition similar to the one-step self-etch adhesive systems with the addition of 10-MDP. The universal adhesives present superior chemical stability and less phase separation than one-step self-etch adhesive systems and can be marketed in a single bottle. In addition, polyalkenoic acid, also found in other adhesive systems, is incorporated in some brands of universal adhesives, providing more molecules to bind the calcium in the hydroxyapatite. Silane has also been incorporated into some commercially available universal adhesives [66].

However, in 2016 research showed interaction in the components of single bottle universal adhesives [67]. The best example in this situation is the use of silane. Manufacturers have added silane in the composition of these adhesive systems to eliminate the need for silane pretreatment, as both would be contained in a single universal bottle. Surprisingly, research has shown an interaction between the silane and water (necessary for the ionization of the acid) in the flask, leading to silane dehydration and reducing its bonding ability to ceramics. Furthermore, despite manufacturers selling the material as if it could be applied with or without prior phosphoric acid treatment, research shows that selective acid etching on enamel improves clinical performance [68].

10.4 Understanding the Challenges of Dental Adhesives Inside the Mouth

The intraoral environment is conducive to microbial growth since it is warm with a temperature range from 0° C to 70° C, moist, and rich in nutrients [69]. In addition, it is subjected to a fluctuating pH, for instance, by consuming acidic food. Finally, the intraoral environment withstands different mechanical forces ranging from rest to mastication, drinking, and cleaning [50].

Oral biofilms are known for their pliability and adaptability, which favor microorganisms growth and survival [70]. However, a shift from a healthy state to a pathogenic one known as dysbiosis can lead to several oral diseases, such as dental caries, periodontal diseases, and implant infection and failure [71].

The monomer infiltration contributes to 1) Dentin bonding to the demineralized intertubular matrix, 2) Formation of hybridized intratubular resin tags, and 3) The chemical bonding to the dentin structure [72]. During the function inside the mouth, dentin adhesives are subjected to chemical and mechanical stresses, which can alter the adhesive's mechanical properties over time. Dentin bonding is challenged by many factors that determine its performance, such as tooth-restoration interface degradation, microleakage, and pulpal reaction. From the following phenomena, interface degradation is the most common failure mode occurring due to the following:

10.4.1 Host-Derived Proteolytic Enzymes

Demineralized dentin collagen matrix acts as a platform for resin infiltration during the resin-dentin bonding procedure, creating the hybrid layer essential to dentin bonding strength [73]. Therefore, the degradation of collagen matrices by matrix metalloproteinases (MMPs) and cysteine cathepsins is among the significant reasons for adhesive breakdown [74, 75]. MMPs and cysteine cathepsins attack type I collagen, the predominant organic element of the dentin matrix [76]. MMPs and cysteine cathepsins are stimulated by proteinases, chemical mediators, low pH, heat treatment, and mechanical stress [76, 77].

10.4.2 Adverse Chemical/Biochemical Interactions

Methacrylate dental adhesives contain ester bonds which can be subjected to chemical and enzymatic hydrolysis, affecting the bond durability [32, 78, 79]. Unfortunately, the cholesterol esterase and pseudocholinesterase present in human saliva cause a synergistic effect in degrading these methacrylate adhesives [32]. Some *in vitro* studies indicated that aging adhesive-dentin samples in artificial saliva containing cholesterol esterase significantly reduced the microtensile bond strength [80]. In addition, the hydrophilic nature of dentin makes it challenging to use a hydrophobic adhesive, but this hydrophilicity promotes water sorption that can accelerate polymers' plasticization and diminish their mechanical properties [42, 81].

10.4.3 Mechanical Loading

Changes in mechanical properties can result either from the proliferation of surface and subsurface flaws or changes in the chemical nature of the polymer by strain hardening, embrittlement, crystallization, or plasticization [31]. Masticatory forces have been identified as the leading stresses affecting the bonding interface negatively, causing the bending of tooth structure and leading to gap formation and leakage around the restoration [82, 83]. Mechanical behavior of dental adhesives under function, i.e., when the patients are biting over the restorations, is described only for a few materials and most of the time describe only flexural strength and linear elasticity. Different adhesive materials are subjected to an extraordinary range of stresses during function which, in turn, causes a drop in elastic modulus leading to extra stress [84, 85]. The stress may stimulate the dentin collagen matrix containing bioactive molecules and signaling factors responsible for mineralization, tissue regeneration, cell differentiation, and other biological functions [86]. Moreover, incomplete infiltration of dental adhesives leads to creep or cycling fatigue during intraoral functions [74].

Oral fluids and bacteria can penetrate the adhesive-dentin interface through marginal gaps resulting from polymerization shrinkage, facilitating bond degradation [87, 88]. The bacterial infiltration has been confirmed using transmission electron microscopy (TEM), where bacterial collagenases had increased nanoleakage at the tooth-dentin interface [80].

During the caries attack, the dentin-pulp complex promotes remineralization as a barrier to bacterial infiltration by depositing a mineral known as reactionary dentin, which is less favorable to adhesive penetration.

This scenario creates an additional challenge to the dentin-adhesive interface [88].

The biosafety of the dental adhesive is also a concern as the dental adhesive's monomers such as bisphenol-glycidylmethacrylate (BisGMA), triethylene glycol dimethacrylate (TEGDMA), and 2-hydroxyethyl methacrylate (HEMA) are also the critical toxic components in dental restorative materials [89, 90]. In addition, the amount of non-polymerized monomers, dentin permeability, and duration of perfusion dictate the toxic effect on pulpal tissue as they weaken the dentin-pulp complex's regenerative and reparative abilities[88, 89, 91, 92].

10.5 New Approaches Targeting Longevity of Adhesive-Dentin Interfaces

Adhesive systems and resin composites have been modified regarding the composition of the organic matrix, fillers, and photoinitiation systems to improve the longevity of adhesive interfaces and dental restorations. The main issues for resin composites are focused on developing materials for insertion in large amounts, or "bulk-fill", self-adhesive, with less hydrolytic degradation, and antimicrobial and bioactive properties. In addition, resins with self-repair potential and "smart materials" that respond to stimuli from the intraoral environment are also being researched [15, 93]. At the same time, the main changes for adhesive systems are based on their simplification with improved adhesion due to more stable acid monomers, such as 10-MDP. Furthermore, this chapter focuses on adhesive systems with less hydrolytic degradation and antimicrobial and bioactive properties.

Bulk-fills are composites with lower polymerization shrinkage and can be polymerized in large increments (about 4 to 5 mm thick). Thus, commercial brands have invested, for example, in additional fragmentation monomers and aromatic urethane dimethacrylate. In addition, some bulk fills present modifications in the initiators, such as the mixture of different initiators with camphorquinone and amine to increase the composite's degree of conversion, modifications in the shape, color, refractive index of the fillers, and factors that can improve polymerization of the composite [94]. So far, bulk fill resins have demonstrated, both in class I and in class II, low values of annual failure rate, around 1 to 2%, and no difference from conventional resin composites in 10 years of follow-up studies [95].

Other areas of development have incorporated antibacterial and remineralizing agents into resin composites. Examples of materials that have been added to resin composites to reduce biofilm formation at the composite-tooth interface, to reduce demineralization processes or induce remineralization of dental tissue are fluoride, chlorhexidine, zinc oxide nanoparticles, methacrylates derived from ammonium compounds, and particles derived from calcium phosphate [93].

Other modifications in the organic matrix currently include using acidic monomers to create self-adhesive resins, such as monomers based on glycerol phosphate dimethacrylate (GPDM) [15]. In addition, the synthesis and application of acrylamide-based monomers without ester groups have also been explored to enhance the hydrolytic stability of dental resins [93].

Furthermore, resins capable of self-repair have been studied, primarily through capsules that carry the repairing agent. Capsules containing initiators and TEGDMA have already been formulated to fill spaces in the restorative material when cracks occur. Finally, "smart materials" are being developed. These materials are formulated to release remineralizing, antimicrobial, or repairing agents when needed [93].

10.6 Dental Adhesives Endowed With Antibacterial Properties

The antibacterial polymers added to dental materials can be categorized into three groups:

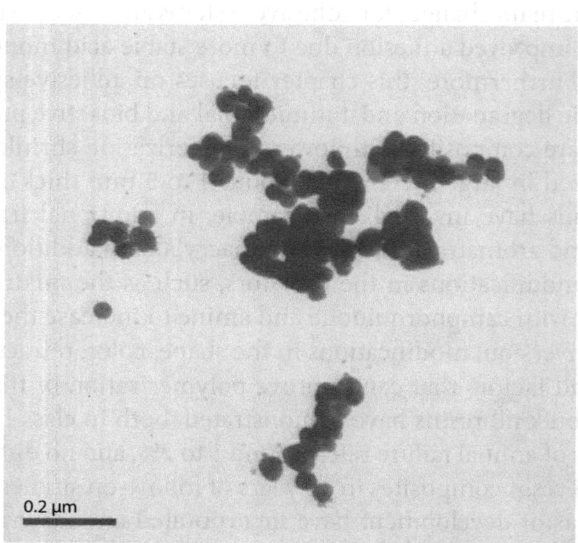

Figure 10.1 Transmission electron microscopy (TEM) image of silver nanoparticles [97].

(1) methyl methacrylate-based organic compounds containing cationic quaternary ammonium; (2) organic compounds possessing properties that stimulate microorganism death, such as chlorhexidine; and (3) inorganic compounds such as metallic particles (silver (Figure 10.1), zinc oxide (Figure 10.2)). Cationic quaternary ammonium compounds are identified as the most added antibacterial agents, while the inorganic additives are the least used among these three groups. On the other hand, dental materials are used as parental materials (carriers). Antibacterial resin materials are divided into primers, adhesives, and resin composites. Composite resins are the dominant parental material, followed by bonding agents, while flowable composites are the least used material [96].

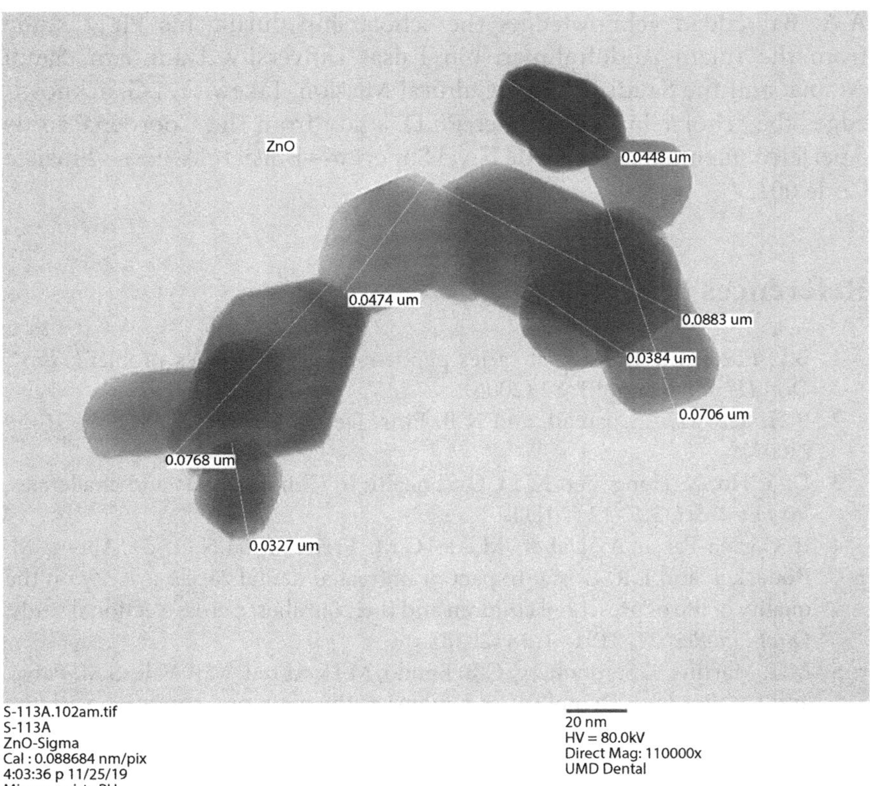

Figure 10.2 Transmission electron microscopy (TEM) image of zinc oxide nanoparticles [98].

10.7 Summary

Bioactive dental adhesives offer a promising approach to combat caries-related biofilm formation. In addition to their antibacterial property, improvements in mechanical properties were also clearly noticed. Understanding dental adhesives' composition and challenges have guided the research for developing and improving dental adhesives. Bioactive functionality was achieved by incorporating different polymers, such as quaternary ammonium compounds. Most of the investigations are in an early stage of development representing *in vitro* and pre-clinical setting taking in consideration the evidence for standard patient's care.

10.8 Acknowledgments

A.A. Balhaddad acknowledges the scholarship during his Ph.D. study from the Imam Abdulrahman bin Faisal University, Dammam, Saudi Arabia, and the Saudi Arabian Cultural Mission. Likewise, I.G. acknowledges the scholarship during her Ph.D. study from the Coordenação de Aperfeiçoamento de Pessoal de Nível Superior—Brasil (CAPES)—Finance Code 001.

References

1. B.L. Edelstein, The dental caries pandemic and disparities problem, *BMC Oral Health*, 6 Suppl 1, S2 (2006).
2. R.H. Selwitz, A.I. Ismail, and N.B. Pitts, Dental caries, *Lancet*, 369, 51–59 (2007).
3. D.-Y. Hu, X. Hong, and X. Li, Oral health in China--trends and challenges, *Int J Oral Sci*, 3, 7–12 (2011).
4. P. Corrêa-Faria, A. Daher, M. do C.M. Freire, M.H.N.G. de Abreu, M. Bönecker, and L.R. Costa, Impact of untreated dental caries severity on the quality of life of preschool children and their families: a cross-sectional study, *Qual Life Res*, 27, 3191–3198 (2018).
5. M.T. Martins, F. Sardenberg, C.B. Bendo, M.H. Abreu, M.P. Vale, S.M. Paiva, and I.A. Pordeus, Dental caries remains as the main oral condition with the greatest impact on children's quality of life, PLoS One, 12, e0185365 (2017).
6. J.D. Featherstone, The continuum of dental caries--evidence for a dynamic disease process, *J Dental Res*, 83, C39-42 (2004).
7. J.D. Featherstone, The science and practice of caries prevention, *J Amer. Dental Assoc*, 131, 887–899 (2000).

8. J. van Houte, Role of micro-organisms in caries etiology, *J Dental Res*, 73, 672–681 (1994).
9. K. Papadimitriou, Á. Alegría, P.A. Bron, M. de Angelis, M. Gobbetti, M. Kleerebezem, J.A. Lemos, D.M. Linares, P. Ross, C. Stanton, F. Turroni, D. van Sinderen, P. Varmanen, M. Ventura, M. Zúñiga, E. Tsakalidou, and J. Kok, Stress physiology of lactic acid bacteria, *Microbiol Mol Biol Rev*, 80, 837–890 (2016).
10. F.E. Dewhirst, T. Chen, J. Izard, B.J. Paster, A.C.R. Tanner, W.-H. Yu, A. Lakshmanan, and W.G. Wade, The human oral microbiome, *J Bacteriol*, 192, 5002–5017 (2010).
11. P.D. Marsh and D.J. Bradshaw, Dental plaque as a biofilm, *J Ind Microbiol*, 15, 169–175 (1995).
12. D.C. Watts, A.S. Marouf, and A.M. Al-Hindi, Photo-polymerization shrinkage-stress kinetics in resin-composites: methods development, *Dental Mater*, 19, 1–11 (2003).
13. J.L. Drummond, Degradation, fatigue, and failure of resin dental composite materials, *J Dental Res*, 87, 710–719 (2008).
14. L. Breschi, A. Mazzoni, A. Ruggeri, M. Cadenaro, R. Di Lenarda, and E. De Stefano Dorigo, Dental adhesion review: aging and stability of the bonded interface, *Dental Mater*, 24, 90–101 (2008).
15. J.L. Ferracane, Resin composite--state of the art, *Dental Mater*, 27, 29–38 (2011).
16. N. Angel Villegas, M.J. Silvero Compagnucci, M. Sainz Ajá, D.M. Rocca, M.C. Becerra, G. Fabián Molina, and S.D. Palma, Novel antibacterial resin-based filling material containing nanoparticles for the potential one-step treatment of caries, *J Healthcare Eng*, 6367919 (2019).
17. B. Van Meerbeek, K. Yoshihara, Y. Yoshida, A. Mine, J. De Munck, and K.L. Van Landuyt, State of the art of self-etch adhesives, *Dental Mater*, 27, 17–28 (2011).
18. D.H. Pashley, F.R. Tay, L. Breschi, L. Tjäderhane, R.M. Carvalho, M. Carrilho, and A. Tezvergil-Mutluay, State of the art etch-and-rinse adhesives, *Dental Mater*, 27, 1–16 (2011).
19. M. Svanberg, and I.A. Mjör, and D. Orstavik, Mutans streptococci in plaque from margins of amalgam, composite, and glass-ionomer restorations, *J Dental Res*, 69, 861–864 (1990).
20. S. Imazato, M. Torii, Y. Tsuchitani, J.F. McCabe, and R.R. Russell, Incorporation of bacterial inhibitor into resin composite, *J Dental Res*, 73, 1437–1443 (1994).
21. N. Beyth, A.J. Domb, and E.I. Weiss, An *in vitro* quantitative antibacterial analysis of amalgam and composite resins, *J Dentistry*, 35, 201–206 (2007).
22. P.A. da Rosa Rodolpho, and M.S. Cenci, T.A. Donassollo, A.D. Loguércio, and F.F. Demarco, A clinical evaluation of posterior composite restorations: 17-year findings, *J Dentistry*, 34, 427–435 (2006).

23. I.A. Mjör, and F. Toffenetti, Secondary caries: a literature review with case reports, *Quintessence Int*, 31, 165–179 (2000).
24. R.L. Sakaguchi, Review of the current status and challenges for dental posterior restorative composites: clinical, chemistry, and physical behavior considerations Summary of discussion from the Portland Composites Symposium (POCOS) June 17-19, 2004, Oregon Health and Science University, Portland, Oregon, *Dental Mater*, 21, 3–6 (2005).
25. T. Beazoglou, S. Eklund, D. Heffley, J. Meiers, L.J. Brown, and H. Bailit, Economic impact of regulating the use of amalgam restorations, *Public Health Reports*, 122, 657–663 (2007).
26. N.J.M. Opdam, F.H. van de Sande, E. Bronkhorst, M.S. Cenci, P. Bottenberg, U. Pallesen, P. Gaengler, A. Lindberg, M.C.D.N.J.M. Huysmans, and J.W. van Dijken, Longevity of posterior composite restorations: a systematic review and meta-analysis, *J Dental Res*, 93, 943–949 (2014).
27. M. Bourbia, D. Ma, D.G. Cvitkovitch, and J.P. Santerre, and Y. Finer, Cariogenic bacteria degrade dental resin composites and adhesives, *J Dental Res*, 92, 989–994 (2013).
28. R.M. Carvalho, A.P. Manso, S. Geraldeli, F.R. Tay, and D.H. Pashley, Durability of bonds and clinical success of adhesive restorations, *Dental Mater*, 28, 72–86 (2012).
29. N. Nakabayashi, M. Nakamura, and N. Yasuda, Hybrid layer as a dentin-bonding mechanism, *J Esthet Dentistry*, 3, 133–138 (1991).
30. J. Krithikadatta, Clinical effectiveness of contemporary dentin bonding agents, *J Conserv Dentistry*, 13, 173–183 (2010).
31. P. Spencer, Q. Ye, J. Park, E.M. Topp, A. Misra, O. Marangos, Y. Wang, B.S. Bohaty, V. Singh, F. Sene, J. Eslick, K. Camarda, and J.L. Katz, Adhesive/dentin interface: the weak link in the composite restoration, *Annals Biomed Eng*, 38, 1989–2003 (2010).
32. Y. Liu, L. Tjäderhane, L. Breschi, A. Mazzoni, N. Li, J. Mao, D.H. Pashley, and F.R. Tay, Limitations in bonding to dentin and experimental strategies to prevent bond degradation, *J Dental Res*, 90, 953–968 (2011).
33. M.G. Brackett, N. Li, W.W. Brackett, R.J. Sword, Y.P. Qi, L.N. Niu, C.R. Pucci, A. Dib, D.H. Pashley, and F.R. Tay, The critical barrier to progress in dentine bonding with the etch-and-rinse technique, *J Dentistry*, 39, 238–248 (2011).
34. S. Imazato, A. Kuramoto, Y. Takahashi, S. Ebisu, and M.C. Peters, *In vitro* antibacterial effects of the dentin primer of Clearfil Protect Bond, *Dental Mater*, 22, 527–532 (2006).
35. D.K. Ratledge, E.A. Kidd, and D. Beighton, A clinical and microbiological study of approximal carious lesions Part 2: efficacy of caries removal following tunnel and class II cavity preparations, *Caries Res*, 35, 8–11 (2001).
36. M. Hannig and C. Hannig, Nanomaterials in preventive dentistry, *Nat Nanotechnol*, 5, 565–569 (2010).
37. R.P. Allaker and G. Ren, Potential impact of nanotechnology on the control of infectious diseases, *Trans R Soc Trop Med Hyg*, 102, 1–2 (2008).

38. S. Qayyum, and A.U. Khan, Nanoparticles vs biofilms: a battle against another paradigm of antibiotic resistance, *Med. Chem. Commun.*, 7, 1479–1498 (2016).
39. A.A. Balhaddad, A.A. Kansara, D. Hidan, M.D. Weir, H.H.K. Xu, and M.A.S. Melo, Toward dental caries: Exploring nanoparticle-based platforms and calcium phosphate compounds for dental restorative materials, *Bioact Mater*, 4, 43–55 (2019).
40. M.A.S. Melo, S.F.F. Guedes, H.H.K. Xu, and L.K.A. Rodrigues, Nanotechnology-based restorative materials for dental caries management, *Trends Biotechnol*, 31, 459–467 (2013).
41. K.R. Raghupathi, R.T. Koodali, and A.C. Manna, Size-dependent bacterial growth inhibition and mechanism of antibacterial activity of zinc oxide nanoparticles, *Langmuir*, 27, 4020–4028 (2011).
42. A. Tezvergil-Mutluay, M.M. Mutluay, L.-S. Gu, K. Zhang, K.A. Agee, R.M. Carvalho, A. Manso, M. Carrilho, F.R. Tay, L. Breschi, B.-I. Suh, and D.H. Pashley, The anti-MMP activity of benzalkonium chloride, *J Dentistry*, 39, 57–64 (2011).
43. C. Sabatini, and D.H. Pashley, Aging of adhesive interfaces treated with benzalkonium chloride and benzalkonium methacrylate, *Eur J Oral Sci*, 123, 102–107 (2015).
44. X. Du, X. Huang, C. Huang, Y. Wang, and Y. Zhang, Epigallocatechin-3-gallate (EGCG) enhances the therapeutic activity of a dental adhesive, *J Dentistry*, 40, 485–492 (2012).
45. S.M.L.C. de Vasconcelos, M.A.S. de Melo, J.P.M.S. Wenceslau, I.C.J. Zanin, H.C.P. Beltrao, C.A.O. Fernandes, P.C. de Almeida, and L.K.A. Rodrigues, *In situ* assessment of effects of the bromide- and fluoride-incorporating adhesive systems on biofilm and secondary caries, *J Contemp Dental Pract*, 15, 142–148 (2014).
46. F.G. de Carvalho, R.M. Puppin-Rontani, L.E.S. Soares, A.M.E. Santo, A.A. Martin, and F.H. Nociti-Junior, Mineral distribution and CLSM analysis of secondary caries inhibition by fluoride/MDPB-containing adhesive system after cariogenic challenges, *J Dentistry*, 37, 307–314 (2009).
47. I.X. Yin, J. Zhang, I.S. Zhao, M.L. Mei, Q. Li, and C.H. Chu, The antibacterial mechanism of silver nanoparticles and its application in dentistry, *Int J Nanomedicine*, 15, 2555–2562 (2020).
48. I.M. Garcia, A.A. Balhaddad, M.S. Ibrahim, M.D. Weir, H.H.K. Xu, F.M. Collares, and M.A.S. Melo, Antibacterial response of oral microcosm biofilm to nano-zinc oxide in adhesive resin, *Dental Mater*, 37, e182–e193 (2021).
49. J.P.Matinlinna and K.L. Mittal (Eds.), *Adhesion Aspects in Dentistry*, CRC Press, Boca Raton, FL (2009).
50. K.-J.M. Söderholm, Dental adhesives . how it all started and later evolved, *J Adhesive Dentistry*, 9, 227–230 (2007).
51. M.G. Buonocore, A simple method of increasing the adhesion of acrylic filling materials to enamel surfaces, *J Dental Res*, 34, 849–853 (1955).

52. A.J. Gwinnett and A. Matsui, A study of enamel adhesives The physical relationship between enamel and adhesive, *Arch Oral Biol*, 12, 1615–1620 (1967).
53. T. Fusayama, M. Nakamura, N. Kurosaki, and M. Iwaku, Non-pressure adhesion of a new adhesive restorative resin, *J Dental Res*, 58, 1364–1370 (1979).
54. N. Nakabayashi, K. Kojima, and E. Masuhara, The promotion of adhesion by the infiltration of monomers into tooth substrates, *J Biomed Mater Res*, 16, 265–273 (1982).
55. J. Kanca, Improving bond strength through acid etching of dentin and bonding to wet dentin surfaces, *J Amer Dental Assoc*, 123, 35–43 (1992).
56. B. Van Meerbeek, J. De Munck, Y. Yoshida, S. Inoue, M. Vargas, P. Vijay, K. Van Landuyt, P. Lambrechts, and G. Vanherle, Buonocore memorial lecture Adhesion to enamel and dentin: current status and future challenges, *Operative Dentistry*, 28, 215–235 (2003).
57. A.Reis and A.D Loguercio, Current status and future challenges, *Operative Dentistry*, 29, 216–239 (2013).
58. K.L. Van Landuyt, J. Snauwaert, J. De Munck, M. Peumans, Y. Yoshida, A. Poitevin, E. Coutinho, K. Suzuki, P. Lambrechts, and B. Van Meerbeek, Systematic review of the chemical composition of contemporary dental adhesives, *Biomaterials*, 28, 3757–3785 (2007).
59. F.A. Rueggeberg, M. Giannini, C.A.G. Arrais, and R.B.T. Price, Light curing in dentistry and clinical implications: a literature review, *Braz Oral Res*, 31, e61 (2017).
60. F.R. Tay, R. Frankenberger, I. Krejci, S. Bouillaguet, D.H. Pashley, R.M. Carvalho, and C.N.S. Lai, Single-bottle adhesives behave as permeable membranes after polymerization I *In vivo* evidence, *J Dentistry*, 32, 611–621 (2004).
61. F.R. Tay, and D.H. Pashley, Water treeing--a potential mechanism for degradation of dentin adhesives, *Amer J Dentistry*, 16, 6–12 (2003).
62. F.R. Tay, C.N.S. Lai, S. Chersoni, D.H. Pashley, Y.F. Mak, P. Suppa, C. Prati, and N.M. King, Osmotic blistering in enamel bonded with one-step self-etch adhesives, *J Dental Res*, 83, 290–295 (2004).
63. Y. Yoshida, K. Yoshihara, N. Nagaoka, S. Hayakawa, Y. Torii, T. Ogawa, A. Osaka, and B.V. Meerbeek, Self-assembled Nano-layering at the Adhesive interface, *J Dental Res*, 91, 376–381 (2012).
64. F.-C. Tian, X.-Y. Wang, Q. Huang, L.-N. Niu, J. Mitchell, Z.-Y. Zhang, C. Prananik, L. Zhang, J.-H. Chen, L. Breschi, D.H. Pashley, and F.R. Tay, Effect of nanolayering of calcium salts of phosphoric acid ester monomers on the durability of resin-dentin bonds, *Acta Biomater*, 38, 190–200 (2016).
65. M. Peumans, J. De Munck, A. Mine, and B. Van Meerbeek, Clinical effectiveness of contemporary adhesives for the restoration of non-carious cervical lesions A systematic review, *Dental Mater*, 30, 1089–1103 (2014).

66. C. Chen, L.-N. Niu, H. Xie, Z.-Y. Zhang, L.-Q. Zhou, K. Jiao, J.-H. Chen, D.H. Pashley, and F.R. Tay, Bonding of universal adhesives to dentine--Old wine in new bottles?, *J Dentistry*, 43, 525–536 (2015).
67. K. Yoshihara, N. Nagaoka, A. Sonoda, Y. Maruo, Y. Makita, T. Okihara, M. Irie, Y. Yoshida, and B. Van Meerbeek, Effectiveness and stability of silane coupling agent incorporated in "universal" adhesives, *Dental Mater*, 32, 1218–1225 (2016).
68. W.L. de O. da Rosa, E. Piva, and A.F. da Silva, Bond strength of universal adhesives: A systematic review and meta-analysis, *J Dentistry*, 43, 765–776 (2015).
69. C.W. Barclay, and D. Spence, and W.R.E. Laird, Intra-oral temperatures during function, *J Oral Rehabil*, 32, 886–894 (2005).
70. P. Stoodley, K. Sauer, D.G. Davies, and J.W. Costerton, Biofilms as complex differentiated communities, *Annu Rev Microbiol*, 56, 187–209 (2002).
71. N.J. Kassebaum, E. Bernabé, M. Dahiya, B. Bhandari, C.J.L. Murray, and W. Marcenes, Global burden of untreated caries: a systematic review and metaregression, *J Dental Res*, 94, 650–658 (2015).
72. A.C.R.L. Caiado, M.F. de Goes, F.J. de Souza-Filho, and F.A. Rueggeberg, The effect of acid etchant type and dentin location on tubular density and dimension, *J Prosthet Dentistry*, 103, 352–361 (2010).
73. M. Fang, R. Liu, Y. Xiao, F. Li, D. Wang, R. Hou, and J. Chen, Biomodification to dentin by a natural crosslinker improved the resin-dentin bonds, *J Dentistry*, 40, 458–466 (2012).
74. A. Tezvergil-Mutluay, K.A. Agee, A. Mazzoni, R.M. Carvalho, M. Carrilho, I.L. Tersariol, F.D. Nascimento, S. Imazato, L. Tjäderhane, L. Breschi, F.R. Tay, and D.H. Pashley, Can quaternary ammonium methacrylates inhibit matrix MMPs and cathepsins?, *Dental Mater*, 31, e25-32 (2015).
75. S. Islam, N. Hiraishi, M. Nassar, C. Yiu, M. Otsuki,and J. Tagami, Effect of natural cross-linkers incorporation in a self-etching primer on dentine bond strength, *J Dentistry*, 40, 1052–1059 (2012).
76. C. Chaussain-Miller, F. Fioretti, M. Goldberg, and S. Menashi, The role of matrix metalloproteinases (MMPs) in human caries, *J Dental Res*, 85, 22–32 (2006).
77. D. Wang, L. Zhang, J. Fan, F. Li, K. Ma, P. Wang, and J. Chen, Matrix metalloproteinases in human sclerotic dentine of attrited molars, *Archives Oral Biol*, 57, 1307–1312 (2012).
78. J. Lin, C. Mehl, B. Yang, and M. Kern, Durability of four composite resin cements bonded to dentin under simulated pulpal pressure, *Dental Mater*, 26, 1001–1009 (2010).
79. J.-G. Park, Q. Ye, E.M. Topp, E.L. Kostoryz, Y. Wang, S.L. Kieweg, and P. Spencer, Preparation and properties of novel dentin adhesives with esterase resistance, *J Appl Polym Sci Symp*, 107, 3588–3597 (2008).

80. Y.-J. Jung, H.-K. Hyun, Y.-J. Kim, and K.-T. Jang, effect of collagenase and esterase on resin-dentin interface: a comparative study between a total-etch adhesive and a self-etch adhesive, *Amer J Dentistry*, 22, 295–298 (2009).
81. M. Diolosà, I. Donati, G. Turco, M. Cadenaro, R. Di Lenarda, L. Breschi, and S. Paoletti, Use of methacrylate-modified chitosan to increase the durability of dentine bonding systems, *Biomacromolecules*, 15, 4606–4613 (2014).
82. R. Pilo and A. Ben-Amar, Comparison of microleakage for three one-bottle and three multiple-step dentin bonding agents, *J Prosthet Dentistry*, 82, 209–213 (1999).
83. C.L. Davidson andA.I. Abdalla, Effect of thermal and mechanical load cycling on the marginal integrity of Class II resin composite restorations, *Amer J Dentistry*, 6, 39–42 (1993).
84. A. Misra, P. Spencer, O. Marangos, Y. Wang, and J.L. Katz, Micromechanical analysis of dentin/adhesive interface by the finite element method, *J Biomed Mater Res B Appl Biomater*, 70, 56–65 (2004).
85. A. Misra, P. Spencer, O. Marangos, Y. Wang, and J.L. Katz, Parametric study of the effect of phase anisotropy on the micromechanical behaviour of dentin-adhesive interfaces, *J R Soc Interface*, 2, 145–157 (2005).
86. M. Toledano, F.S. Aguilera, M. Yamauti, M.E. Ruiz-Requena, and R. Osorio, *In vitro* load-induced dentin collagen-stabilization against MMPs degradation, *J Mech Behav Biomed Mater*, 27, 10–18 (2013).
87. K. Khosravi, S.-M. Mousavinasab, and M.S. Samani, Comparison of microleakage in Class II cavities restored with silorane-based and methacrylate-based composite resins using different restorative techniques over time, *Dental Res J (Isfahan)*, 12, 150–156 (2015).
88. G. Bergenholtz, Evidence for bacterial causation of adverse pulpal responses in resin-based dental restorations, *Crit Rev Oral Biol Med*, 11, 467–480 (2000).
89. A. Bogović, J. Nižetić, N. Galić, D. Zelježić, V. Micek, and M. Mladinić, The effects of hyaluronic acid, calcium hydroxide, and dentin adhesive on rat odontoblasts and fibroblasts, *Arh Hig Rada Toksikol*, 62, 155–161 (2011).
90. N.D. Chandwani, M.G. Pawar, J.V. Tupkari, and M. Yuwanati, Histological evaluation to study the effects of dental amalgam and composite restoration on human dental pulp: an *in vivo* study, *Med Princ Pract*, 23, 40–44 (2014).
91. A. Sengün, M. Yalçın, H.E. Ülker, B. Öztürk, and S.S. Hakkı, Cytotoxicity evaluation of dentin bonding agents by dentin barrier test on 3-dimensional pulp cells, *Oral Surg Oral Med Oral Pathol Oral Radiol Endod*, 112, e83-88 (2011).
92. S. Krifka, C. Seidenader, K.-A. Hiller, G. Schmalz, and H. Schweikl, Oxidative stress and cytotoxicity generated by dental composites in human pulp cells, *Clin Oral Investig*, 16, 215–224 (2012).
93. A.P.P. Fugolin and C.S. Pfeifer, New resins for dental composites, *J Dental Res*, 96, 1085–1091 (2017).

94. J.W.V. van Dijken, and U. Pallesen, Bulk-filled posterior resin restorations based on stress-decreasing resin technology: a randomized, controlled 6-year evaluation, *Eur J Oral Sci*, 125, 303–309 (2017).
95. A. Van Ende, J. De Munck, D.P. Lise, and B. Van Meerbeek, Bulk-fill composites: a review of the current literature, *J Adhesive Dentistry*, 19, 95–109 (2017).
96. M.S. Ibrahim, I.M. Garcia, A. Kensara, A.A. Balhaddad, F.M. Collares, M.A. Williams, A.S. Ibrahim, N.J. Lin, M.D. Weir, H.H.K. Xu, and M.A.S. Melo, How we are assessing the developing antibacterial resin-based dental materials? A scoping review, *J Dentistry*, 99 (2020).
97. M.A.S. de Melo, *Designing Bioactive Polymeric Materials For Restorative Dentistry*, CRC Press (2020).
98. F.W. Degrazia, V.C.B. Leitune, I.M. Garcia, R.A. Arthur, S.M.W. Samuel, and F.M. Collares, Effect of silver nanoparticles on the physicochemical and antimicrobial properties of an orthodontic adhesive, *J Appl Oral Sci*, 24, 404–410 (2016).

11

Role of Adhesive-Based Systems for Diagnostic Imaging and Theranostic Applications

Aishee Dey and Sudarsan Neogi*

Department of Chemical Engineering, Indian Institute of Technology Kharagpur, Kharagpur, India

Abstract

Adhesives have been popularly used in the biomedical industry for various applications including wound healing, drug delivery, etc. Recently, adhesives have also found their use in the diagnostic imaging field which comprises of manufacturing of proper adhesive-aided instruments meant to endure harsh conditions, along with instruments engineered for better comfort of patients. It also includes the use of bioadhesives as image contrast agents for better image acquisition. Further, bioadhesives find their application in the theranostic industry which uses adhesive-based systems for improved diagnosis, image acquisition, and targeted therapeutics. This chapter discusses in detail the various uses of adhesives in the fields of diagnostic imaging and theranostics.

Keywords: Adhesive, diagnostic imaging, theranostics, image acquisition, targeted therapeutics

11.1 Introduction

Adhesives are known to attach and hold materials together due to the intermolecular forces and certain other mechanical effects [1]. They can be divided into bioadhesives and synthetic adhesives. Bioadhesives can be defined as naturally occurring polymeric materials that are biocompatible in nature and generally used for various biomedical applications. The most

**Corresponding author*: sneogi@che.iitkgp.ac.in

K.L. Mittal and S. Neogi (eds.) Adhesives in Biomedical Applications, (279–312) © 2023 Scrivener Publishing LLC

popular theories of adhesion include mechanical interlocking, diffusion, electrostatic, acid-base, and adsorption/surface reaction [2–5]. Mechanical interlocking theory suggests that adhesion occurs due to the penetration of adhesives into surface irregularities. Adhesion according to diffusion theory is explained by inter-diffusion of adhesive and adherend materials. Electrostatic forces between adhesive and adherend describe the electrostatic theory. Wetting theory suggests that surface forces and molecular contact between two materials result in adhesion [6].

While many publications clearly state the role of adhesives in revolutionizing the field of wound closure, therapeutics, and drug delivery applications [7], there are only a few publications stating their involvement in other biological applications such as diagnostic imaging and theranostics. This chapter describes the use of adhesives in various diagnostic imaging methods including x-ray, magnetic resonance imaging (MRI), ultrasonography, endoscopy, fluorescence imaging, etc. Also, it provides insights into their contribution in theranostics (combination of diagnosis and targeted therapeutics) predominantly in the detection and treatment of cancers.

11.2 Role of Adhesives in Diagnostic Imaging

Medical diagnosis involves instruments and imaging devices to aid in the diagnosis of diseases and further helps in the evaluation of method of treatment with great accuracy. These devices are required to provide impeccable diagnosis and imaging while being exposed to a harsh environment with repeated disinfecting protocols including autoclaving. Hence, adhesives like epoxies and silicones have found their way into the manufacturing of these diagnostic medical devices. Also, bioadhesives have been used as couplants/contrast agents for better image acquisition while being biocompatible with human tissues. The various roles of adhesives have been explained in detail in the next sections.

11.2.1 X-Ray Imaging

An x-ray imaging system consists mainly of an x-ray film placed between a set of screen supports and fluorescent screens. When x-rays pass through the desired sample, they hit the fluorescent screens and an imaging pattern is produced [8].

An x-ray system for dental examinations is mainly comprised of an image receptor bonded to a holder using of an adhesive layer. The flexibility of the dental x-ray positioning system gives the clinician the liberty of adjustment, hence providing superior quality x-ray images along with comfort to patients. Hence, Bratslavsky *et al.* [9] invented a dental x-ray positioning system where the positioning device is flexibly connected to the receptor through an epoxy adhesive system with no uncomfortable protrusions, holding tabs, or other protruding edges associated with the receptor. The image sensor was also bonded to the holder with an epoxy adhesive system [9]. Similarly, Steck *et al.* [10] established an x-ray positioning system using a low-tack, non-toxic adhesive. Further, in an effort to mitigate the discomfort faced by patients during intraoral x-ray imaging caused by the peripheral portion of the flat recording medium, a dental adhesive gel plate bonded to the x-ray recording material was devised. Several advantages of the adhesive gel plate include ease in operation, protection provided by the plate to the x-ray film which negates the requirement of any further protection, and most importantly the whole x-ray film and plate can be repeatedly subjected to disinfection using a chemical solution. Due to repeated use, when the adhesive strength decreased, a soap wash helped in the restoration of adhesion strength proving its use for long-term purposes [11]. A barrier and cushioning apparatus for improved comfort of patients and a method to avoid contamination has been designed by Strong and Vonda [12]. The oral cushion accommodated an adhesive section for securing various accessories which helped in aligning the radiographic detection device for ease of patients. The better cushioning further helped in avoiding any tissue injury. Another similar invention related to increased comfort of patients suggested means of flexibly affixing an imaging plate and digital sensor to the backing plate using a releasable or pressure-sensitive adhesive so that the bite block can be moved within the channels to orient accordingly [13]. Finally, an adhesive system based on polyvinylpyrrolidone and silver nanoparticle x-ray markers was employed to allow visualization of interfaces and intermediate layers of hard tooth tissues and restorative materials using the x-ray method [14].

11.2.2 Magnetic Resonance Imaging

Magnetic Resonance Imaging is the most prevalent technique for diagnosis and imaging of soft tissues in the human body. Human body is made up of water molecules which consist of hydrogen and oxygen. A uniform magnetic field with a suitable radiofrequency passes through the desired region in the patient's body that holds the protons of the hydrogen atoms and is aligned in the direction of magnetic field. When radiowaves are sent to the particular location, they knock the protons out of alignment. On turning off the radiowaves, the protons realign which are picked up by the receivers to produce MR images. Magnetic contrast agents like gadolinium or manganese are used to detect minute changes in the human body [8, 15].

Open MRI systems include 'U' or 'C' shaped magnets, where two magnets are separated from each other. These systems consist of multilayer structures for support to help achieve high permeability of magnetic flux. These support layers are typically connected through adhesives like epoxy, fasteners, weldments, or a combination of all three [16–18]. In an MRI apparatus, the field gradient is generated due to superimposition of Gx, Gy and Gz static magnetic fields, which helps to specify the particular position of the object to be examined. This field gradient core arrangement has been assembled using epoxy resin to obtain high positional accuracy of coils [19].

Parkinson's disease is characterized by continuous degeneration of dopaminergic neurons which are the main source of dopamine in the mammalian nervous system, leading to cognitive deficits. The absence of dopamine can lead to visual impairments commonly found in Parkinson's disease. Hence, manganese oxide nanoparticles tagged with 3,4-dihydroxy-L-phenylalanine (L-DOPA) were developed by McDonagh *et al.* [15]. Mn^{2+} ions can enter the neurons through the Ca^{2+} ion channels due to their analogous nature which helps in magnetic resonance imaging. When injected in the body, dopamine tagged manganese oxide nanoparticles disintegrate in water and Mn^{2+} ions are released resulting in a positive contrast for MRI. Thus, the resultant system was successfully used for enhanced imaging of Parkinson's disease with the help of the magnetic resonance imaging technique [15].

Conventional MRI technique in biological systems is primarily based on contrast agents which work by modulating magnetic effects on ^{1}H nuclei of water. However, to obtain more complementary information, current research has focused on other capable imaging contrast agents containing heteronuclei. Hence, Orlandi *et al.* [20] have suggested fluorinated compounds due to their high signal sensitivity and negligible presence in living organisms. Perfluoro-*tert*-butoxy conjugated 3,4-dihydroxy-L-phenylalanine was synthesized, where the presence of fluorine atoms helped in obtaining prominent and intense ^{19}F NMR singlet signal. The ^{19}F NMR singlet signal was easy to detect, making it apt as a probe in ^{19}F magnetic resonance imaging applications [20].

11.2.3 Ultrasound Imaging

Real-time quantitative physiological and anatomical data acquisition through ultrasound imaging is a well-established diagnostic tool. In addition to providing high-resolution images, the lack of any radiation along with low purchase and maintenance costs make it popular for long-term monitoring of patients [21]. Typically, the probe used for ultrasound imaging comprises a piezoelectric element for transmission and receiving of ultrasound waves, an acoustic lens, an acoustic matching layer for adjustment of acoustic impedance, and a backing layer. An epoxy resin adhesive or silicone resin adhesive has been suggested for the integration of the above-mentioned units [22].

For superior ultrasonic images, conductive gel-based couplants are used between transducers and patients to focus on the required region of interest [23]. Andersen and Olsen [24] proposed the use of silicone gel in place of traditional hydrogel couplants for better adhesion to skin [24]. Conventional couplants do not possess sufficient adhesive property, hence acting as a disadvantage. On the other hand, the proposed silicone gel with better adhesive property is capable of transmitting ultrasound pulses without any significant interference which is highly desirable for a clearer image of the region of interest. These gels reduce/remove echoes and provide acoustic impedance matching with skin or probe surface [24, 25]. When applied intraorally, the couplants are expected to be biocompatible, possess excellent deformability to suit the contour of the gums, and be stable in water [26]. Aquaflex, a commercially used couplant, although suitable for external use, cannot be used for intraoral purposes due to lack of the above-mentioned qualities. Hence, Yi *et al.* [27] reported a mussel-inspired poly(vinyl alcohol)-polyacrylamide-polydopamine (PVA-PAM-PDA) double network hydrogel as the ultrasound couplant candidate for dento-periodontal tissue imaging and periodontal disease diagnosis [27]. The synthesized PVA-PAM-PDA hydrogel supersedes the commercial Aquaflex by showing better mechanical properties, higher stability in water, superior adhesive property, and most importantly significantly good image quality suitable for diagnostic purposes. Thus adhesive-based biocompatible ultrasound couplants help in gaining new insights into interfaces between human tissues and medical devices for disease diagnosis applications [27]. Similarly, while designing a coupling agent for use for wearable ultrasound patches some factors need to be considered to establish the desired acoustic coupling. These include the longevity of the couplant along with its ability not to interfere with the adhesion of the patch. Beckers *et al.* [28] proposed an ultrasound transducer on a wearable patch usable for long-term purposes. Polyvinylpyrrolidone (PVP) or poly(vinyl alcohol) (PVA) was used as an adhesive which was applied to the ultrasound transducer. PVA dissolved in water has a specific advantage of providing acoustic impedance like on a patient's skin, hence resulting in a coupling fluid which facilitates superior quality ultrasound images and hence was preferred [28].

Conventional ultrasound diagnosis although effective, faces the threat of bacterial transmission between patients. Generally, the ultrasound probes are wiped with a paper towel to remove the coupling gels, which not only lead to cross-contamination but also pose damage to the acoustic lens which is the most crucial acoustic part. Some other commonly used disinfecting methods include the use of UV rays, alcohol, Tristel wipes, quaternary ammonium compounds, or glutaraldehyde-based disinfectants [29–31]. Although effective against some bacteria, these are not effective against all pathogens and also may lead to probe cracking or peeling, hence reducing their service life. Thus, acrylate pressure-sensitive adhesive (PSA) tapes have been suggested by Qin *et al.* [32] which act as a covering to resist cross-infections. These PSA tapes are easy to adhere to and do not leave any adhesive residue on the patients. Secondly, the PSA tapes are a replacement for conventional coupling gels used for obtaining good ultrasound images, thus eliminating the requirement of wiping using paper towels or other methods leading to a reduction of cross-infection. Further, a PSA tape used as a cover on the probe increases the life of the acoustic lens due to reduced usage of chemical disinfection. Finally, the PSA tapes provide a great clarity of ultrasonic images and thus have a huge prospect for ultrasonic examinations as shown in Figure 11.1 [32].

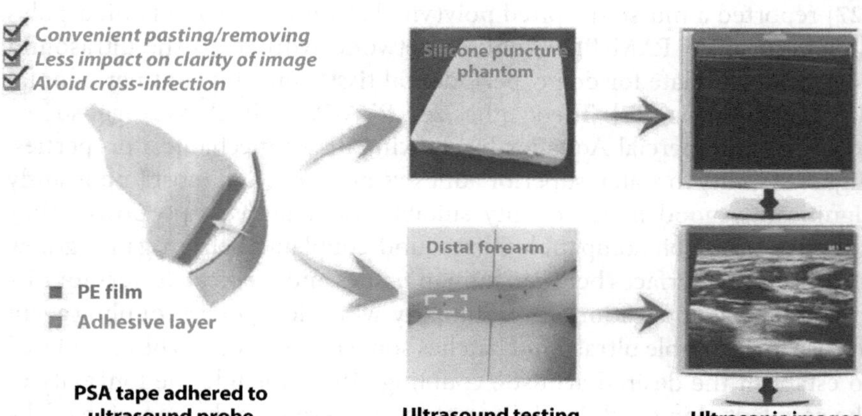

Figure 11.1 Use of PSA tapes for ultrasound testing with good image quality and avoiding cross-infections [32]. Reprinted from *Surfaces and Interfaces*., 27, Y. Qin, A. Zhang, X. Wang, P. Mao, Y. Xia, S. Oliver, T. Wang, L. Lei, B. Fan, and S. Shi, Acrylate pressure-sensitive adhesives tape as cover membrane for preventing ultrasound probes from cross-infections, 101503, Copyright 2021, with permission from Elsevier.

Current fiber-optic ultrasound transmitters used for biological imaging have a bandwidth of tens of MHz, preventing to obtain images at μm level. Hence, Fan *et al.* [33] proposed an ultrasound transmitter with an ultraviolet cured adhesive waveguide for an ultrasound generator. The aforementioned arrangement generates ultrasound by optical pulse excitation and thermal expansion of the adhesive, increasing the contrast by almost tenfold [33]. Existing ultrasound contrast agents are known to break down easily during circulation in the blood. Hence, an improved microcapsule contrast agent was proposed by Makuta and Tamakawa [34] which was fabricated from cyanoacrylate adhesive where polymerization took place at the gas-liquid interface. Thus, the obtained hollow microcapsule contrast agent enhanced the acoustic signals leading to better images.

11.2.4 Endoscopy

Endoscopy is a process where a thin, long tube is inserted directly into the body to image an internal organ or for treatment and minor surgery [35].

As an endoscope requires to be inserted into infectious sites such as bronchi, intestine, etc, extraordinary cleanliness is recommended to avoid cross-infection. Due to the previously mentioned reason, disinfection and sterilization using hot water/saturated water vapor, chemical solutions, active oxygen of peracetic acid, ethylene oxide gas, strong acidic water, or hydrogen peroxide plasma treatment are vastly prevalent [36–41]. Conventionally used adhesives, when exposed to such sterilization repetitively, might peel off, change color, get dissolved, and develop cracks resulting in a deteriorated appearance, depreciated performance, and overall short lifetime [36–41]. Thus, adhesives used in endoscope devices should exhibit high resilience towards these severe sterilization conditions. Such publications using adhesive compositions for resistance to sterilization have been listed in Table 11.1.

Current research on endoscopy focuses on miniature endoscopic capsule robot which offers advantages like higher convenience, virtually non-invasive diagnosis, fewer complications, and most importantly enables imaging and screening of small intestine which could not be imaged by conventional endoscopes [43]. Although an endoscopic microcapsule robot possesses many advantages, its inaccuracy and inability to control its position, orientation, and speed pose a problem in biomedical diagnostic imaging [44]. Karagozler *et al.* [45] suggested poly(dimethylsiloxane) micro-patterned adhesives to control the locomotion of the robot in the intestinal tract inspired by beetles and inchworms [45].

Table 11.1 Adhesive compositions used in an endoscope for sterilization resistance.

Main component of adhesive	Curing agent of adhesive	Sterilization procedure	Reference
Epoxy resin belonging to phenol novolac type epoxy resin, bisphenol-F-type epoxy resin, or bisphenol-A-type epoxy resin	Tertiary amine compound	Hydrogen peroxide plasma treatment	Furukawa *et al.* [36]
Epoxy resin belonging to phenol novolac type epoxy resin, bisphenol-F-type epoxy resin, or bisphenol-A-type epoxy resin	Polyamine compound	Ethylene oxide gas/hydrogen peroxide plasma	Furukawa *et al.* [37]
Epoxy resin belonging to phenol novolac type epoxy resin, bisphenol-F-type epoxy resin, or bisphenol-A-type epoxy resin	m-Xylylenediamine acrylic rubber and alumina	Low temperature plasma using hydrogen peroxide	Kobayashi *et al.* [38]

(*Continued*)

Table 11.1 Adhesive compositions used in an endoscope for sterilization resistance. (*Continued*)

Main component of adhesive	Curing agent of adhesive	Sterilization procedure	Reference
Epoxy resin belonging to phenol novolac type epoxy resin, bisphenol-F-type epoxy resin, or bisphenol-A-type epoxy resin	m-Xylylenediamine and a silica filler	Aldehyde sterilants	Yokoyama *et al.* [39]
Epoxy resin belonging to phenol novolac type epoxy resin, bisphenol-F-type epoxy resin, or bisphenol-A-type epoxy resin	Polyamine compound, and alcohol	Hydrogen peroxide plasma	Kazushi and Yoshihiro [40]
Epoxy resin and inorganic zwitterion exchanger from bismuth, antimony, zirconium, magnesium, and aluminum	Polyamine, m-Xylylenediamine, tertiary amine	High temperature, high pressure steam	Hayashi *et al.* [41]
Bisphenol-A-epoxy resin, phenol novolac epoxy resin, and acrylic rubber	m-Xylylenediamine and silica filler	Aldehyde sterilants. The adhesive composition also restricted water vapour permeation.	Hirano *et al.* [42]

11.2.5 Fluorescence Imaging

Fluorescence imaging in the field of imaging techniques uses fluorescence emitting substances to gain insight into the desired subject [46]. A light source of a specific wavelength is used for exciting the fluorophores which are then detected using appropriate instruments like fluorescence microscope, confocal microscope, etc. [47–49].

Modern orthodontics uses a light-cured adhesive for bracket bonding which although provides improved bonding but is difficult to remove [50]. The residual resin on the tooth is difficult to identify and causes unaesthetic staining [51]. Hence, addition of an agent providing colour to the adhesive would make its removal easier. Thus, an adhesive paste made of dimethacrylates, photoinitiators, and fillers along with coumarin-derived fluorescent dye (excitable at visible light) was used to distinguish the adhesive from the teeth resulting in negligible aesthetic impairment but with sufficient bond strength of brackets [52, 53].

Similarly, fluorescent dye-tagged adhesive systems allow easy and non-destructive inspection of imperfections or tears in intraluminal medical devices such as catheter assembly or stent delivery systems [54, 55]. Due to lack of experience of operators or human errors along with the absence of color in adhesives, the ultraviolet (UV) excitable fluorescent agents (derivative of benzoxazole, pyrene, or naphthalene) help in identifying the exact location for application of adhesives. Yang *et al.* [54] displayed the bonding of various parts (elongated tubular member and inflated balloon member) of an intraluminal medical device using epoxy, cyanoacrylate, or polyurethane adhesives which fluoresce upon exposure to UV light [54]. Further, the use of the previously mentioned fluorescent adhesive system was explored in the stent delivery system where the stent retaining sleeve was attached to the stent delivery catheter using the adhesive system [54].

Human-derived stem/stromal cells are widely used for cartilage tissue engineering due to their remarkable self-renewal property and capability to produce cartilage-specific matrix proteins found in the cartilage extracellular matrix [56]. Additionally, fibrin glue has been popularly used for chondrogenic differentiation of stromal cells [57]. Thus, these human-derived stromal cells along with fibrin and fluorescent nanodiamonds have been used by Wu *et al.* for long-term tracking of *in-vivo* chondrogenesis [58]. These help to overcome difficulties like multiple surgeries, donor site morbidity, and limited cell source while helping in cell proliferation, differentiation, and cellular functions [58].

11.2.6 Other Applications

Multifunctional tantalum oxide/silica core/shell nanoparticles (TSNPs) were synthesized which could be used for multiple imaging techniques including ultrasonography, computed tomography, and x-ray fluoroscopy. The particles are comprised of a radiopaque core made of tantalum oxide along with a tissue adhesive shell made of silica. When compared to commercially available alternatives i.e. SiO_2, iopamidol, and CA-Lp agents for fluoroscopy, ultrasonography, and computed tomography, respectively, TSNPs exhibited better contrast and had excellent adhesion strength. When the signal-to-noise ratio, a measure of signal attenuation in fluoroscopy, was measured, TSNPs showed 15 times higher attenuation compared to SiO_2. Moreover, comparable contrast effects were observed for TSNPs when compared with CA-Lp and iopamidol as the images showed prominent difference among cancer and healthy liver tissues. Also, the synthesized NPs showed significantly less cellular toxicity and inflammation, which is highly desirable in adhesives used for imaging procedures. Thus, due to the atypical properties found in nanomaterials, they have a significant possibility of replacing conventional image contrast agents [59].

Methylene blue-stained N-butyl cyanoacrylate and N-octyl cyanoacrylate glues were used for identification of pulmonary small nodules before thoracoscopic resection using a computed tomography scan. In contrast to the conventional method of using dyes which faces the disadvantage of rapid diffusion and difficulty in visual inspection, the methylene blue dye stains the pleural surface while the solidified glue helps to make the lung parenchyma palpable. Further, with the sealing of the needle track, the frequency of complications related to puncture decreases [60, 61].

Surface enhanced Raman scattering (SERS) is a popularly used method to amplify weak Raman signals obtained from visible light excitation. When laser is excited on roughened surfaces of gold or silver nanostructures, a highly localized plasmonic light field is created, enhancing the Raman signals by several orders of magnitude. Thus, lower concentrations are possible to detect through this method. Au-Ag core-shell nanostructures with an intermediate layer of Poly(3,4-dihydroxy-L-phenylalanine) were fabricated for cellular imaging. The obtained NPs showed stable and enhanced surface-enhanced Raman spectroscopy (SERS) beneficial for imaging purposes. Moreover, functionalization with folic acid helped in targeting and imaging lung cancer cells due to the ample presence of folic acid receptors on the cells. Hence, folic acid tagged Au-Ag core-shell nanostuctures with a poly-DOPA intermediate layer acted as a promising platform for SERS imaging of cancer cells [62].

11.3 Theranostics

Theranostics is the combination of efficient diagnosis and targeted treatment. Nanoparticles intended for theranostics are engineered for targeted and specific treatment of the disease along with diagnostic capability with imaging-based monitoring using systems such as MRI, CT, photoacoustic imaging (PA), or optical imaging [63].

Bioadhesives are characterized by their strong electrostatic interaction and surface free energy properties, which exhibit prominent adhesive nature with tumors, epithelial barriers, and associated mucous membranes [64]. These bioadhesives are specifically tailored to facilitate contact and prolong residence time in targeted areas, ensuring improved image acquisition and better therapeutic activity [64]. Deng *et al.* [65] ensured prolonged stay of bioadhesive nanosystem by interaction with mesothelial cell lining. Similar adhesive capability of bioadhesive particle systems was evaluated by Saltzman and coworkers [66]. In both cases, the bioadhesive systems were suggested to form bonds by Schiff-base interaction with proteins, leading to enhanced particle adhesion and prolonged stay on protein-rich surfaces.

11.3.1 DOPA (3,4-dihydroxyl-L-phenylalanine) and Dopamine

Dopamine is a small molecule mimic of adhesive proteins of mussels. Polydopamine (PDA), which can be obtained by simple self-polymerization of dopamine is known to exhibit excellent biocompatibility and *in-vivo* stability. The indigenous adhesive nature of PDA helps in the use of this material for magnetic resonance (MR), computed tomography (CT), or other forms of imaging [67]. DOPA (3,4-dihydroxyl-L-phenylalanine) and Polydopamine (PDA) have been widely used for magnetic resonance imaging-guided photothermal therapy for the treatment of cancer. Upon introduction of PDA or DOPA on the surface of nanoparticles, water can be trapped leading to better MR contrast signals. Furthermore, due to the strong absorption of these in the near-infrared region, a photothermal effect can be obtained [68, 69]. Table 11.2 lists the uses of DOPA in theranostic applications.

Table 11.2 Polydopamine-based systems for theranostic applications.

Particle description	Therapeutic method	Diagnostic method	Reference
$MnCO_3$ - PDA core-shell NPs	Photothermal therapy (PTT) for ablation of tumors.	MRI	Cheng *et al.* [68]
Metal-organic framework-based nanogels by hybridization of dopamine monomer in the skeleton of MnCo, functionalized with poly(ethylene glycol).	PTT of tumors. Poly(ethylene glycol) helps in increasing blood circulation time whereas thiol terminal cyclic arginine–glycine–aspartic acid peptide helps in achieving higher tumor accumulation leading to better therapeutic efficacy	MRI	Wang *et al.* [70]
Hollow manganese oxide was surface functionalized with DOPA. DOPA helped in the anchoring of organic molecules or polymers on the surface of metal oxides.	Nanoparticles attached to Herceptin, a therapeutic monoclonal antibody that specifically binds to human epidermal growth factor receptor 2, which is found to be present in breast and ovarian cancers	MRI	Bae *et al.* [71]
PDA-iron NPs with indocyanine green to increase the NIR absorption	PTT of tumor	MRI	Hu *et al.* [67]

(*Continued*)

Table 11.2 Polydopamine-based systems for theranostic applications. (*Continued*)

Particle description	Therapeutic method	Diagnostic method	Reference
Fe_3O_4-PDA core-shell nanocomposites	PTT	Used for mRNA detection (single-stranded ribonucleic acids, which are markers for disease and stage of disease)	Lin *et al.* [69]
$CaCO_3$–PDA composite NPs loaded with chlorin e6.	Photodynamic therapy (PDT) is a treatment procedure known to combine light energy and a photosensitizer tailored to kill cancerous cells post light activation. The particles showed increased degradation in the acidic environment of the tumor, increasing the photoactivity of the photosensor, finally leading to high singlet oxygen generation.		Dong *et al.* [72]

(*Continued*)

Table 11.2 Polydopamine-based systems for theranostic applications. (*Continued*)

Particle description	Therapeutic method	Diagnostic method	Reference
PDA/ human serum albumin/ doxorubicin coated bismuth selenide NPs	Bismuth is widely known as a therapeutic agent with a high x-ray attenuation coefficient justifying its use for CT imaging, whereas selenium is an essential element known to restrict the growth of various cancers occurring in the lung, liver, and prostate.	CT and infrared thermal imaging	Li *et al.* [73]
Nd^{3+}-doped $NaYF_4$ nanophosphors/ $NaLuF_4$/PDA	PTT for tumors	CT scan and photoluminescence from second layer of $NaLuF_4$.	Dai *et al.* [74]
PDA-coated gold NPs	PDA facilitates successful loading of cisplatin, arginine-glycine-aspartic acid (RGD) peptide, and labeling with chelator-free iodine-125. The particles showed high selectivity towards $\alpha v\beta_3$ integrin, and subsequently could accumulate in tumors	CT and photoacoustic imaging	Zhang *et al.* [75]

(*Continued*)

Table 11.2 Polydopamine-based systems for theranostic applications. (*Continued*)

Particle description	Therapeutic method	Diagnostic method	Reference
PDA-coated gold NPs.	Dual PTT and PDT therapeutic effects for killing cancer cells via an apoptotic pathway with a mild increase in temperature at low laser power over a short exposure time	SERS probes were used for monitoring the process	Kumar *et al.* [76]

11.3.2 Gelatin

Gelatin is a natural, biocompatible, and biodegradable polyampholyte protein [77]. It is a versatile material used in the biomedical industry due to its ability to be designed as nanoparticles, microparticles, fibers, and hydrogels for delivery of bioactive molecules or intravenous delivery of the drug. Thus, engineered gelatin has been used for targeted spatiotemporal release in response to demands in the body [78].

Iron oxide tagged with gelatin has been used for theranostic applications. Amphiphilic gelatin/iron oxide coated with doxorubicin (DOX) containing calcium phosphate outer shell was used to treat cancer. The outer shell acted as a pH-responsive drug release system which ensured selective delivery of the drug at low pH found in cancer cells [79]. Similarly, ease of iron oxide NPs to be tagged by gelatin due to amphiphilic segments present in gelatin has been reported by Cheng *et al.* [80]. Furthermore, the NPs were tagged with fluorescein isothiocyanate (FITC) and Pt(IV). While the iron oxide acted as an MRI contrast agent, FITC helped in the monitoring of the anticancerous effect through fluorescence imaging. The conversion of Pt(IV) to Pt(II) in an intracellular environment acted as an anticancer agent [80]. Another analogous theranostic system includes silica-coated iron oxide functionalized with oleic acid and gelatin. The magnetic core helped in providing prominent MR contrast for imaging, whereas the oleic acid-gelatin shell helped in loading of anticancer Taxol® drug [81].

Gold and gelatin-based theranostic systems have equally gained importance in the biomedical industry. A pH and temperature-sensitive gelatin-coated gold NP system functionalized with doxorubicin (DOX) has been suggested by Suarasan *et al.* [82] for the treatment of breast cancer. The accumulation of NPs and release of DOX were successfully monitored using fluorescence techniques and confocal Raman microscopy [82]. The gold-gelatin-DOX system when further functionalized with a tandem peptide RRGD was used to treat glioma. RGD is known to target $\alpha_v\beta_3$ (a transmembrane receptor responsible for cell-cell or cell-extracellular matrix adhesion), found abundantly on glioma cells. Thus, the system obtained was able to deliver doxorubicin to glioma cells while monitoring the process through fluorescence *in vivo* imaging [83]. The fluorescent property of gold NPs with adhesive characteristic of gelatin combined, along with the cytotoxic effect of anticancer drugs provided multimodal applications for cancer therapy and imaging.

Melanin-gelatin nanoparticles were found as a promising diagnostic and therapeutic platform for tumor targeting [84]. The particles accumulated in tumors by enhanced permeability and retention (EPR) effect and exhibited excellent photoacoustic imaging and MR imaging when injected intravenously in tumor-bearing mice. Additionally, the particles were cleared completely due to the shrinkable nature of the particles and no toxicity was observed in major organs [84].

Triple-negative breast cancer (TNBC) is one of the most life-threatening cancers. Angiopep-2 was anchored with gelatin/doxorubicin/dendrigraft poly-L-lysine NPs to target lipoprotein- receptor-related protein found on TNBC. This helped in achieving high accumulation in breast cancer cells along with deep penetration within them leading to a tumor growth inhibition rate of 74.1% which proved to be effective for the treatment of TNBC as monitored through fluorescence imaging [85].

11.3.3 Chitosan

Chitosan is a natural carbohydrate widely popular due to its mucoadhesive, biodegradable, and biocompatible nature and is known for its enhanced penetration property [86]. Furthermore, its ability to be easily derivatized with photosensitizers, imaging agents, biomarkers, and therapeutic agents has a large potential to be utilized in real-time cancer therapy after the systematic evaluation of their cytotoxicity and immunotoxicity and completion of preclinical studies [87]. Table 11.3 lists the chitosan-based systems for theranostic applications.

Table 11.3 Chitosan-based systems for theranostic applications.

Particle description	Therapeutic method	Diagnosis method	Reference
Chitosan NPs loaded with photosensitizer chlorin e6	Iodinated chitosan-chlorin e6 conjugate was found to accelerate singlet oxygen production due to the intraparticle heavy-atom effect, which was used to treat breast cancer cells.	Fluorescence Imaging	Lim *et al.* [88]
Glycol chitosan particles loaded with chlorin e6	Particles showed effective accumulation in tumor cells along with high photodynamic therapy efficiency.	Fluorescence Imaging	Lee *et al.* [89]
Folic acid-functionalized CuS-chitosan nanosystem	The presence of chitosan and folic acid ensured good biocompatibility and higher cellular uptake. Resultant particles exhibited near-infrared absorption with high PTT efficiency.	Photoacoustic Imaging	Yu *et al.* [90]

(Continued)

Table 11.3 Chitosan-based systems for theranostic applications. (*Continued*)

Particle description	Therapeutic method	Diagnosis method	Reference
IR 820 dye-tagged glycol chitosan-coated polycaprolactone NPs.	Hyperthermic cell death was observed in breast cancer cells via apoptosis and necrosis due to increased uptake of the particles.	Fluorescent Imaging	Kumar and Srivastava [91]
Chitosan-polypyrrole nanocomposites	Particles showed photothermal ablation of drug-resistant cancer.	Photoacoustic Imaging	Manivasagan *et al.* [92]
Chitosan oligosaccharide lactate and the ZW800-1 NIR fluorophore NPs	Photothermal treatment showed no recurrence of tumors upon a single dose of the NPs.	Fluorescence Imaging	Lee *et al.* [93]
Palladium-tagged chitosan oligosaccharide particles functionalized with RGD peptide	Palladium facilitated the PTT, whereas RGD helped in targeting the cancer cells.	Photoacoustic Imaging	Bharathiraja *et al.* [94]

(*Continued*)

Table 11.3 Chitosan-based systems for theranostic applications. (*Continued*)

Particle description	**Therapeutic method**	**Diagnosis method**	**Reference**
Natural fluorescent protein R-phycoerythrin conjugated chitosan-silver nanoclusters	The NPs were found to be effective in the treatment of triple-negative breast cancer (MDA-MB-231) cells. A stable fluorescent nanocomposite resulted from ionic interaction between the protein and chitosan-silver complex, which helped in real-time monitoring of the therapeutic activity of the nanocomposite. The obtained composite showed selective toxicity towards cancer cells while being biocompatible towards healthy cells.	Fluorescent Imaging	Thangam *et al.* [95]

(*Continued*)

Table 11.3 Chitosan-based systems for theranostic applications. (*Continued*)

Particle description	Therapeutic method	Diagnosis method	Reference
Magnetic graphene NPs functionalized with chitosan	The particles were used for the delivery of genes and super magnetic iron oxide to tumor cells. The particles proved to be effective in the delivery of genes and chemotherapeutic drugs in cancer cells while monitoring the real-time process through magnetic resonance imaging.	MRI	Wang *et al.* [96]

11.3.4 Poly(lactic-co-glycolic acid)

US Food and Drug Administration (FDA) and the European Medicine Agency (EMA) approved poly(lactic-co-glycolic acid) (PLGA) for use for drug delivery and other biomedical applications. The ratio of glycolic acid to lactic acid plays a major role in their *in vivo* distribution. Moreover, the flexibility of PLGA controlled with peptides, imaging probes, or proteins offers an advantage for theranostic purposes [97].

Mn^{2+} coordinated on the surface of PDA modified DOX-PLGA NPs were developed for chemo-photothermal therapy of tumors. While PDA acted as a photothermal agent, DOX was used for chemotherapy. Furthermore, the presence of Mn^{2+} produced high MR contrast, where the production of Mn^{2+} was sensitive to temperature and pH. The particles obtained produced a synergistic effect to inhibit tumor growth monitored through MR imaging [98]. Biocompatible Prussian blue NPs were coated with poly(lactic-co-glycolic acid) followed by conjugation with folic acid. The system obtained was useful for MRI as well as photoacoustic imaging which helped in monitoring the PTT treatment of breast cancer [99].

Similarly, PLGA coated DOX tagged upconversion NPs (particles with unique optical property which can convert two lower energy photons into one higher energy photon) with dual imaging characteristics were reported by Zhao *et al.* [100]. The upconversion core ($NaYF_4$:Yb,Er-$NaGdF_4$ core-shell structure) helped in providing bright upconversion fluorescence as well as acted as a contrast agent for MR imaging. These particles delivered

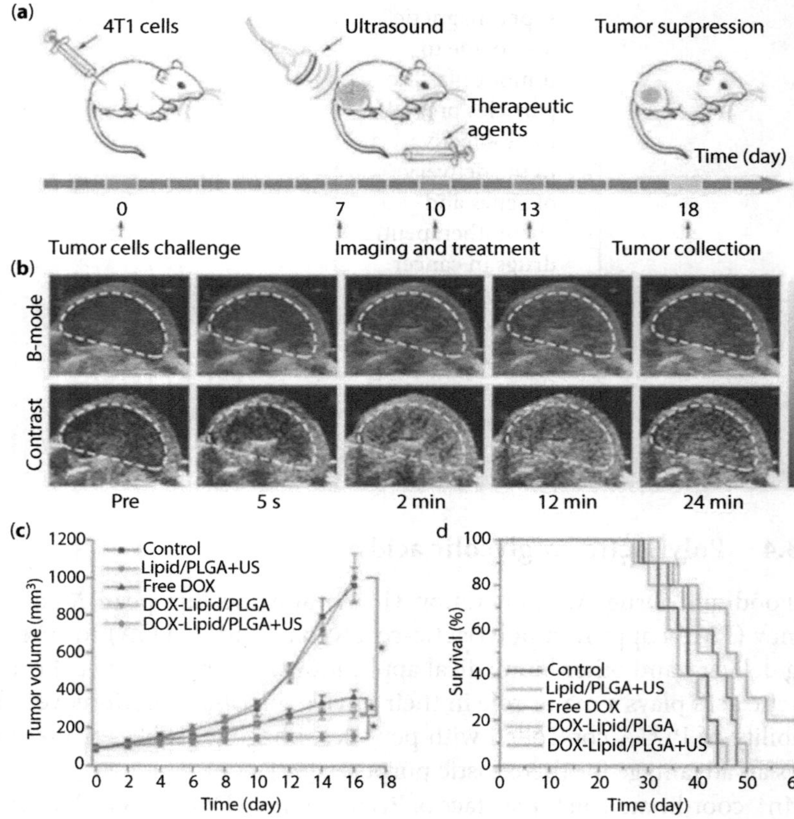

Figure 11.2 *In vivo* antitumor efficacy of DOX-lipid/PLGA microbubbles combined with ultrasound. (a) Schematic illustration showing treatment procedure and outcome of therapy by using DOX-lipid/PLGA microbubbles with ultrasound in 4T1 tumor. (b) Ultrasound contrast and B-mode imaging of DOX-lipid/PLGA microbubbles accumulation in tumors. (c) Tumor growth curves of the mice with different treatments. (d) Survival curves of 4T1 tumor-bearing mice after various treatments indicated [101]. Reprinted with permission from Y. Chen, Y. Liang, P. Jiang, F. Li, B. Yu, and F. Yan, Lipid/PLGA hybrid microbubbles as a versatile platform for noninvasive image-guided targeted drug delivery. *ACS Appl. Mater. Interfaces.* 11, 41842–41852 (2019). Copyright 2019 American Chemical Society

DOX in response to acidic pH, hence treating lung cancer while being biocompatible to healthy cells.

Tumor cells are characterized by irregular vascularization due to the production of proteins that belong to the vascular endothelial growth factor family. Thus, in an attempt to target the cancer cells, PLGA NPs were functionalized with vascular endothelial growth factor (VEGF)-A165, where A165 refers to an inflammatory cytokine known to promote pathological angiogenesis in tumour growth. This combined with a near-infrared fluorophore allowed deep tissue penetration imaging which helped in investigating the treatment of tumor cells [97].

Microbubbles have gained popularity as ultrasound imaging contrast agents with drug delivery capability. These drug-loaded microbubbles when irradiated with ultrasound energy release the drugs in the targeted spots. The conventional lipid microbubbles face the disadvantages of poor drug loading and weak contrast imaging capability. Hence, Chen *et al.* [101] developed lipid/PLGA hybrid microbubbles with soft shells and porous bubble structures able to generate strong signals leading to excellent ultrasound images and ultrasound-triggered drug delivery as shown in Figure 11.2. Thus, the microbubble-based theranostic system showed promising ultrasound imaging with precise spatiotemporal drug delivery [101]. A major problem faced today is multiple drug-resistant tumors, which can reduce the efficacy of chemotherapy. PLGA nanobubbles functionalized with DOX and P-glycoprotein were used for ultrasonic imaging-guided therapy of breast cancer. These were further found to overcome the multiple drug resistance seen in tumours [102].

11.3.5 Poly(acrylic acid)

Poly(acrylic acid) (PAA), a biodegradable and biocompatible polymer, can be chemically modified for various biomedical applications. Its stimuli-responsive nature makes it ideal for drug delivery, antimicrobial, and theranostics [103].

Metal oxide-loaded PAA has been widely used for theranostics guided through MR imaging. PAA functionalized $MnCo_2O_4$ nanodots were constructed for PTT guided through MRI. The magnetic properties of the particles made them outstanding contrast agents for MRI which helped in gaining information regarding tumor diagnosis. Moreover, while they acted as PTT agents owing to strong photoabsorption in the near-infrared window, their small size ensured deep penetration in tumors, restricting their reoccurrence [104]. Similarly, $CoFe_2O_4$-PAA-Folic Acid Doxorubicin NPs were designed for the pH-sensitive drug release which ensured cytotoxicity to cancer cells while being biocompatible to healthy cells. The whole process was monitored through MRI due to the contrast by metal ion release [105]. $MnCO_3/Mn_3O_4$-PAA coated gold nanoparticles were used for CT scans and MRI contrast agents. When functionalized with folic acid and loaded with DOX, the system was used for the targeted treatment of folate receptor overexpressed breast cancer. The affinity of particles towards cancer cells was 5.9 times higher than normal cells, making them effective anticancer agents [106].

PAA/silica encapsulated gold NPs loaded with DOX were synthesized as chemotherapy agents for use in dual fluorescence and computed tomography techniques. The particles led to efficient tumor ablation, proving their usefulness as theranostic agent [107]. Similarly, PAA/calcium phosphate encapsulated gold nanoclusters were used for fluorescence contrast imaging and CT-guided efficient chemotherapy of hepatocarcinoma [108]. PAA functionalized bismuth fluoride nanocarriers were synthesized to encapsulate DOX. These acted as pH-sensitive drug release nanosystems used for successful treatment of cancer imaged through CT scan [109]. Next-generation theranostic agents include superparamagnetic iron oxide/PAA nanobubbles which can be monitored through ultrasound and magnetic resonance. Combined magnetic targeting and ultrasound-triggered drug release ensured tumor therapy [110].

11.4 Summary

In the past years, adhesives have found their extensive use in the biomedical industry. The prominent use of epoxy and silicone adhesives has been found in the manufacturing of x-ray, MRI, ultrasound, endoscopes, fluorescent microscopes, and CT scan-based instruments. These adhesives ensure proper assembly of the instruments while making them resistant to harsh autoclave conditions. Also, adhesives make the instruments flexible, hence aiding in their operation and increasing patient comfort. Next, adhesive-based systems have been commonly used as contrast agents for MRI, fluorescence imaging, or CT imaging, while being successfully used as couplants in ultrasound imaging. Finally, bioadhesives like PDA, chitosan, PLGA, gelatin, PAA, etc. have been widely used for better attachment of the theranostic agents to the targeted tissues, ensuring better drug delivery or therapeutic agents, while being biocompatible to other healthy tissues. Given the need for various adhesives in the biomedical industry, this chapter provides additional insights into adhesive-based systems designed for the intended purpose of diagnostic imaging and theranostics primarily for the treatment of various cancers.

References

1. A. Dey, P. Bhattacharya, and S. Neogi, Bioadhesives in biomedical applications: A critical review, in: *Progress in Adhesion and Adhesives*, Vol 6, K.L. Mittal (Ed.) pp. 131–153, Wiley-Scrivener, Beverly, MA (2021).
2. F.M. Fowkes, Acid-base interactions in polymer adhesion, in: *Physicochemical Aspects of Polymer Surfaces*, Vol 2, K.L. Mittal (Ed.) pp. 583-603, Plenum Press, New York (1983).
3. M.M. Chehimi, A. Azioune, and E. Cabet-Deliry, Acid-base interactions: relevance to adhesion and adhesive bonding, in: *Handbook of Adhesive Technology*, second edition, A. Pizzi and K.L. Mittal (Eds.) pp. 95-144, CRC Press, Boca Raton, FL (2003).
4. D. E. Packham, The mechanical theory of adhesion, in: *Handbook of Adhesive Technology*, second edition, A. Pizzi and K.L. Mittal (Eds.) pp. 69-93, CRC Press, Boca Raton, FL (2003).
5. J. Schultz and M. Nardin, Theories and mechanisms of adhesion, in: *Handbook of Adhesive Technology*. A. Pizzi and K.L. Mittal (Eds.) pp. 19-34, Marcel Dekker, New York (1994).
6. S. Ebnesajjad, Introduction and adhesion theories, in *Handbook of Adhesives and Surface Preparation,* S. Ebnesajjad (Ed.), second Edition, pp. 3-13, William Andrew, Norwich, NY (2011).
7. M. Mehdizadeh and J. Yang, Design strategies and applications of tissue bioadhesives, *Macromol. Biosci.* 13, 271-288 (2013).
8. S.S. Ilangovan, B. Mahanty, and S. Sen, Biomedical imaging techniques. in: *Emerging Technologies in Intelligent Applications for Image and Video Processing.* V. Santhi, D. P. Acharjya and M. Ezhilarasan (Eds.), pp. 401-421, IGI Global, Hershey, Pennsylvania (2016).
9. A. Bratslavsky, A. Royzen, and S. Mandelkern, Dental x-ray positioning using adhesives, US Patent 6811312B2, assigned to Sirona Dental Inc (2003).
10. J. Steck, K. Brown, and N. Sullivan, Method and apparatus for supporting a dental x-ray sensor, US Patent 7261463B2, assigned to Darrell A Becht (2007).
11. 眞正 山川, Dental adhesive gel plate, Japanese Patent 3130398U, assigned to 眞正 山川 (2007).
12. V. Strong and M. Vonda, Barrier and cushioning apparatus for dental radiography, US Patent 7004627B2, assigned to V. Strong (2006).
13. C. Steward, Jr., P. McDonough, T. Hartlaub, N. Sullivan, and B. Kunz, Adjustable dental x-ray image media holder, US Patent 20100177875A1, assigned to Dentsply Sirona Inc. (2010).
14. N.O. Bessudnova, D.I. Bilenko, S.B. Venig, V.S. Atkin, and A.M. Zacharevich, A new x-ray adhesive system with embedded nanoparticulate silver markers for dental applications, in *Proc. Saratov Fall Meeting 2012: Optical Technologies in Biophysics and Medicine XIV*; Laser Physics and Photonics XIV. 8699, 86990R (2013).

15. B.H. McDonagh, G. Singh, S. Hak, S. Bandyopadhyay, I.L. Augestad, D. Peddis, I. Sandvig, A. Sandvig, and W.R. Glomm, L -DOPA-coated manganese oxide nanoparticles as dual MRI contrast agents and drug-delivery vehicles. *Small.* 12, 301–306 (2016).
16. J. Mao, J. Wang, H. Liu, and A. Wu, Support structure for open MRI and method and system thereof, US Patent 20080201930A1, assigned to General Electric Company (2008).
17. J. Mao, J. Wang, H. Liu, and A. Wu, Method of manufacturing support structure for open MRI, US Patent 7802354B2, assigned to General Electric Company (2009).
18. J. Schuster, Gradient coil system for MRI machine is made up of two or more units fixed into tubular casing using adhesive, DE10246310A1, assigned to Siemens AG (2004).
19. R. Takahashi, Y. Yoshida, and D. Tsuruno, Method of manufacturing a gradient magnetic field coil assembly of an MRI apparatus, US Patent 5409558A, assigned to Toshiba Corp. (1995).
20. S. Orlandi, M. Cavazzini, S. Capuani, A. Ciardello, and G. Pozzi, Synthesis and 19F NMR parameters of a perfluoro-tert-butoxy tagged L-DOPA analogue, *J. Fluorine Chem.* 237, 109596 (2020).
21. C.M. Moran and A.J.W. Thomson, Preclinical ultrasound imaging—a review of techniques and imaging applications, *Frontiers Phys.* 8, 124 (2020).
22. K. Morita, Method for manufacturing ultrasound probe and ultrasound dagnostic imaging apparatus, US Patent 2014/0249419 A1, assigned to Konica Minolta, Inc. (2014).
23. S. Afzal, M. Zahid, Z.A. Rehan, H.M. Fayzan Shakir, H. Javed, M.M.H. Aljohani, S.K. Mustafa, M. Ahmad, and M.M. Hassan, Preparation and evaluation of polymer-based ultrasound gel and its application in ultrasonography. *Gels.* 8, 42-67 (2022).
24. N. Andersen and G. Olsen, Coupling an ultrasound probe to the skin, US Patent 20130165005A1, assigned to Neorad AS (2011).
25. L. Smith, Adhesive hydrophilic membranes as couplants in ultrasound imaging applications, US Patent 6719699B2, assigned to Next Medical Products Co. (2004).
26. J. Yi, K.C.T. Nguyen, W. Wang, W. Yang, M. Pan, E. Lou, P.W. Major, L.H. Le, and H. Zeng, Polyacrylamide/Alginate double-network tough hydrogels for intraoral ultrasound imaging, *J. Colloid Interface Sci.* 578, 598–607 (2020).
27. J. Yi, K.C.T. Nguyen, W. Wang, W. Yang, M. Pan, E. Lou, P.W. Major, L.H. Le, and H. Zeng, Mussel-inspired adhesive double-network hydrogel for intraoral ultrasound imaging, *ACS Appl. Bio Mater.* 3, 8943–8952 (2020).
28. L. Beckers, F. Hakkens, E. Berben, and T. Grob, Wearable ultrasound patch and application method of such a patch, European Patent 3684262B1, assigned to Philips Intellectual Property & Standards (2021).

29. P. Bhattacharya, A. Dey, and S. Neogi, An insight into the mechanism of antibacterial activity by magnesium oxide nanoparticles, *J. Mater. Chem. B.* 9, 5329–5339 (2021).
30. P. Bhattacharya, I. Sengupta, A. Dey, S. Chakraborty, and S. Neogi, Antibacterial effect of ciprofloxacin loaded reduced graphene oxide nanosheets against *Pseudomonas aeruginosa* strain. *Colloids Interface Sci. Commun.* 40, 100344 (2021).
31. I. Sengupta, P. Bhattacharya, M. Talukdar, S. Neogi, S.K. Pal, and S. Chakraborty, Bactericidal effect of graphene oxide and reduced graphene oxide: Influence of shape of bacteria. *Colloids Interface Sci. Commun.* 28, 60–68 (2019).
32. Y. Qin, A. Zhang, X. Wang, P. Mao, Y. Xia, S. Oliver, T. Wang, L. Lei, B. Fan, and S. Shi, Acrylate pressure-sensitive adhesives tape as cover membrane for preventing ultrasound probes from cross-infections. *Surfaces and Interfaces.* 27, 101503 (2021).
33. H. Fan, L. Chen, and X. Bao, Fiber-optic ultrasound transmitter based on multi-mode interference in curved adhesive waveguide. *IEEE Photonics Technol. Lett.* 32, 325–328 (2020).
34. T. Makuta and Y. Tamakawa, Ultrasound contrast agent fabricated from microbubbles containing instant adhesives, and its ultrasound imaging ability. *Proc. Nanosensors, Biosensors and Info-Tech Sensors and Systems* 2012. 8344, 83441O (2012).
35. Endoscopy: Types, preparation, procedure & risks, https://www.medicalnewstoday.com/articles/153737 (2022).
36. K. Furukawa, T. Yoshitani, and Y. Nakai, Endoscope adhesive, cured material, endoscope, and method for manufacturing endoscope, US Patent 10973392B2, assigned to Fujifilm Corp (2021).
37. K. Furukawa, Y. Nakai, and S. Uehira, Adhesive for endoscope, cured product, endoscope, and method for producing endoscope, US Patent 20210007579A1, assigned to Fujifilm Corp (2021).
38. K. Kobayashi, R. Niino, M. Nakamura, and J. Matsumoto, Adhesive composition for medical apparatus and endoscopic device, European Patent 2765171A1, assigned to Olympus Corporation (2012).
39. H. Yokoyama, K. Kobayashi, J. Matsumoto, and M. Nakamura, Adhesive composition and endoscope device, US Patent 9822287B2, assigned to Olympus Medical Systems Corp (2017).
40. F. Kazushi and N. Yoshihiro, Adhesive for endoscope and cured product thereof, and endoscope and method for producing the same, US Patent 20210189203A1, assigned to Fujifilm Corp (2021).
41. T. Hayashi, K. Kobayashi, and H. Yokoyama, Adhesive composition, ultrasonic transducer, endoscopic device, and ultrasonic endoscopic device, US Patent 20190082937A1, assigned to Olympus Corp, assignee (2018).

42. Y. Hirano, M. Nakamura, and J. Matsumoto, Adhesive composition and endoscope, European Patent 0879946A2, assigned to Olympus Medical Systems Corp. (2012).
43. Y.P. Lee, B. Kim, M.G. Lee, and J.O. Park, Locomotive mechanism design and fabrication of biomimetic micro robot using shape memory alloy, in *Proceedings - IEEE International Conference on Robotics and Automation* 2004 pp. 5007–5012 (2004).
44. A. Menciassi, J.H. Park, S. Lee, S. Gorini, P. Dario, and J.O. Park, Robotic solutions and mechanisms for a semi-autonomous endoscope, in *Proc. EEE/RSJ International Conference on Intelligent Robots and Systems*, pp. 1379–1384 (2002).
45. M.E. Karagozler, E. Cheung, J. Kwon, and M. Sitti, Miniature endoscopie capsule robot using biomimetic micro-patterned adhesives, in *Proceedings of The First IEEE/RAS-EMBS International Conference on Biomedical Robotics and Biomechatronics*, 2006. BioRob 2006. pp. 105–111 (2006).
46. V. Ntziachristos, Fluorescence imaging. *in: Encyclopedia of Diagnostic Imaging.* A. L.Baert. (Ed.) pp. 723–726, Springer, Heidelberg. (2008).
47. A. Dey, P. Guha Ray, S. Dhara, and S. Neogi, Optically engineered ZnO nanoparticles: excitable at visible wavelength and lowered cytotoxicity towards bioimaging applications. *Appl. Surf. Sci.* 592, 153303 (2022).
48. A. Dey, S. Gare, S. Swain, P. Bhattacharya, V. Dhyani, L. Giri, and S. Neogi, 3D imaging and quantification of PLL coated fluorescent ZnO NP distribution and ROS accumulation using laser scanning confocal microscopy. *AIChE J.* e17801 (2022).
49. P. Bhattacharya, S. Swain, L. Giri, and S. Neogi, Fabrication of magnesium oxide nanoparticles by solvent alteration and their bactericidal applications. *J. Mater. Chem. B.* 7, 4141–4152 (2019).
50. N.R. Krishnaswamy and C. Sunitha, Light-emitting diode vs halogen light curing of orthodontic brackets: A 15-month clinical study of bond failures. *Amer. J. Orthod. Dentofacial Orthop.* 132, 518–523 (2007).
51. H.W. Fields, Bonded resins in orthodontics, *Pediatric Dentistry*, 4, 51-60 (1982).
52. Y. Namura, T. Tsuruoka, C. Ryu, M. Kaketani, and N. Shimizu, Usefulness of orthodontic adhesive-containing fluorescent dye. *Eur. J. Orthodontics* 32, 620–626 (2010).
53. P.H. Rossato, E.N. Kaneshima, F. Domingues, T.M.F. Fernandes, S.B. Berger, and P.V.P. Oltramari, Do fluorescent agents alter the mechanical strength of orthodontic adhesives? An *in vitro* and clinical study. *Prog. Orthodontics* 21, 1–6 (2020).
54. D. Yang, J. Stanslaski, L. Tang, and J. Scanlon, Fluorescent dyed adhesive for bonding various components in a medical device, US Patent 6461326B1, assigned to Boston Scientific Scimed Inc (2002).
55. D. Yang, L. Tang, and D. Lang, Fluorescent dyed lubricant for medical devices, US Patent 7052512B2, assigned to Boston Scientific Scimed Inc (2006).

56. P.A. Zuk, M. Zhu, H. Mizuno, J. Huang, J.W. Futrell, A.J. Katz, P. Benhaim, H.P. Lorenz, and M.H. Hedrick, Multilineage cells from human adipose tissue: implications for cell-based therapies. *Tissue Eng.* 7, 211–228 (2004).
57. M. Sheykhhasan, R.T. Qomi, and M. Ghiasi, Fibrin scaffolds designing in order to human adipose-derived mesenchymal stem cells differentiation to chondrocytes in the presence of TGF-β3. *Int. J. Stem Cells.* 8, 219–227 (2015).
58. Y.-C. Wu, Y.-C. Wang, W.-T. Wang, H.-M.D. Wang, H.-H. Lin, L.-J. Su, Y.-R. Kuo, C.-S. Lai, M.-L. Ho, and J. Yu, Fluorescent nanodiamonds enable long-term detection of human adipose-derived stem/stromal cells in an *in vivo* chondrogenesis model using decellularized extracellular matrices and fibrin glue polymer., *Polymers (Basel).* 11, 1391 (2019).
59. K. Shin, J.W. Choi, G. Ko, S. Baik, D. Kim, O.K. Park, K. Lee, H.R. Cho, S.I. Han, S.H. Lee, D.J. Lee, N. Lee, H.C. Kim, and T. Hyeon, Multifunctional nanoparticles as a tissue adhesive and an injectable marker for image-guided procedures. *Nature Commun.* 8, 1–12 (2017).
60. J. Wang, L. Bin Gao, H. Zhang, J. Wang, L. Liu, and W.Y. Shi, Preresection stained glue injection to localize pulmonary small nodules and ground-glass opacities. *J. Thoracic Imaging.* 35, 260–264 (2020).
61. G. Tao, Y. Jingying, G. Tan, D. Xiaotao, and C. Min, A novel CT-guided technique using medical adhesive for localization of small pulmonary ground-glass nodules and mixed ground-glass nodules (≤20 mm) before video-assisted thoracoscopic surgery. *Diagnostic Intervention Radiology* 24, 209–212 (2018).
62. H. Wen, P. Jiang, Y. Hu, and G. Li, Synthesis of Au@Ag core-shell nanostructures with a poly(3,4-dihydroxy-L-phenylalanine) interlayer for surface-enhanced Raman scattering imaging of epithelial cells. *Microchim. Acta.* 185, 1–8 (2018).
63. F. Chen, E.B. Ehlerding, and W. Cai, Theranostic nanoparticles, *J. Nucl. Med.* 55, 1919–1922 (2014).
64. M.K. Viswanadh and M.S. Muthu, Targeted bioadhesive nanomedicine: An effective approach for synergistic drug delivery to cancers. *Nanomedicine.* 13, 1401–1403 (2018).
65. Y. Deng, F. Yang, E. Cocco, E. Song, J. Zhang, J. Cui, M. Mohideen, S. Bellone, A.D. Santin, and W.M. Saltzman, Improved i.p. drug delivery with bioadhesive nanoparticles, *Proc. Natl. Acad. Sci. (U. S. A.)* 113, 11453–11458 (2016).
66. M. Mohideen, E. Quijano, E. Song, Y. Deng, G. Panse, W. Zhang, M.R. Clark, and W.M. Saltzman, Degradable bioadhesive nanoparticles for prolonged intravaginal delivery and retention of elvitegravir. *Biomaterials.* 144, 144–154 (2017).
67. D. Hu, C. Liu, L. Song, H. Cui, G. Gao, P. Liu, Z. Sheng, and L. Cai, Indocyanine green–loaded polydopamine–iron ions coordination nanoparticles for photoacoustic/magnetic resonance dual-modal imaging-guided cancer photothermal therapy, *Nanoscale.* 8, 17150–17158 (2016).

68. Y. Cheng, S. Zhang, N. Kang, J. Huang, X. Lv, K. Wen, S. Ye, Z. Chen, X. Zhou, and L. Ren, Polydopamine-coated manganese carbonate nanoparticles for amplified magnetic resonance imaging-guided photothermal therapy. *ACS Appl. Mater. Interfaces*. 9, 19296–19306 (2017).
69. L.-S. Lin, Z.-X. Cong, J.-B. Cao, K.-M. Ke, Q.-L. Peng, J. Gao, H.-H. Yang, G. Liu, and X. Chen, Multifunctional Fe_3O_4 @ polydopamine core A shell nanocomposites for intracellular RNA detection and imaging-guided photothermal therapy, *ACS Nano*. 8, 3876–3883 (2014).
70. D. Wang, H. Wu, J. Zhou, P. Xu, C. Wang, R. Shi, H. Wang, H. Wang, Z. Guo, Q. Chen, D. Wang, P. Xu, C. Wang, R. Shi, Q. Chen, H. Wang, H. Wu, J. Zhou, and Z. Guo, *In situ* one-pot synthesis of MOF–polydopamine hybrid nanogels with enhanced photothermal effect for targeted cancer therapy. *Advanced Science* 5, 1800287 (2018).
71. K.H. Bae, K. Lee, C. Kim, and T.G. Park, Surface functionalized hollow manganese oxide nanoparticles for cancer targeted siRNA delivery and magnetic resonance imaging. *Biomaterials*. 32, 176–184 (2011).
72. Z. Dong, L. Feng, Y. Hao, M. Chen, M. Gao, Y. Chao, H. Zhao, W. Zhu, J. Liu, C. Liang, Q. Zhang, and Z. Liu, Synthesis of hollow biomineralized $CaCO_3$-polydopamine nanoparticles for multimodal imaging-guided cancer photodynamic therapy with reduced skin photosensitivity. *J. Amer. Chem. Soc.* 140, 2165–2178 (2018).
73. Z. Li, Y. Hu, K.A.A. Howard, T. Jiang, X. Fan, Z. Miao, Y. Sun, F. Besenbacher, and M. Yu, Multifunctional bismuth selenide nanocomposites for antitumor thermo-chemotherapy and imaging. *ACS Nano*. 10, 984–997 (2016).
74. Y. Dai, D. Yang, D. Yu, C. Cao, Q. Wang, S. Xie, L. Shen, W. Feng, and F. Li, Mussel-inspired polydopamine-coated lanthanide nanoparticles for NIR-II/CT dual imaging and photothermal therapy. *ACS Appl. Mater. Interfaces*. 9, 26674–26683 (2017).
75. L. Zhang, H. Su, J. Cai, D. Cheng, Y. Ma, J. Zhang, C. Zhou, S. Liu, H. Shi, Y. Zhang, and C. Zhang, A multifunctional platform for tumor angiogenesis-targeted chemo-thermal therapy using polydopamine-coated gold nanorods. *ACS Nano*. 10, 10404–10417 (2016).
76. A. Kumar, S. Kumar, W.K. Rhim, G.H. Kim, and J.M. Nam, Oxidative nanopeeling chemistry-based synthesis and photodynamic and photothermal therapeutic applications of plasmonic core-petal nanostructures, *J. Amer. Chem. Soc.* 136, 16317–16325 (2014).
77. Y. Gou, D. Miao, M. Zhou, L. Wang, H. Zhou, and G. Su, Bio-inspired protein-based nanoformulations for cancer theranostics. *Frontiers Pharmacol.* 9, 1-16 (2018).
78. M. Foox and M. Zilberman, Drug delivery from gelatin-based systems. *Expert Opin. Drug. Deliv.* 12, 1547–1563 (2015).
79. W.M. Li, S.Y. Chen, and D.M. Liu, *In situ* doxorubicin–CaP shell formation on amphiphilic gelatin–iron oxide core as a multifunctional drug delivery

system with improved cytocompatibility, pH-responsive drug release and MR imaging. *Acta Biomater.* 9, 5360–5368 (2013).

80. Z. Cheng, Y. Dai, X. Kang, C. Li, S. Huang, H. Lian, Z. Hou, P. Ma, and J. Lin, Gelatin-encapsulated iron oxide nanoparticles for platinum (IV) prodrug delivery, enzyme-stimulated release and MRI. *Biomaterials.* 35, 6359–6368 (2014).
81. T.T.D. Tran, P.H.L. Tran, T.J. Yoon, and B.J. Lee, Fattigation-platform theranostic nanoparticles for cancer therapy. *Mater. Sci. Eng. C.* 75, 1161–1167 (2017).
82. S. Suarasan, M. Focsan, M. Potara, O. Soritau, A. Florea, D. Maniu, and S. Astilean, Doxorubicin-incorporated nanotherapeutic delivery system based on gelatin-coated gold nanoparticles: Formulation, drug release, and multimodal imaging of cellular internalization, *ACS Appl. Mater. Interfaces.* 8, 22900–22913 (2016).
83. S. Ruan, Q. He, and H. Gao, Matrix metalloproteinase triggered size-shrinkable gelatin-gold fabricated nanoparticles for tumor microenvironment sensitive penetration and diagnosis of glioma. *Nanoscale.* 7, 9487–9496 (2015).
84. T. Li, Y. Yang, W. Jing, Z. Yan, J. Che, H. Xu, X. Hu, and R. Zhang, Melanin-gelatin nanoparticles with both EPR effect and renal clearance for PA/MRI dual-modal imaging of tumors. *Biomater. Advances* 134, 112718 (2022).
85. G. Hu, X. Chun, Y. Wang, Q. He, H. Gao, G. Hu, X. Chun, Y. Wang, Q. He, and H. Gao, Peptide mediated active targeting and intelligent particle size reduction-mediated enhanced penetrating of fabricated nanoparticles for triple-negative breast cancer treatment. *Oncotarget.* 6, 41258–41274 (2015).
86. M.C. Bonferoni, E. Gavini, G. Rassu, M. Maestri, and P. Giunchedi, Chitosan nanoparticles for therapy and theranostics of hepatocellular carcinoma (HCC) and liver-targeting. *Nanomaterials.* 10, 870 (2020).
87. A.S. Soubhagya, and M. Prabaharan, Chitosan-based theranostics for cancer therapy. *Adv. Polym. Sci.* 288, 271–292 (2021).
88. C.K. Lim, J. Shin, I.C. Kwon, S.Y. Jeong, and S. Kim, Iodinated photosensitizing chitosan: Self-assembly into tumor-homing nanoparticles with enhanced singlet oxygen generation. *Bioconjugate Chem.* 23, 1022–1028 (2012).
89. S.J. Lee, H. Koo, H. Jeong, M.S. Huh, Y. Choi, S.Y. Jeong, Y. Byun, K. Choi, K. Kim, and I.C. Kwon, Comparative study of photosensitizer loaded and conjugated glycol chitosan nanoparticles for cancer therapy. *J. Controlled Release.* 152, 21–29 (2011).
90. W. Yu, N. Yu, Z. Wang, X. Li, C. Song, R. Jiang, P. Geng, M. Li, S. Yin, and Z. Chen, Chitosan-mediated green synthesis and folic-acid modification of CuS quantum dots for photoacoustic imaging guided photothermal therapy of tumor. *J. Colloid Interface Sci.* 555, 480–488 (2019).
91. P. Kumar and R. Srivastava, IR 820 stabilized multifunctional polycaprolactone glycol chitosan composite nanoparticles for cancer therapy. *RSC Advances* 5, 56162–56170 (2015).

92. P. Manivasagan, N. Quang Bui, S. Bharathiraja, M. Santha Moorthy, Y.O. Oh, K. Song, H. Seo, M. Yoon, and J. Oh, Multifunctional biocompatible chitosan-polypyrrole nanocomposites as novel agents for photoacoustic imaging-guided photothermal ablation of cancer. *Sci. Reports.* 7, 1–14 (2017).
93. S. Lee, G. Jo, J.S. Jung, D.H. Yang, and H. Hyun, Near-infra-red fluorescent chitosan oligosaccharide lactate for targeted cancer imaging and photothermal therapy. *Artificial Cells Nanomed. Biotechnol.* 48, 1144–1152 (2020).
94. S. Bharathiraja, N.Q. Bui, P. Manivasagan, M.S. Moorthy, S. Mondal, H. Seo, N.T. Phuoc, T.T. Vy Phan, H. Kim, and K.D. Lee, and J. Oh, Multimodal tumor-homing chitosan oligosaccharide-coated biocompatible palladium nanoparticles for photo-based imaging and therapy. *Sci. Reports.* 8, 1–16 (2018).
95. R. Thangam, S. Sundarraj, R. Vivek, V. Suresh, S. Sivasubramanian, M. Paulpandi, S.V. Karthick, A.S. Ragavi, and S. Kannan, Theranostic potentials of multifunctional chitosan–silver–phycoerythrin nanocomposites against triple negative breast cancer cells. *RSC Adv*ances 5, 12209–12223 (2015).
96. C. Wang, S. Ravi, U.S. Garapati, M. Das, M. Howell, J. Mallela, S. Alwarappan, S.S. Mohapatra, and S. Mohapatra, Multifunctional chitosan magnetic-graphene (CMG) nanoparticles: a theranostic platform for tumor-targeted co-delivery of drugs, genes and MRI contrast agents. *J. Mater. Chem. B.* 1, 4396–4405 (2013).
97. M. Varani, F. Galli, G. Capriotti, M. Mattei, R. Cicconi, G. Campagna, F. Panzuto, and A. Signore, Theranostic designed near-infrared fluorescent poly (lactic-co-glycolic acid) nanoparticles and preliminary studies with functionalized VEGF-nanoparticles. *J. Clinical Med.* 9, 1–13 (2020).
98. J. Xi, L. Da, C. Yang, R. Chen, L. Gao, L. Fan, and J. Han, Mn^{2+}-coordinated PDA@DOX/PLGA nanoparticles as a smart theranostic agent for synergistic chemo-photothermal tumor therapy, *Int. J. Nanomedicine.* 12, 3331–3345 (2017).
99. T. Shang, J. Liu, Y. Chen, Z. Hu, L. Deng, H. Ran, P. Li, Y. Zheng, D. Wang, Z. Wang, and Y. Sun, In vivo targeted cancer theranostics by core/shell-structured multifunctional Prussian blue/PLGA "Nanococktails," *Particle Syst. Charact.* 35, 1700306 (2018).
100. J. Zhao, H. Yang, J. Li, Y. Wang, and X. Wang, Fabrication of pH-responsive PLGA(UCNPs/DOX) nanocapsules with upconversion luminescence for drug delivery. *Sci. Reports.* 7, 1–11 (2017).
101. Y. Chen, Y. Liang, P. Jiang, F. Li, B. Yu, and F. Yan, Lipid/PLGA hybrid microbubbles as a versatile platform for noninvasive image-guided targeted drug delivery. *ACS Appl. Mater. Interfaces.* 11, 41842–41852 (2019).
102. H. Yang, X. Shen, J. Yan, X. Xie, Z. Chen, T. Li, S. Li, X. Qin, C. Wu, and Y. Liu, Charge-reversal-functionalized PLGA nanobubbles as theranostic agents for ultrasonic-imaging-guided combination therapy. *Biomater. Sci.* 6, 2426–2439 (2018).

103. H. Arkaban, M. Barani, M.R. Akbarizadeh, N.P.S. Chauhan, S. Jadoun, M.D. Soltani, and P. Zarrintaj, Polyacrylic acid nanoplatforms: Antimicrobial, tissue engineering, and cancer theranostic applications. *Polymers (Basel).* 14, 1259 (2022).
104. Y. Zhao, Y. Liu, Y. Wang, B. Xu, S. Zhang, J. Liu, T. Zhang, L. Jin, S. Song, and H. Zhang, Rapidly clearable MnCo2O4@PAA as novel nanotheranostic agents for T1/T2 bimodal MRI imaging-guided photothermal therapy, *Nanoscale.* 13, 16251–16257 (2021).
105. H. Arkaban, A. Khajeh Ebrahimi, A. Yarahmadi, P. Zarrintaj, and M. Barani, Development of a multifunctional system based on CoFe2O4@polyacrylic acid NPs conjugated to folic acid and loaded with doxorubicin for cancer theranostics, *Nanotechnology.* 32, 305101 (2021).
106. H. Arkaban, R. Karimi Shervedani, F. Yaghoobi, and A. Kefayat, A nanocomposite theranostic system, consisting of AuNPs@MnCO3/Mn3O4 coated with PAA and integrated with folic acid, doxorubicin, and propidium iodide: Synthesis, characterization and examination for capturing of cancer cells, *Inorg. Chem. Commun.* 128, 108566 (2021).
107. X. Wu, L. Li, L. Zhang, T. Wang, C. Wang, and Z. Su, Multifunctional spherical gold nanocluster aggregate@polyacrylic acid@mesoporous silica nanoparticles for combined cancer dual-modal imaging and chemo-therapy. *J. Mater. Chem. B.* 3, 2421–2425 (2015).
108. L. Li, L. Zhang, T. Wang, X. Wu, H. Ren, C. Wang, and Z. Su, Facile and scalable synthesis of novel spherical Au nanocluster assemblies@polyacrylic acid/calcium phosphate nanoparticles for dual-modal imaging-guided cancer chemotherapy, *Small.* 11, 3162–3173 (2015).
109. S. Zhao, R. Tian, B. Shao, Y. Feng, S. Yuan, L. Dong, L. Zhang, Z. Wang, and H. You, One-pot synthesis of Ln3+-doped porous BiF3@PAA nanospheres for temperature sensing and pH-responsive drug delivery guided by CT imaging. *Nanoscale.* 12, 695–702 (2020).
110. H.Y. Huang, S.H. Hu, S.Y. Hung, C.S. Chiang, H.L. Liu, T.L. Chiu, H.Y. Lai, Y.Y. Chen, and S.Y. Chen, SPIO nanoparticle-stabilized PAA-F127 thermosensitive nanobubbles with MR/US dual-modality imaging and HIFU-triggered drug release for magnetically guided *in vivo* tumor therapy. *J. Controlled Release.* 172, 118–127 (2013).

Index

Also of Interest

Check out these published and forthcoming in the Adhesion and Adhesives series

Polymer Surface Modification to Enhance Adhesion: Techniques and Applications
Edited by K.L. Mittal and Anil Netravali
Forthcoming 2024. ISBN 978-1-394-23100-3

Progress in Adhesion and Adhesives, Volume 7
Edited by K.L. Mittal
Forthcoming 2023. ISBN 978-1-394-19810-8

Adhesives in Biomedical Applications
Edited by K.L. Mittal and S. Neogi
Published 2023. ISBN 978-1-394-20920-0

Biobased Adhesives
Sources, Characteristics and Applications
Edited by Manfred Dunky and K.L. Mittal
Published 2023. ISBN 978-1-394-17463-8

Structural Adhesives
Properties, Characterization and Applications
Edited by K.L. Mittal and S.K. Panigrahi
Published 2023. ISBN 978-1-394-17472-0

Progress in Adhesion and Adhesives, Volume 6
Edited by K.L. Mittal
Published 2021. ISBN 978-1-119-84665-9

Rubber to Rubber Adhesion
By Dinesh K. Kotnees and Anil K. Bhowmick
Published 2021. ISBN 978-1-119-76889-0

Surface Science and Adhesion in Cosmetics
Edited by K.L. Mittal and H.S. Bui
Published 2021. ISBN 978-1-119-65482-7

Ice Adhesion
Mechanism, Measurement and Mitigation
Edited by. K.L. Mittal and Chang-Hwan Choi
Published 2020. ISBN 978-1-119-64037-0

Structural Adhesive Joints
Design, Analysis and Testing
Edited by K.L. Mittal and S. K. Panigrahi
Published 2020. ISBN 978-1-119-73643-1

Progress in Adhesion and Adhesives, Volume 5
Edited by K.L. Mittal
Published 2020. ISBN 978-1-119-74806-9

Bioadhesives in Drug Delivery
Edited by K. L. Mittal, I, S. Bakshi and J.K. Narang
Published 2020. ISBN 978-1-119-64019-6

Advances in Contact Angle, Wettability and Adhesion
Volume 4
Edited by K.L. Mittal
Published 2019. ISBN 978-1-119-59254-9

Adhesives for Wood and Lignocellulosic Materials
By R.N. Kumar and A. Pizzi
Published 2019. ISBN 978-1-119-60543-0

Progress in Adhesion and Adhesives, Volume 4
Edited by K.L. Mittal
Published 2019. ISBN 978-1-119-62525-4

Advances in Contact Angle, Wettability and Adhesion
Volume 3
Edited by K.L. Mittal
Published 2018. ISBN 978-1-119-45994-1

Laser Technology: Applications in Adhesion and Related Areas
K.L. Mittal and Wei-Sheng Lei
Published 2018. ISBN 978-1-119-18493-5

Progress in Adhesion and Adhesives, Volume 3
Edited by K.L. Mittal
Published 2018. ISBN 978-1-119-52629-2

Textile Finishing: Recent Developments and Future Trends
Edited by K.L. Mittal and Thomas Bahners
Published 2017. ISBN 978-1-119-42676-9

Progress in Adhesion and Adhesives, Volume 2
Edited by K.L. Mittal
Published 2017. ISBN 978-1-119-40638-9

Adhesion in Pharmaceutical, Biomedical and Dental Fields
Edited by K.L. Mittal and F.M. Etzler
Published 2017. ISBN: 978-1-119-32350-1

Interface/Interphase in Polymer Nanocomposites
Edited by Anil Netravali and K.L. Mittal
Published. 2016. ISBN 978-1-119-18491-1

Progress in Adhesion and Adhesives
Edited by K.L. Mittal
Published 2015. ISBN 978-1-119-16219-3

Advances in Contact Angle, Wettability and Adhesion
Volume 2
Edited by K.L. Mittal
Published 2015. ISBN 978-1-119-11698-1

Particle Adhesion and Removal
Edited by K.L. Mittal and Ravi Jaiswal
Published 2014. ISBN 978-1118-83153-3

Laser Surface Modification and Adhesion
Edited by K.L. Mittal and Thomas Bahners
Published 2014. ISBN 978-1-118-83163-2

Adhesion in Microelectronics
Edited by K.L. Mittal and Tanweer Ahsan
Published 2014. ISBN 978-1-118-83133-5

Advances in Contact Angle, Wettability and Adhesion
Volume 1
Edited by K.L. Mittal
Published 2013. ISBN 978-1-118-47292-7

Advances in Modeling and Design of Adhesively Bonded Systems
Edited by S. Kumar and K.L. Mittal
Published 2013. ISBN 978-1-118-68637-9

Atmospheric Pressure Plasma Treatment of Polymers
Edited by Michael Thomas and K.L. Mittal
Published 2013. ISBN 978-1-118-59621-0

Printed in the USA/Agawam, MA
September 22, 2023

852010.012